**Catalytic Asymmetric
Friedel–Crafts Alkylations**

*Edited by
Marco Bandini and
Achille Umani-Ronchi*

Further Reading

E.M. Carreira, L. Kvaerno

Classics in Stereoselective Synthesis

2009

ISBN: 978-3-527-32452-1 (Hardcover)
ISBN: 978-3-527-29966-9 (Softcover)

M. Christmann, S. Bräse

Asymmetric Synthesis – The Essentials

Second, Completely Revised Edition

2007
ISBN: 978-3-527-32093-6

T. Hudlicky, J.W. Reed

The Way of Synthesis

Evolution of Design and Methods for Natural Products

2007
ISBN: 978-3-527-32077-6

T. Toru, C. Bolm

Organosulfur Chemistry in Asymmetric Synthesis

2008
ISBN: 978-3-527-31854-4

M. Hiersemann, U. Nubbemeyer (Eds.)

The Claisen Rearrangement

Methods and Applications

2007
ISBN: 978-3-527-30825-5

B. Cornils, W.A. Herrmann, M. Muhler, C.-H. Wong (Eds.)

Catalysis from A to Z

A Concise Encyclopedia

Third, Completely Revised and Enlarged Edition

2007
ISBN: 978-3-527-31438-6

A. Berkessel, H. Gröger

Asymmetric Organocatalysis

From Biomimetic Concepts to Applications in Asymmetric Synthesis

2005
ISBN: 978-3-527-30517-9

J. Christoffers, A. Baro, S.V. Ley

Quaternary Stereocenters

Challenges and Solutions for Organic Synthesis

2005
ISBN: 978-3-527-31107-1

Catalytic Asymmetric Friedel–Crafts Alkylations

Edited by
Marco Bandini and Achille Umani-Ronchi

With a Foreword by George A. Olah

WILEY-VCH Verlag GmbH & Co. KGaA

The Editors

Prof. Marco Bandini
Università di Bologna
Dipartamento di Chimica "G. Ciamician"
Via Selmi 2
40126 Bologna
Italy

Prof. Achille Umani-Ronchi
Università di Bologna
Dipartamento di Chimica "G. Ciamician"
Via Selmi 2
40126 Bologna
Italy

All books published by Wiley-VCH are carefully produced. Nevertheless, authors, editors, and publisher do not warrant the information contained in these books, including this book, to be free of errors. Readers are advised to keep in mind that statements, data, illustrations, procedural details or other items may inadvertently be inaccurate.

Library of Congress Card No.: applied for

British Library Cataloguing-in-Publication Data
A catalogue record for this book is available from the British Library.

Bibliographic information published by the Deutsche Nationalbibliothek
The Deutsche Nationalbibliothek lists this publication in the Deutsche Nationalbibliografie; detailed bibliographic data are available on the Internet at http://dnb.d-nb.de

© 2009 WILEY-VCH Verlag GmbH & Co. KGaA, Weinheim

All rights reserved (including those of translation into other languages). No part of this book may be reproduced in any form – by photoprinting, microfilm, or any other means – nor transmitted or translated into a machine language without written permission from the publishers. Registered names, trademarks, etc. used in this book, even when not specifically marked as such, are not to be considered unprotected by law.

Cover design   Grafik-Design Schulz, Fußgönheim
Typesetting   Thomson Digital, Noida, India
Printing   Strauss GmbH, Mörlenbach
Binding   Litges & Dopf Buchbinderei GmbH, Heppenheim

Printed in the Federal Republic of Germany
Printed on acid-free paper

ISBN: 978-3-527-32380-7

Foreword

Charles Friedel, a professor at the Sorbonne in Paris, and James Mason Crafts, his long standing American collaborator from MIT (who later became the President of his institution and founded its renowned graduate school), in 1877 disclosed in a series of papers the discovery of aluminum chloride catalyzed conversion of hydrocarbons. These included alkylation with alkyl halides and other related reactions. Rapid development of the Friedel–Craft-type reaction using a large variety of catalysts made it one of the most versatile and useful fields of organic chemistry.

Marco Bandini has now undertaken to edit, and in no small part write, a most valuable monograph on one of the more recent and important applications of Friedel–Crafts chemistry to catalytic asymmetric alkylations. This clearly is one of the most fascinating and also practical applications of the heritage of Friedel and Crafts applied to the increasingly significant synthesis of asymmetric, optically active compounds. Each chapter was written by leading investigators in the area, covering topics such as Michael Addition, Addition to Carbonyl Compound, Alkylic Akylation, Nucleophilic Substitution on Csp^3 Carbons, Unactivated Alkenes, Catalytic Asymmetric Alkylations in Total Synthesis and Application in Industrial Friedel–Crafts Chemistry.

It is a real pleasure to introduce and recommend to the chemical community, including not only academic and industrial researchers but also other interested chemists, this most useful and stimulating book. It will be I am sure not only of well-deserved interest but also of great practical value to those involved in this significant area of asymmetric synthesis.

Los Angeles, California, October 2008   *George A. Olah*

Catalytic Asymmetric Friedel–Crafts Alkylations. Edited by M. Bandini and A. Umani-Ronchi
Copyright © 2009 WILEY-VCH Verlag GmbH & Co. KGaA, Weinheim
ISBN: 978-3-527-32380-7

Contents

Foreword *V*
Preface *XIII*
List of Contributors *XV*

1	**General Aspects and Historical Background** *1*	
	Marco Bandini	
1.1	Introduction *1*	
1.2	General Aspects and Historical Background *3*	
1.3	Catalytic Enantioselective FC Reactions: An Introduction *5*	
1.3.1	Alkylating Agents *6*	
1.3.2	Privileged Catalysts *9*	
1.3.3	Aromatic Compounds and Reaction Conditions *13*	
	References *14*	
2	**Michael Addition** *17*	
2.1	Chelating α,β-Unsaturated Compounds *17*	
	Jesús M. García, Mikel Oiarbide and Claudio Palomo	
2.1.1	Chelating Compounds *18*	
2.1.1.1	Alkylidene Malonates *19*	
2.1.1.2	β,γ-Unsaturated α-Ketoesters *25*	
2.1.1.3	α′-Hydroxy Enones *28*	
2.1.1.4	α′-Phosphonate Enones *31*	
2.1.1.5	α,β-Unsaturated Acyl Compounds *32*	
2.1.1.5.1	α,β-Unsaturated Acyl Phosphonates *32*	
2.1.1.5.2	α,β-Unsaturated 2-Acyl Imidazoles *34*	
2.1.1.6	α,β-Unsaturated Thioesters *39*	
2.1.1.7	Nitroacrylates *40*	
2.1.1.8	Experimental: Selected Procedures *42*

Catalytic Asymmetric Friedel–Crafts Alkylations. Edited by M. Bandini and A. Umani-Ronchi
Copyright © 2009 WILEY-VCH Verlag GmbH & Co. KGaA, Weinheim
ISBN: 978-3-527-32380-7

2.2	Simple α,β-Unsaturated Substrates 49
	Giuseppe Bartoli and Paolo Melchiorre
2.2.1	Introduction 49
2.2.1.1	α,β-Unsaturated Aldehydes 50
2.2.1.1.1	Organocatalysis 50
2.2.1.1.2	Organocatalytic Domino Reactions 55
2.2.1.1.3	Experimental: Selected Procedures 57
2.2.1.2	α,β-Unsaturated Ketones 60
2.2.1.2.1	Organometallic Catalysis 60
2.2.1.2.2	Organocatalysis 61
2.2.1.2.3	Experimental: Selected Procedures 64
2.2.2	Intramolecular Approach 65
2.3	Nitroalkenes 67
	Luca Bernardi and Alfredo Ricci
2.3.1	Introduction 68
2.3.2	Organocatalytic Enantioselective Reactions 68
2.3.2.1	Friedel–Crafts Alkylation of Indoles 68
2.3.2.1.1	Thiourea Catalysts 69
2.3.2.1.2	Bissulfonamide Catalysts 72
2.3.2.1.3	Phosphoric Acid Catalysts 73
2.3.2.1.4	Synthetic Applications 74
2.3.3	Friedel–Crafts Reactions of Naphthols 75
2.3.3.1	Thiourea Catalysts 75
2.3.4	Addition of 1,2,3-Triazcles 76
2.3.4.1	Cinchona Alkaloids Catalysts 77
2.3.5	Conclusion and Outlook 78
2.3.6	Experimental: Selected Procedures 79
2.3.7	Lewis Acid Catalyzed Enantioselective Reactions 80
2.3.7.1	Friedel–Crafts Reactions of Indoles 81
2.3.7.1.1	Zinc Catalysts 81
2.3.7.1.2	Copper and Aluminum Catalysts 84
2.3.7.2	Friedel–Crafts Reactions of Furans and Pyrroles 88
2.3.7.2.1	Zinc Catalysts 88
2.3.7.3	Conclusion and Outlook 90
2.3.7.4	Experimental: Selected Procedures 91
	References 94
3	**Addition to Carbonyl Compounds** 101
3.1	Aldehydes/Ketones 101
	Jia-Rong Chen and Wen-Jing Xiao
3.1.1	Introduction 101
3.1.2	Organometallic Catalysis 102
3.1.2.1	Fundamental Examples and Mechanism 102
3.1.2.1.1	Chiral Titanium (IV) Catalysis 103

3.1.2.1.2	Chiral Bisoxazoline-Cu(II) Catalysis	108
3.1.2.2	Experiments: Selected Representative Procedures	114
3.1.3	Organocatalysis	116
3.1.3.1	Fundamental Examples and Mechanism	116
3.1.3.2	Experiments: Selected Representative Procedures	118
3.1.4	Heterogeneous Catalysis	118
3.2	Imines	119
	Shu-Li You	
3.2.1	Catalytic Asymmetric Friedel–Crafts Reaction of Imines	120
3.2.1.1	Intermolecular Approach	120
3.2.1.1.1	Organometallic Catalysis	120
3.2.1.1.2	Organocatalysis	123
3.2.1.2	Pictet–Spengler Reaction	132
3.2.2	Selected Procedures	138
	References	140
4	**Nucleophilic Allylic Alkylation and Hydroarylation of Allenes**	**145**
	Marco Bandini and Achille Umani-Ronchi	
4.1	Introduction	145
4.2	Allylic Alkylations	147
4.2.1	Intermolecular Approach	147
4.2.1.1	Benzene-Like Compounds	147
4.2.1.2	Indoles	150
4.2.1.3	Asymmetric Allylations	152
4.2.1.4	Experimental: Selected Procedures	155
4.2.2	Intramolecular Approaches	156
4.2.2.1	Introduction and Fundamental Examples	156
4.2.2.2	Experimental: Selected Procedures	159
4.3	Metallo-Catalyzed Hydroarylation of Allenes	160
4.3.1	Introduction and Fundamental Examples	160
4.3.2	Experimental: Selected Procedures	163
	References	164
5	**Nucleophilic Substitution on Csp^3 Carbon Atoms**	**167**
	Marco Bandini and Pier Giorgio Cozzi	
5.1	Ring-Opening of Epoxides	168
5.1.1	Introduction	168
5.1.2	Enantiomerically Pure Epoxides	169
5.1.2.1	Introduction	169
5.1.2.2	Indium(III) Catalysis	172
5.1.2.3	Mechanism of Indium(III) Catalysis	175
5.1.2.4	Gold(III) Catalysis	176
5.1.2.5	Mechanism of Gold(III) Catalysis	177
5.1.2.6	Experiments: Selected Procedures	177

5.1.3	Asymmetric Ring-Opening of Racemic and *meso* Epoxides	179
5.1.3.1	Introduction	179
5.1.3.2	Salen-Chromium-Catalyzed Kinetic Resolution of Epoxides with Indoles	179
5.1.3.3	Experiments: Selected Procedures	180
5.2	Direct Activation of Alcohols	181
5.2.1	Introduction	181
5.2.2	Diastereoselective Reactions	182
5.2.2.1	BF_3-Mediated Reactions	183
5.2.2.2	Gold(III) Catalysis	184
5.2.3	Enantioselective Reactions	187
5.2.3.1	Ruthenium Catalysis	187
5.2.3.2	Brønsted Acid Catalysis	191
5.2.3.3	Experiments: Selected Procedures	192
5.2.3.4	FC Reactions with Chiral Ferrocenyl Compounds	193
5.2.3.4.1	Indium-Promoted Nucleophilic Substitution with Ferrocene	193
5.2.3.4.2	"On Water" FC Reactions	194
5.2.3.4.3	Experiments: Selected Procedures	196
	References	198

6	**Unactivated Alkenes**	203
	Ross A. Widenhoefer	
6.1	Introduction	203
6.2	Early Studies	204
6.2.1	Enantioselective Hydroarylation of Norbornene	204
6.2.2	Atropselective Alkylation of Biaryls	206
6.3	Rh(I)-Catalyzed Enantioselective Hydroarylation of Iminoarenes	207
6.3.1	Catalyst Control	207
6.3.1.1	Alkylation of Aromatic Ketimines	207
6.3.1.2	Alkylation of Aromatic Aldimines	209
6.3.2	Directing Group Control	211
6.3.2.1	Synthesis of (+)-Lithospermic Acid	211
6.3.2.2	Optimization and Scope	213
6.4	Pt(II)-Catalyzed Enantioselective Hydroarylation of Alkenylindoles	214
6.5	Au(I)-Catalyzed Enantioselective Hydroarylation of Allenylindoles	217
6.6	Conclusions and Outlook	219
6.7	Experimental: Selected Procedures	220
	References	221

7	**Catalytic Asymmetric Friedel–Crafts Alkylations in Total Synthesis**	223
	Gonzalo Blay, José R. Pedro and Carlos Vila	
7.1	Introduction	223
7.2	Total Synthesis of Indole-Containing Compounds	224
7.2.1	Synthesis of β-Indolyl-Propanoic Acids	224
7.2.2	Synthesis of Tryptamine and Tryptophan Analogs	225

7.2.3	Synthesis of Polycyclic Indoles 228
7.2.4	Synthesis of 2-Aminomethyl Indoles 229
7.2.5	Experiments: Selected Procedures 232
7.3	Total Synthesis of Pyrrole-Containing Compounds 240
7.3.1	Synthesis of (+)-Heliotridane 240
7.3.2	Synthesis of the Indolizidine Alkaloids Tashiromine, epi-Tashiromine, Razhinal, Rhazinilam, Leuconolam and epi-Leuconalam 241
7.3.3	Synthesis of Pyrrolo[1,2-a]pyrazines 242
7.3.4	Experiments: Selected Procedures 244
7.4	Friedel–Crafts Alkylation of Furan Derivatives in Total Synthesis 247
7.4.1	Synthesis of Aminobutenolides 248
7.4.2	Experiments: Selected Procedures 248
7.5	Friedel–Crafts Alkylation of Arenes in Total Synthesis 249
7.5.1	Synthesis of Optically Active Mandelic Acid Derivatives and Aromatic α-Amino Acids 250
7.5.2	Synthesis of Optically Active Chromanes 250
7.5.3	Experiments: Selected Procedures 251
7.6	Asymmetric Synthesis of Natural Products Based on Diastereoselective Friedel–Crafts Reactions 253
7.6.1	Synthesis of Hapalindole Alkaloids 253
7.6.2	Synthesis of Acremoauxin A and Oxazinin 3 255
7.6.3	Synthesis of (S)-Ketoralac 256
7.6.4	Synthesis of (−)-Lintetralin 257
7.6.5	Synthesis of (+)-Erogorgiaene 258
7.6.6	Synthesis of (+)-Bruguierol C 260
7.6.7	Synthesis of (−)-Talaumidin 261
7.6.8	Experiments: Selected Procedures 262
	References 268
8	**Industrial Friedel–Crafts Chemistry** 271
	Duncan J. Macquarrie
8.1	Introduction 271
8.2	Green Chemistry and the Friedel–Crafts Reaction 272
8.2.1	Green Chemical Assessment of Friedel–Crafts Processes 272
8.3	Heterogeneous Catalysts for the Friedel–Crafts Reaction 275
8.3.1	Zeolites 275
8.3.2	Clays and Other Solid Acids 277
8.4	Large Scale Hydrocarbon Processing 278
8.4.1	Ethylbenzene and Cumene 278
8.4.2	Linear Alkyl Benzenes 279
8.4.3	Dialkylated Biaryls 280
8.4.4	Alkylanilines and Related Compounds 280
8.4.5	Alkylphenol Production 281
8.4.6	Diarylmethanes 282
8.4.7	Hydroxyalkylation of Aromatics 282

8.4.8	Pechmann Syntheses	*284*
8.4.9	Use of Epoxides as Alkylating Agents	*284*
8.5	Conclusions and Perspectives	*286*
	References	*286*

Index *289*

Preface

Friedel–Crafts (FC) alkylation is one of the cornerstones in organic chemistry. Since the pioneering discovery by Charles Friedel and James Mason Crafts in 1887, an impressive number of applications of such a process have appeared for the synthesis of challenging aromatic compounds. Numerous comprehensive treatises on the general aspects of the process, mechanistic details and scope have already been published. In 1963, George A. Olah edited the well referred *Friedel–Crafts and Related Reactions* (1963), a decade later Professor Olah updated his original treatise, producing a monograph entitled *Friedel–Crafts Chemistry*. Almost simultaneously, Royston M. Roberts and Ali A. Khalaf edited their monograph entitled *Friedel–Crafts Alkylation Chemistry*. Other more general monographs focusing on general aspects of aromatic substitutions have appeared later (e.g., *Electrophilic Aromatic Substitution* edited by Roger Taylor in 1990), however, a comprehensive overview addressing strategies to perform catalytic enantioselective FC alkylation is still lacking. With *Catalytic Asymmetric Friedel–Crafts Alkylations* we wish to fill this gap.

Because of the exceptional number of asymmetric FC transformations, we deliberately decided to focus mainly on catalytic enantioselective reactions only (up to July 2008) with a collection of more representative diastereoselective approaches being reported in Chapter 7.

We decided to consider the nature of the electrophilic species as the guideline to creating the chapters of the book. Moreover, the single chapters present further subdivisions for an independent treatment of intramolecular and intermolecular approaches or organometallic and organocatalytic reactions. In our opinion, this choice should help the reader to easily find the necessary information in the articulate scenario of enantioselective FC processes. In doing so, five chapters originated (Chapters 2 to 6) focusing on: conjugate addition to α,β-unsaturated compounds (Chapter 2), direct condensation with carbonyl compounds (Chapter 3), allylic alkylations (Chapter 4), nucleophilic substitution on Csp^3 carbon centers (Chapter 5), and hydroarylation of unactivated carbon–carbon double bonds (Chapter 6).

In Chapter 1, after an introduction to general aspects and the historical background of FC alkylation, a summary of guidelines is reported concerning general trends in alkylating agents, chiral catalysts and aromatic systems adopted in catalytic enantioselective aromatic functionalizations.

Chapter 7 has a more target-oriented content. In particular, the authors succeeded in the difficult task of collecting and organizing the plethora of synthetic applications in which asymmetric FC-alkylation is employed as the key step in the total syntheses of natural compounds.

Finally, the tremendous impact of FC processes on actual worldwide chemical industry production is reviewed in Chapter 8. Catalytic enantioselective versions of this class of transformations has not yet been applied to large scale production. However, the ongoing use of innovative catalysts to drive industrial Friedel–Crafts processes towards cleaner and safer production, represents a strong *viaticum* for new developments in the near future.

All the chapters are integrated with a collection of experimental procedures and analytical characterization of model substrates, critically selected by the authors. The presence of these sections will expand the scope of the book to a wide readership such as undergraduate students, graduate students, and researchers both in academia and in industry.

We personally feel indebted to all the authors (we like to call them *friends*) who enthusiastically joined us in the project, providing outstanding contributions. Special mention goes to Professor George A. Olah who kindly agreed to write the foreword.

I (MB) would also like to thank my wife Manuela for putting up with my prolonged absence from the family scene during the editing of this text.

We hope that readers will enjoy reading of the developments in catalytic enantioselective Friedel–Crafts processes presented herein as much as the Editors did during the assembly of the chapters.

Bologna, May 2009

Marco Bandini
Achille Umani-Ronchi

List of Contributors

Marco Bandini
Università di Bologna
Alma Mater Studiorum
Dipartimento di Chimica
"G. Ciamician"
Via Selmi 2
40126 Bologna
Italy

Giuseppe Bartoli
Università di Bologna
Alma Mater Studiorum
Dipartimento di Chimica Organica
"A. Mangini"
Viale Risorgimento 4
40136 Bologna
Italy

Gonzalo Blay
Universidad de València
Facultat de Química
Departament de Química Orgànica
C/ Dr. Moliner 50
46100 Burjassot (València)
Spain

Jia-Rong Chen
Central China Normal University
College of Chemistry
Key Laboratory of Pesticide &
Chemical Biology
Ministry of Education
152 Luoyu Road
Wuhan, Hubei 430079
China

Pier Giorgio Cozzi
Università di Bologna
Alma Mater Studiorum
Dipartimento di Chimica
"G. Ciamician"
Via Selmi 2
40126 Bologna
Italy

Jesús M. García
Universidad Pública de Navarra
Campus de Arrosadía
Departamento de Química Aplicada
31006 Pamplona
Spain

Duncan J. Macquarrie
University of York
Centre of Excellence in Green
Chemistry
Department of Chemistry
Heslington, York YO10 5DD
UK

Paolo Melchiorre
Università di Bologna
Alma Mater Studiorum
Dipartimento di Chimica Organica
"A. Mangini"
Viale Risorgimento 4
40136 Bologna
Italy

Mikel Oiarbide
Universidad del País Vasco
Facultad de Química
Departamento de Química Orgánica I
Manuel Lardizabal 3
20018 San Sebastián
Spain

José R. Pedro
Universidad de València
Facultat de Química
Departamento de Química Orgànica
C/ Dr. Moliner 50
46100 Burjassot (València)
Spain

Claudio Palomo
Universidad del País Vasco
Facultad de Química
Departamento de Química Orgánica I
Manuel Lardizabal 3
20018 San Sebastián
Spain

Achille Umani-Ronchi
Università di Bologna
Alma Mater Studiorum
Dipartimento di Chimica
"G. Ciamician"
Via Selmi 2
40126 Bologna
Italy

Carlos Vila
Universidad de València
Facultat de Química
Departamento de Química Orgànica
C/ Dr. Moliner 50
46100 Burjassot (València)
Spain

Ross A. Widenhoefer
Duke University
French Family Science Center
Durham, NC 27708-0346
USA

Wen-Jing Xiao
Central China Normal University
College of Chemistry
Key Laboratory of Pesticide &
Chemical Biology
Ministry of Education
152 Luoyu Road
Wuhan, Hubei 430079
China

Shu-Li You
Chinese Academy of Sciences
Shanghai Institute of Organic
Chemistry
State Key Laboratory of
Organometallic Chemistry
354 Fenglin Lu
Shanghai 200032
China

1
General Aspects and Historical Background
Marco Bandini

Summary

The scope of catalytic enantioselective Friedel–Crafts alkylations is expanding rapidly and since the seminal papers appeared in the mid 1980s, numerous examples featuring enantioselectivities higher than 90% have been published. At present, nearly all the organic compounds displaying electrophilic character have been reacted with aromatic systems in FC-type alkylation reactions. However, the typology of reagents becomes slightly narrower if we limit the survey to approaches that employ chiral catalysts capable of traducing stereochemistry in the final products. Activated as well as unactivated carbon–carbon double bonds and C=X frameworks characterize the most used classes of electrophilic agents, that are generally combined with privileged chiral organometallic and organic catalysts. It is also worth mentioning the actual distribution of enantioselective FC processes based on the type of aromatic system employed. Interestingly, highly reactive electron-rich arenes (pyrrole and indole) still constitute almost 80% of catalytic enantioselective FC-processes, while asymmetric transformations of benzene-like compounds are quite undeveloped.

1.1
Introduction

The Friedel–Crafts (FC) alkylation of aromatic compounds is one of the cornerstones of organic chemistry. Since it was first reported (three consecutive notes appeared in *Comptes Rendus de l'Académie des Sciences* in 1877) [1] by Charles Friedel and James Mason Crafts, countless versions of this process have been reported. In this context, it is worth mentioning that "... one third of worldwide organic chemical production involves aromatic compounds..." [2] with the consequent synthetic interest in their chemical manipulation.

The reaction introduced by the European (CF, Strasbourg, France, 1832–1889) and the youngest American (JMC, Boston, MA, USA, 1839–1917) researchers [3], has always been the subject of lively scientific debate, and one of the most controversial aspects is the definition of the process.

Catalytic Asymmetric Friedel–Crafts Alkylations. Edited by M. Bandini and A. Umani-Ronchi
Copyright © 2009 WILEY-VCH Verlag GmbH & Co. KGaA, Weinheim
ISBN: 978-3-527-32380-7

It can be seen from the more than one hundred papers rapidly published by Crafts and Friedel on the topic that there is not only one but numerous organic transformations that can be listed under the name of FC-alkylation. In particular, Olah and Dear concluded in their outstanding treatise on FC-reactions (1963) that "... *today we consider Friedel–Crafts type reactions to be any isomerization, elimination, cracking, polymerization, or addition reactions taking place under the catalytic effect of Lewis acid type ... or protic acids*" [3].

Nowadays, this view has probably altered, and although the original experiments addressed the breaking of a C–H bond in aliphatic compounds, the actual definition of FC processes is restricted to the specific functionalization of aromatic systems, namely alkylation and acylation reactions.

Most importantly, FC processes are probably the oldest organic transformations requiring metal halides (known as Lewis acids (LAs), e.g., aluminum trichloride, zinc chloride, boron trifluoride, ferric chloride etc.), as chemical promoters (catalysts).

Over more than 130 years, the original scope of the reaction has been significantly enlarged. Chemical aspects (e.g., reactivity, selectivity) combined with environmental concerns associated with the production/use of poorly manageable catalysts and the disposal of hazardous wastes have prompted many chemists to re-address their research toward the discovery of *greener* and economically viable alternatives. At present, a FC alkylation reaction constitutes an essential synthetic step in a number of commercial processes in the bulk chemical industry. From a restricted and merely indicative survey of the SciFinder Scholar (ACS), CARPLUS, and MEDLINE databases for the term "Friedel–Crafts alkylation" (1928–2007) emerged a striking increase in the use of FC-processes in organic synthesis (Figure 1.1).

These numbers would be even more impressive if FC-acylation reactions were also included.

Due to the intrinsic industrial interest in this transformation, most of these reports still focus exclusively on the use of recoverable solid Lewis acid promoters, however, since the late 1990s, an expanding volume of effort facing challenging issues such as catalysis and stereocontrol in alkylations of aromatic compounds has begun. Here, before moving to a more specific description of the actual landmarks regarding chiral organic and organometallic catalysis for the construction of enantiomerically

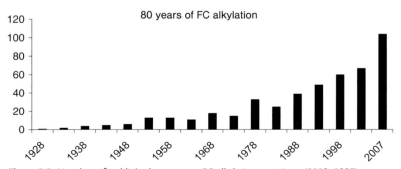

Figure 1.1 Number of published papers on FC-alkylation reactions (1928–2007).

enriched benzylic stereocenters, a brief overview of general aspects of the Friedel–Crafts reaction is provided in the following sections.

1.2
General Aspects and Historical Background

Friedel–Crafts alkylation processes involve the replacement of a C—H atom of an aromatic ring by an alkyl group "R$^+$" in the presence of a Lewis or Brønsted acid catalyst (Scheme 1.1).

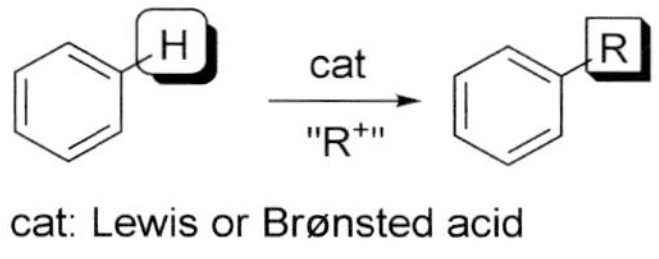

cat: Lewis or Brønsted acid

Scheme 1.1 Pictorial representation of the Friedel–Crafts alkylation.

Textbooks of basic organic chemistry usually mention exclusively the use of reactive alkyl halides in combination with aromatics. However, as partially outlined in the following chapters, activated and unactivated alkenes, alkynes, paraffins, alcohols, ethers, carbonyl compounds and so on, can also be effectively employed as alkylation agents.

A range of molecular Lewis acids with order of catalytic power: $AlBr_3 > AlCl_3 > GaCl_3 > FeCl_3 > SbCl_5$ $TiCl_4$, $ZnCl_2 > SnCl_4 > BCl_3$, BF_3 [4], modified solid LAs and Brønsted–Lowry acids (HF, H_3PO_4, H_2SO_4) have proved efficient in accelerating this process. The right choice of the additive to be employed is often a matter of trial and error, because a narrow correlation between type of alkylation agent and reactivity of aromatic compounds frequently occurs.

The reactivity of aromatics is another key aspect that soon emerged from the seminal studies of FC alkylation reactions. As known, by considering benzene as an *electron-neutral* arene, substitution with electron-donating groups (EDGs) usually increases the nucleophilic character of the aromatic compound and causes substitution to occur predominantly at the *ortho* and *para* positions. In contrast, electron-withdrawing groups (EWGs) deactivate the arene toward alkylation processes with a concomitant *meta*-oriented regiochemistry.

There are of course many exceptions to these trends. For instance, the substitution of aromatic compounds with strongly coordinating electron-releasing substituents such as, hydroxy, alkoxy, amino, and so on, quenches the catalytic performance of the catalyst by deactivating interactions and, additionally, the original electron-donating group is transformed into a deactivating substituent (Scheme 1.2).

Despite the great volume of effort devoted to gaining insight into the role of the acid catalyst in the mechanism of the reactions, an unequivocal and clear answer is still lacking. It should be mentioned that, although organometallic intermediates like "$C_6H_5 \cdot Al_2Cl_5$" were proposed by the authors in the original papers, here Friedel and Crafts finally stated "*Nous n'avons donc encore aucune preuve décisive à apporter en*

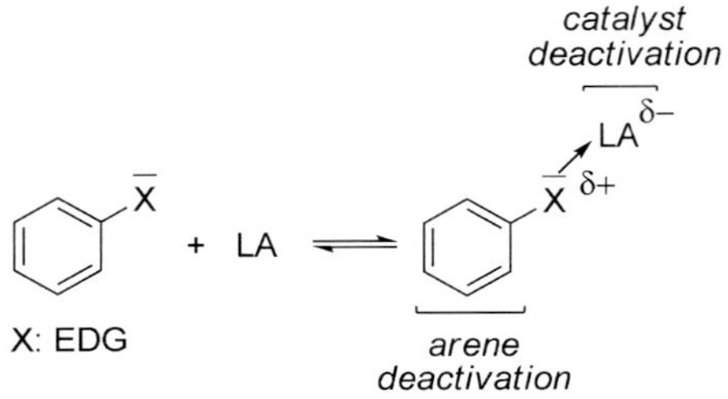

Scheme 1.2 Trapping of Lewis acids by coordinating heteroatomic substituents.

faveur de l'hypothèse que nous faisons sur le mécanisme de la réaction..." (we do not have any decisive proof to support the hypothesis of the reaction mechanism).

In this context, the requirement of acid additives for optimal reaction outcomes suggests that an interaction of the promoting agent (A) with a donor species (D), present in solution, is operating in the process. Among all the possible combinations, the catalyst-alkylating agent (1) and the catalyst–substrate interactions (2) are considered essential for the process. On the contrary, the complexation of the catalyst by the product/s (3) will preclude irremediably its availability for the reaction course.

By considering the alkylation of arenes with alkyl halides (R–X) as the model reaction, it is largely accepted that the Lewis/Brønsted acid initially interacts with the reagent, through an *n*-donor interaction (Scheme 1.3a) [5]. This complexation weakens the R–X bond with the formation of highly reactive carbonium ion [R$^+$] feasible during the reaction course. The existence of such an interaction has been proved experimentally through numerous chemical and physical investigations [6], however, in no case was it possible to shed light on the existence of alkyl cation complexes as reaction intermediates.

A breakthrough in the field was achieved by the Nobel Laureate in Chemistry, George A. Olah (1994), who proved the existence of "long-lived" long alkyl chain cations by employing "superacids" (HF·SbF$_5$) in combination with alkyl fluorides [7].

However, the possibility that the acid species could interact directly with the π-rich aromatic counterpart, via either π-complexes (all the six electrons of the arenes interact with the LA, Scheme 1.3b) or σ-complexes (formation of arenonium species, Scheme 1.3c) cannot be ruled out [8,9]. Here, the intermediate originated from a π-donor interaction between arene and catalyst is generally characterized by low

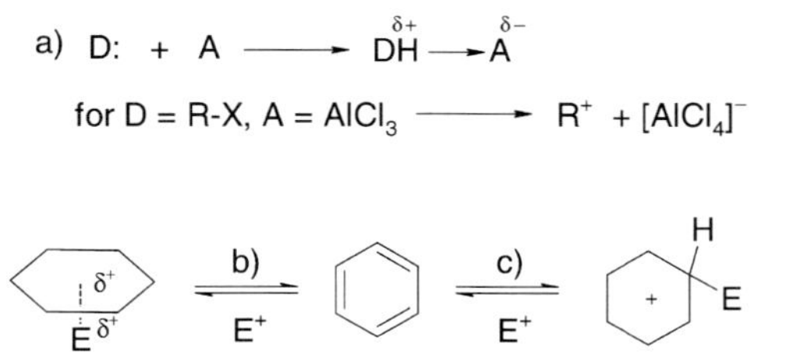

Scheme 1.3 Types of intermediate complexes in FC alkylations: (a) carbonium ion, (b) π-complexes, (c) σ-complexes.

stability with a marginal role in the FC mechanistic cycle. On the contrary, with σ-donor-like contacts, a new covalent σ-bond (strong interaction) is formed.

A typical example of spontaneous reaction between aromatic systems and metal salts is the auration process that has long been recognized to involve reaction between anhydrous gold(III) chloride and benzene [8]. The corresponding arylauric intermediate is highly unstable leading to Au(I) chloride and PhCl, but, under particular circumstances, the auration products have been isolated as air- and moisture-stable crystalline solids [10].

A major outcome of these catalyst–arene interactions is that the carbon atoms of the benzene system become highly nucleophilic. This has been quantified in a theoretical investigation (*ab initio* study) focusing on the role of $AlCl_3$– and BCl_3–benzene interactions in Friedel–Crafts reactions in the gas-phase [11]. Here, for the first time, a tight Al–C contact (2.35 Å) was found with consequent marked pyramidalization of $AlCl_3$ (Cl–Al $\cdots$ Cl $\approx 98°$) and loss of the benzene nodal plane. Such an interaction should be even more pronounced with electron-rich arenes or with late-transition metal-based catalysts in which the back-donation of charge should strengthen the interaction [12].

1.3
Catalytic Enantioselective FC Reactions: An Introduction

Despite the fact that the original studies date back more than 130 years, and the large volume of research effort devoted to FC-alkylation reactions, it has taken more than a century for asymmetric catalytic versions of this process to be developed [13]. In fact, before 1999 the examples of enantioselective catalytic FC processes were sporadic [14]. The use of chiral Lewis acids (stoichiometric amount) in the alkylation of phenols with chloral **2** was first developed by an Italian team in 1985 [15]. Here, menthol-Al Lewis acid **3** promoted the hydroxyalkylation reaction with high *ortho*-regiocontrol and enantioselectivity up to 80% (Scheme 1.4a). The simultaneous coordination of phenol and alkylating agent to the catalyst was postulated to account for the high regio- and stereochemical outcomes.

An enantioselective intramolecular alkylation of indoles starting from N-hydroxy-tryptamine and aldehydes (Pictet–Spengler condensation) was also reported, furnishing enantiomerically enriched tetrahydro-β-carbolines (*ee* up to 91%) with a stoichiometric amount of (+)-Ipc_2BCl as the chiral promoter [16].

A few years later the use of chiral LAs, in catalytic amounts, for the construction of benzylic stereocenters was first reported [17]. The study considers the condensation of naphthol and ethyl pyruvate **6** with $ZrCl_3$-dibornacyclopentadienyl **7** (5 mol%), leading to the desired *ortho*-hydroxyalkylated compound **8** in moderate conversion (70%) and good enantiomeric excess (89%, Scheme 1.4b). However, the methodology suffered from a quite narrow scope.

Nowadays, asymmetric FC-processes have been greatly improved with the possibility of frequently isolating the products with enantioselectivities $\geq 90\%$. This is partially due to the continued efforts in the detailed investigation of mechanistic

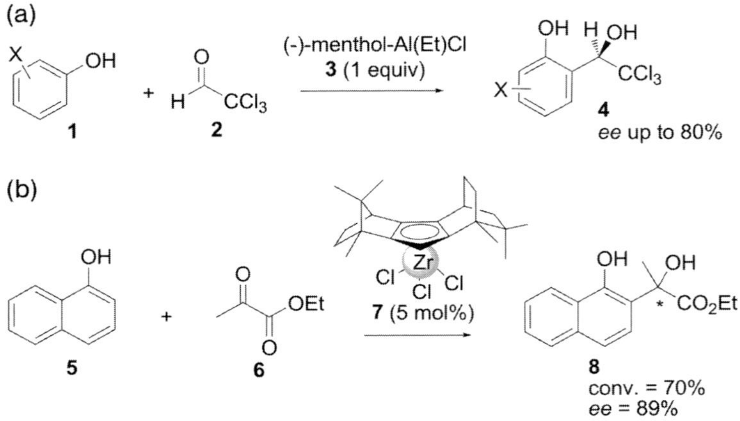

Scheme 1.4 (a) Chiral aluminum-based complexes in the ortho-hydoxyalkylation of phenols; (b) first example of catalytic enantioselective FC-type alkylation arenes.

aspects, such as specific catalyst–substrate interactions, that leads to the rapid development of ever more efficient catalytic systems.

In this scenario, asymmetric organocatalysis deserves a special mention. In fact, despite its young age, it has already had an impressive impact in enantioselective aromatic functionalization, widening dramatically the scope of such methodology [14b].

In the following sections, a summary of guidelines, concerning general trends in electrophilic agents, catalysts and aromatic systems, is given in order to help the reader become oriented in the complex scenario of catalytic enantioselective aromatic functionalizations. In doing this, comprehensive literature, such as reviews and monographs, will be preferably cited, leaving more detailed descriptions of specific applications to the other contributors in this book.

1.3.1
Alkylating Agents

Nowadays, nearly all organic compounds with electrophilic character have been reacted with aromatics in FC-type alkylation reactions under suitable activation conditions. Starting from the most inert alkanes and cycloalkanes, that are split into reactive olefins and smaller paraffins via rearrangements, up to highly reactive carbonyl-containing compounds, a very high number of alkylating agents have been employed, among them: alkyl halides, alkenes, alkynes, epoxides, alcohols and ethers.

However, the typology of the reagents becomes slightly narrower if we limit the survey to approaches that employ chiral promoting agents in catalytic amounts able to traduce stereochemistry in the final products. Here, for instance, conventional alkylating agents for aromatic functionalization, (i.e., alkyl halides), have not found use in stereocontrolled catalytic reactions due to the intrinsic difficulties in stereodifferentiating the enantiotopic faces of the highly reactive prochiral carbocation

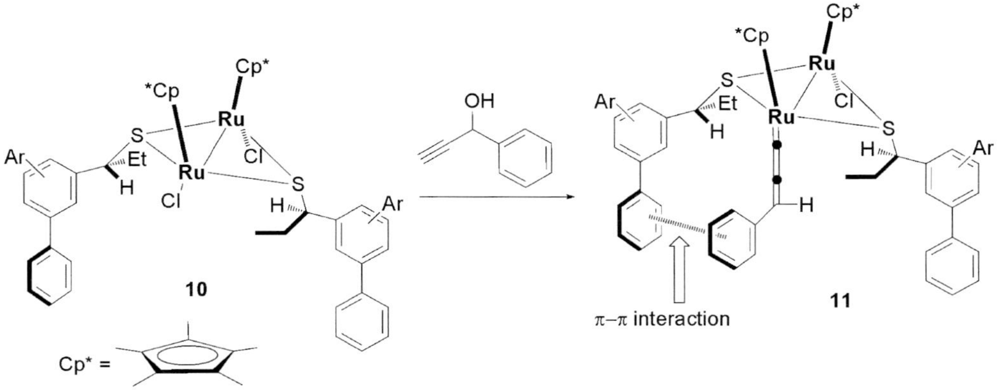

Scheme 1.5 Chiral Ru-carbenium intermediate (**11**) in the catalyzed enantioselective propargylic FC alkylation.

species formed during the reaction mechanism. In fact, the premature leaving of the chiral catalyst from the reactive center, with respect to the stereodiscriminating carbon–carbon forming event, precludes chiral translations in the final product.

An analogous scenario concerns the use of alcohols. They are stronger coordinating species than the analogous halides, and the direct activation of alcohols in FC-alkylations generally requires a high loading of catalyst and harsh reaction conditions. Only very recently, a specific family of chiral Ru-based complexes **10** has been described to be efficient in the catalytic and enantioselective alkylation of 2-methylfuran and N,N-dimethylaniline with propargyl alcohols through the formation of carbenium ion intermediates **11** (Scheme 1.5) [18].

Continuing the survey of Csp^3-based electrophilic agents for enantioselective aromatic substitutions, oxiranes have been considered, but in this case also efficient catalytic examples are sporadic. The use of noncovalent-type activation is the only strategy investigated to date and involves the use of achiral Lewis acids with enantiomerically pure epoxides or the employment of chiral metal-based catalysts with racemic or meso substrates (Scheme 1.6). The challenging searching for a sufficiently active, but simultaneously mild catalyst, to prevent the formation of ionic intermediates, has limited the number of applications to a handful of examples with further restrictions to electron-rich heteroaromatic compounds [19]. It should be mentioned that, at present, no examples of organocatalyzed asymmetric FC-type ring-opening of epoxides have been reported.

Carbon–carbon double bonds have risen to prominence in enantioselective electrophilic aromatic substitutions due to their fine-tunable reactivity, via conjugation with electron-withdrawing groups (i.e., Michael acceptors) and via covalent and noncovalent interactions with late-transition metal complexes.

The Michael-type Friedel–Crafts alkylation is probably the most popular and investigated methodology for the direct construction of benzylic stereocenters in a stereocontrolled fashion. The activation of electron-deficient alkenes (LUMO-activation), with the simultaneous enantiodiscrimination of the arene attack, has

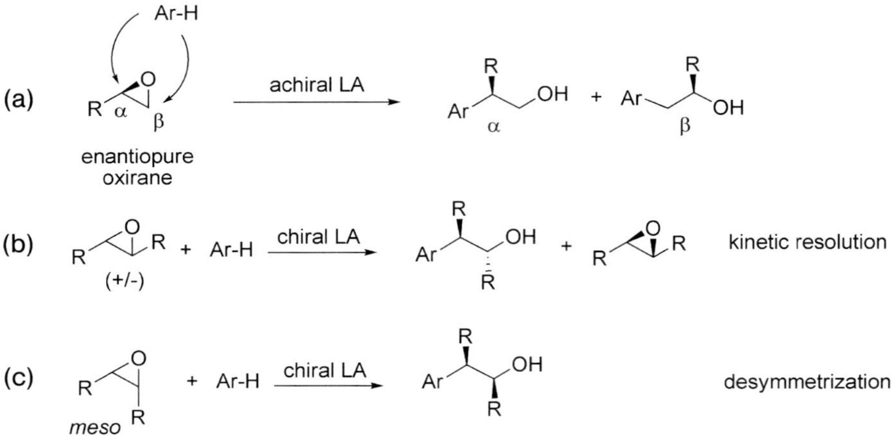

Scheme 1.6 Asymmetric electrophilic aromatic substitutions with epoxides: (a) ring-opening of enantiopure epoxides promoted by achiral catalysts; (b) kinetic resolution of racemic epoxides with chiral LAs; (c) desymmetrization of *meso* epoxides with chiral catalysts.

been obtained with the coordination of Lewis or Brønsted acids to the basic center of the Michael acceptor (ketones and chelating substrates) or via *in situ* formation of reactive iminium intermediates (mainly for α,β-unsaturated aldehydes and ketones) by chiral primary or secondary amines (Scheme 1.7a). It should be mentioned, that simple α,β-unsaturated acid derivatives have not yet found application in this class of reaction. The relatively poor Michael acceptor character of these compounds accounts for their inertness. On the contrary, activated chelating α,β-unsaturated carboxylic acid and nitro alkene derivatives have been largely employed in stereoselective FC condensations, combined with cationic chiral Lewis acids.

Nucleophilic addition of arenes to unactivated alkenes and allenes (hydroarylation of C=C) is a well established procedure that requires coordinative-activation of the double bond by π-acid late metal complexes (i.e., Au, Pt, Ru, etc.). The consequent lowering of the LUMO energy of the C=C framework makes possible the nucleophilic attack of the aromatic system under high atom-economical conditions (Scheme 1.7b).

Despite the high synthetic and practical appeal of the latter approach, the use of unactivated alkenes generally requires harsh reaction conditions: high reaction temperatures, long reaction times, and the reaction scope is usually limited to electron-rich arenes or to properly functionalized aromatic systems carrying *ortho*-directing moieties such as: imines, pyridines, carboxylates, and so on. Finally, carbon–carbon double bonds bearing a leaving group in the allylic position found significant application in the catalytic enantioselective FC-type allylic alkylation (Scheme 1.7c). In this case, the initial insertion of the metal, in a low oxidation state, into the C–LG bond (LG = leaving group) originates in the electrophilic η^3-metal allyl species that undergoes the FC process [20]. Several combinations of chiral metal catalysts (Ir, Pd) and FC partners have been documented.

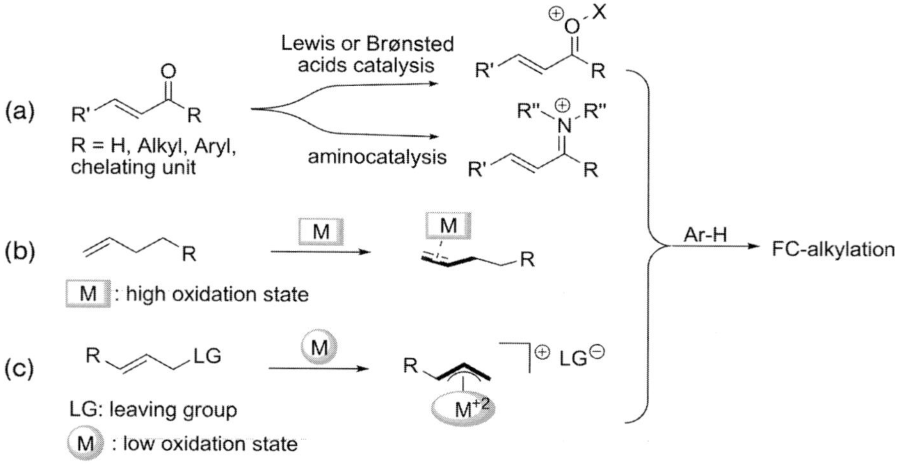

Scheme 1.7 Catalytic strategies for the use of alkenes in enantioselective functionalizations of arenes: (a) low-LUMO activation of electron-deficient carbon–carbon double bonds via covalent and noncovalent interactions; (b) unactivated olefins in FC alkylation; (c) metal-catalyzed allylic alkylation.

Finally, the 1,2-additions of arenes to carbonyl compounds have demonstrated efficiency for the construction of benzylic stereocenters in a stereocontrolled manner, in the presence of noncovalent activation with chiral Lewis and Brønsted acids. Focusing on the electrophilic partners, a high level of stereodiscrimination has been reached with activated ketones (e.g., pyruvate and trifluoropyruvates) and a marginal level with simple aldehydes and ketones. Here, in fact, the concomitant formation of bis-arylmethanes, via a dehydrative mechanism, affects markedly the use of such a type of substrate. More recently, aldo-imines, enamines and enecarbamates have also been successfully employed in the alkylation of electron-rich arenes through hydrogen-bond-type activation [21]. Contrary to the afore-described scenario of enones and enals, no covalent activation (organocatalysis), with simple carbonyl compounds, has proved efficient to date.

1.3.2
Privileged Catalysts

Adopting the general definition of Jacobsen and Yoon of privileged chiral catalysts [22] in Friedel–Crafts alkylations, a distinct separation between organometallic and metal-free catalysts seems convenient.

Asymmetric organometallic catalysis was first employed in the mid-1980s with a remarkable consolidation over the following two decades. A classic composition of a catalytic active metal-containing catalyst involves the presence of a cataliphor (metallic reactive site), exercising a noncovalent activation via Lewis acid catalysis, complexed with a proper chiral non-racemic organic ligand (chiraphor) that is accountable for the stereodiscrimination occurring during the process (Figure 1.2) [23].

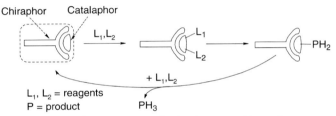

Figure 1.2 Pictorial representation of molecular composition and activity of chiral organometallic catalysts.

A collection of the most efficient organometallic catalytic systems employed in the direct construction of benzylic stereocenters is documented in Chart 1.1. Both soft and hard chiral Lewis acids have been used, ranging from transition metal (early and late) complexes to more conventional chiral systems of Group 13 elements. Among the others, cationic complexes of C2- and C3-symmetric bis-oxazolines [25] with Cu(II), Zn(II) and Sc(III) have found several applications in asymmetric FC chemistry involving bidentate (chelating) alkylating reagents. This prerequisite ensures the formation of rigid conformations between catalyst and electrophilic partner in the transition state with the consequent enhancement in stereodiscrimination during the aromatic species attack.

Binol–titanium and binol–zirconium adducts have also been employed in the 1,2-addition and 1,4-addition of activated aromatic compounds to simple aldehydes and enones, respectively [25]. In these cases, the real nature of the catalytically active organometallic species is unknown. Still in the area of privileged organometallic catalysts for the alkylation of aromatic compounds, Salen-metal (Al(III) and Cr(III)) complexes have risen to prominence for their broad scope in Michael additions and the ring-opening of epoxides.

Worthy of mention is the outstanding catalytic efficiency demonstrated also by complexes of soft late-transition metals (Pt, Au, Ir, Pd, Rh) with chiral P-ligands such as BINAP- and DPPBA-based ligands, ferrocenyl ligands, biphen and phosphoramidites. In this area, while metals in low oxidation states are generally adopted in nucleophilic allylic alkylations, cationic organometallic species featuring metal centers in a high oxidation state (π-acceptors) have found application in stereoselective alkylations of aromatic compounds with unactivated carbon–carbon multiple bonds.

In the scenario of asymmetric organocatalysis, the iminium catalysis [26] was firstly adopted in Friedel–Crafts chemistry in 2001 by MacMillan and coworkers. Here, the outstanding performances of chiral imidazolidinones of first generation were applied in the condensation of pyrroles with enales [27]. This catalytic approach generally requires the use of a strong Brønsted acid as co-catalyst (e.g., TFA, AcOH, HCl, TfOH, MsOH, HClO$_4$) in order to improve the turn-over frequency of the reaction cycle. The pioneering study was subsequently extended by the same team and other groups, creating a large library of imidazolidinone systems with different stereochemistries for FC alkylation of more challenging substrates. Aziridine alcohols were also employed in the alkylation of indoles with α,β-unsaturated aldehydes, obtaining enantiomeric excesses up to 75%. Very recently, the historical

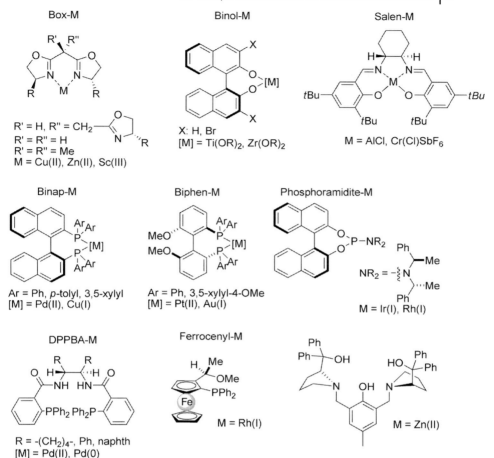

Chart 1.1 Collection of chiral organometallic catalysts utilized in enantioselective aromatic substitutions.

tabu on the use of α,β-unsaturated ketones in organocatalyzed Michael-type FC alkylation was finally solved by using chiral primary amines, such as derivatives of the natural *Cinchona* alkaloids [28a, b]. The hypothesis of why these amines work lies in the favorable sterical hindrance encountered in the condensation of enones and primary amines. Also in this case the role of the acidic co-catalyst was described in almost concomitant papers, with N-Bocphenylglycine [28c] leading to high reaction rates and enantiocontrol.

Hydrogen-bond asymmetric catalysis is the second major effort of organocatalysis in FC processes. Single-site interactions (e.g., *Cinchona* alkaloids, chiral phosphoric acids) and two-site binding catalysts (e.g., chiral ureas and thioureas) expanded dramatically the potential of asymmetric catalysis for the construction of complex molecular targets bearing stereochemically defined benzylic stereocenters, even through intramolecular approaches [29]. In Chart 1.2, some of the most efficient chiral architectures for organocatalytic FC processes are shown.

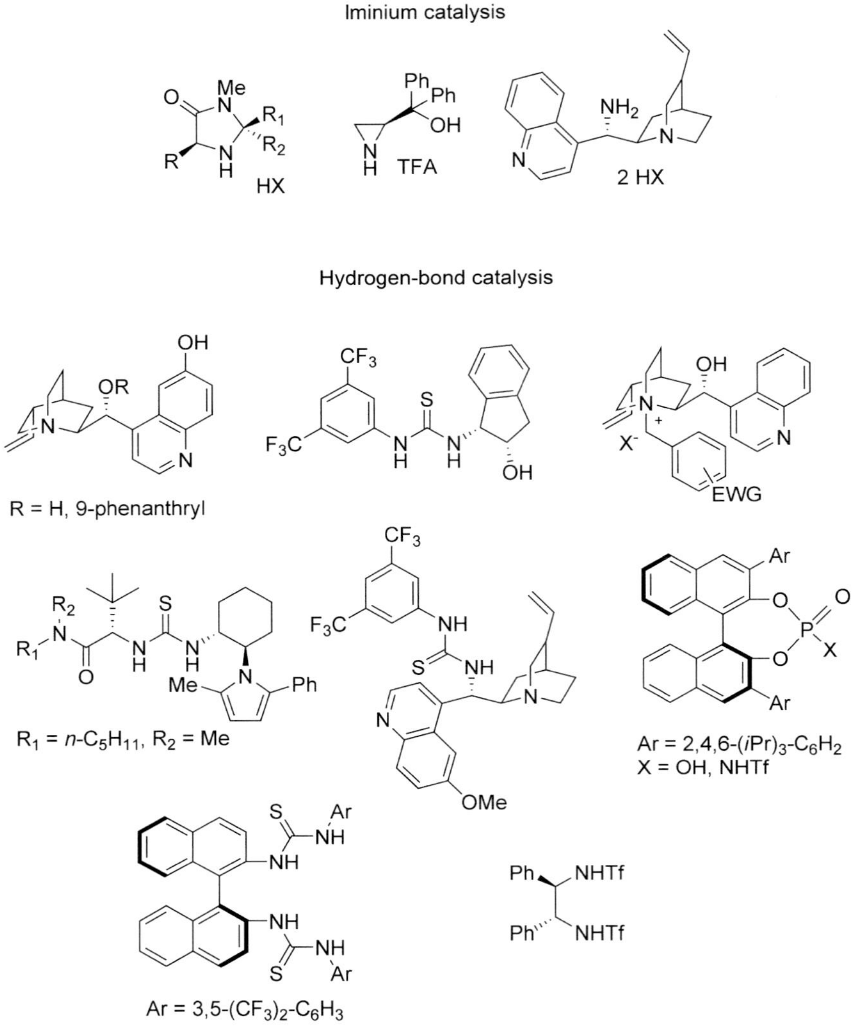

Chart 1.2 Chiral organocatalysts for asymmetric FC alkylations, via covalent (iminium) and noncovalent activation (H-bond).

Very recently, phase transfer conditions, with chiral ammonium salts of *Cinchona* alkaloids, displayed their potential in the enantioselective N(1)-alkylation of indoles via intramolecular aza-Michael addition [30].

The growing demand for mild and *green* synthetic methodologies has prompted chemists to develop innovative strategies enabling the minimization of leaching of catalyst into the reaction products. Among them, immobilization of homogeneous chiral catalysts onto inert matrixes, through covalent and noncovalent attachments (heterogeneization), is among the most promising options for the recovery and reuse

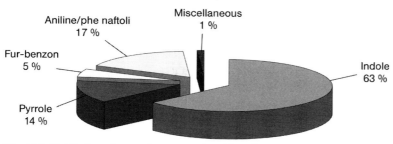

Chart 1.3 Distribution of aromatic compounds subjected to catalytic stereoselective FC-type alkylations.

of active species. Surprisingly, such an approach has been addressed and exploited only marginally in the Friedel–Crafts scenario, with scattered but effective examples in organometallic and organo-catalysis [31].

1.3.3
Aromatic Compounds and Reaction Conditions

Despite the remarkable developments recorded in this relatively young discipline, the most stringent limitations in applicability concern the nature of aromatic compounds. Here, the chemistry of reactive C-5-membered heteroaromatic compounds (mainly, nitrogen atom-containing indole and pyrrole derivatives) has been extensively expanded and accounts for almost 80% of the published methodologies (Chart 1.3). The presence of electron-donating as well as electron-withdrawing substituents on the heteroaromatic skeletons is generally well tolerated in these reactions, with the latter ones generally requiring higher loading of catalyst and prolonged reaction times. The use of indolyl and pyrrolyl derivatives usually enables one to control also the regiochemistry of the aromatic substitution, an important task for benzene-like analogs. In fact, while the use of indoles generally leads to selective C(3)-alkylated compounds, pyrrolyl cores direct the functionalization at the carbon atoms(s) adjacent to the nitrogen atom (C(2)-position). However, examples of different regiochemistries have been recorded, mainly for intramolecular processes or via introduction of specific sterically demanding groups on the nitrogen atom.

On the contrary, simple benzene-like compounds have received less attention due to their intrinsic inertness to reaction under the conventional catalytic reaction conditions used in asymmetric FC chemistry. In fact, the need for low reaction temperatures ($\leq -20\,°C$) and the impracticability of using large excesses of aromatic compounds (intramolecular variants) led to unacceptable kinetics for synthetic purposes.

At present, an exception to this trend is constituted by a few examples of enantioselective FC alkylations exploiting unactivated alkenes in combination with late transition metal complexes. In these cases harsh reaction conditions (refluxing highly boiling solvents, prolonged reaction times) are needed [32]. Benzene-like-arenes have found application also in examples of asymmetric organo-catalyzed transformations, with several restrictions to electron-rich systems such as amino- or alkoxy-substituted benzenes.

Acknowledgments

Financial support was provided by MIUR, Rome, and Alma Mater Studiorum, University of Bologna.

Abbreviations

DPPBA	diphenylphosphino benzoic acid
EDG	electro-donating groups
EWG	electron-withdrawing group
LA	Lewis acid

References

1 (a) Friedel, C. and Crafts, J.-M. (1877) *Comptes Rendus de l'Academie des Sciences Paris*, **84**, 1450–1454; (b) Friedel, C. and Crafts, J.-M. (1877) *Comptes Rendus de l'Academie des Sciences Paris*, **84**, 1392–1395; (c) Friedel, C. and Crafts, J.-M. (1877) *Comptes Rendus de l'Academie des Sciences Paris*, **85**, 74–77.

2 Taylor, R. (1990) *Electrophilic Aromatic Substitution*, Wiley, Chichester.

3 (a) For extensive biographical sketches of Friedel and Crafts see Olah, G.A. and Dear, R.E.A. (1963) *Historical Friedel–Crafts and Related Reactions*, Wiley, New York, pp. 1–24, Chapter I; (b) Bataille, X. and Bram, G. (1998) *Comptes Rendus de l'Academie des Sciences Paris, Série II*, 293–296.

4 (a) Russell, G.A. (1959) *Journal of the American Chemical Society*, **81**, 4834–4838; (b) Olah, G.A., Tashiro, M. and Kobayashi, M. (1970) *Journal of the American Chemical Society*, **92**, 6369–6371; (c) Olah, G.A., Olah, J.A. and Ohyama, T. (1984) *Journal of the American Chemical Society*, **106**, 5284–5290.

5 For a comprehensive treatise see: Olah, G.A. and Meyer, M.W. (1963) *Intermediate Complexes, Friedel–Crafts and Related Reactions*, Wiley, New York, p. 623, Chapter VIII.

6 (a) Brown, H.C., Pearsall, H. and Eddy, L.P. (1950) *Journal of the American Chemical Society*, **72**, 5347–5347; (b) Wertyporoch, E. and Firla, T. (1933) *Annali di Chimica*, **500**, 287; (c) Olah, G.A., Kuhn, S.J. and Olah, J.A. (1957) *Journal of the Chemical Society*, 2174.

7 Olah, G.A. (1994) Nobel Lecture (http://nobelprize.org/nobel_prizes/chemistry/laureates/1994/).

8 (a) Pioneering and representative examples: Kharasch, M.S. and Isbell, H.S. (1931) *Journal of the American Chemical Society*, **53**, 3053–3059; (b) Brown, H.C. and Wikkala, R.A. (1966) *Journal of the American Chemical Society*, **88**, 1447–1452; (c) Olah, G.A., Hashimoto, I. and Lin, H.C. (1977) *Proceedings of the National Academy of Sciences of the United States of America*, **74**, 4121–4125.

9 For comprehensive treatise on the formation of σ-metal-arene bonds during the FC functionalisation of benzenes: Dyker, G.(ed.) (2005) *Handbook of C-H Transformations Applications in Organic Synthesis*, Wiley-VCH, Weinheim.

10 (a) de Graaf, P.W.J., Boersma, J. and van der Kerk, G.J.M. (1976) *Journal of Organometallic Chemistry*, **105**, 399–406; (b) Fuchita, Y., Utsunomiya, Y. and Yasutake, M. (2001) *Journal of The Chemical Society – Dalton Transactions*, 2330–2334; (c) Liddle, K.S. and Parkin, C. (1972) *Journal of the Chemical Society – Chemical Communications*, 26–27; (d) Porter, K.A., Schier, A. and Schmidbaur, H. (2003) *Organometallics*, **22**, 4922–4927.

11 Tarakeshwar, P., Lee, J.Y. and Kim, K.S. (1998) *Journal of Physical Chemistry A*, **102**, 2253–2255.

12 Ottoni, O., Neder, A. de V.F., Dias, A.K.B., Cruz, R.P.A. and Aquino, L.B. (2001) *Organic Letters*, **3**, 1005–1007.

13 (a) Jørgensen, K.A. (2003) *Synthesis*, 1117–1125; (b) Bandini, M. Melloni, A. and Umani-Ronchi, A. (2004) *Angewandte Chemie*, **116**, 560–566; (2004) *Angewandte Chemie – International Edition*, **43**, 550–556.

14 For a recent review on catalytic enantioselective aromatic functionalization see: (a) Poulsen, T.B. and Jørgensen, K.A. (2008) *Chemical Reviews*, **108**, 2903–2915; (b) Sheng, Y.-F., Zhang, A.-J., Zheng, X.-J. and You, S.-L. (2008) *Chinese Journal of Chemistry*, **28**, 605–616.

15 Bigi, F., Casiraghi, G., Casnati, G., Sartori, G., Gasparri Fava, G. and Ferrari Belicchi, M. (1985) *The Journal of Organic Chemistry*, **50**, 5018–5022.

16 Yamada, H., Kawate, T., Matsumizu, M., Nishida, A., Yamaguchi, K. and Nakagawa, M. (1998) *The Journal of Organic Chemistry*, **63**, 6348–6354.

17 Erker, G. and van der Zeijden, A.A.H. (1990) *Angewandte Chemie*, **102**, 562–565; (1990) *Angewandte Chemie – International Edition*, **29**, 512–514.

18 Matsuzawa, H., Miyake, Y. and Nishibayashi, Y. (2007) *Angewandte Chemie – International Edition*, **46**, 6488–6491.

19 For representative examples see: (a) Bandini, M., Cozzi, P.G., Melchiorre, P. and Umani-Ronchi, A. (2002) *The Journal of Organic Chemistry*, **67**, 5386–5389; (b) Bandini, M., Cozzi, P.G., Melchiorre, P. and Umani-Ronchi, A. (2004) *Angewandte Chemie*, **116**, 83–86; (2004) *Angewandte Chemie – International Edition*, **43**, 84–87.

20 (a) Trost, B.M. and van Vranken, D.L. (1996) *Chemical Reviews*, **96**, 395–422; (b) Helmchen, G. and Pfaltz, A. (2000) *Accounts of Chemical Research*, **33**, 336–345; (c) Trost, B.M. and Crawley, M.L. (2003) *Chemical Reviews*, **103**, 2921–2943.

21 (a) Terada, M. and Sorimachi, K. (2007) *Journal of the American Chemical Society*, **129**, 292–293; (b) Jia, Y.-X., Zhong, J., Zhu, S.-F., Zhang, C.-M. and Zhou, Q.-L. (2007) *Angewandte Chemie*, **119**, 5661–5663; (2007) *Angewandte Chemie – International Edition*, **46**, 5565–5567.

22 Yoon, T.P. and Jacobsen, E.N. (2003) *Science*, **299**, 1691–1693.

23 Mulzer, J. (1999) *Comprehensive Asymmetric Catalysis*, Vol. 1 (eds E.N. Jacobsen, A. Pfaltz and H. Yamamoto), Springer, Berlin, pp. 34–94, Chapter 3.

24 For recent reviews see: (a) Ghosh, A.K., Mathivanan, P. and Cappiello, J. (1998) *Tetrahedron: Asymmetry*, **9**, 1–45; (b) Jørgensen, K.A., Johannsen, M., Yao, S., Audrain, H. and Thorhauge, J. (1999) *Accounts of Chemical Research*, **32**, 605–613; (c) Johnson, J.S. and Evans, D.A. (2000) *Accounts of Chemical Research*, **33**, 325–335; (d) McManus, H.A. and Guiry, P.J. (2004) *Chemical Reviews*, **104**, 4151–4202; (e) Desimoni, G., Faita, G. and Jørgensen, K.A. (2006) *Chemical Reviews*, **106**, 3561–3651.

25 (a) Chen, Y., Yekta, S. and Yudin, A.K. (2003) *Chemical Reviews*, **103**, 3155–3211; (b) Brunel, J.M. (2005) *Chemical Reviews*, **105**, 857–897.

26 (a) Erkkila, A., Majander, I. and Pihko, P.M. (2007) *Chemical Reviews*, **107**, 5416–5470; (b) Melchiorre, P., Marigo, M.,

Carlone, A. and Bartoli, G. (2008) *Angewandte Chemie*, **120**, 6232–6265; (2008) *Angewandte Chemie – International Edition*, **47**, 6138–6171.

27 Paras, N.A. and MacMillan, D.W.C. (2001) *Journal of the American Chemical Society*, **123**, 4370–4371.

28 (a) Bartoli, G. and Melchiorre, P. (2008) *Synlett*, 1759–1772; (b) Chen, Y.-C. (2008) *Synlett*, 1919–1930; Martin, N.J.A. and List, B. (2006) *Journal of the American Chemical Society*, **128**, 13368–13369.

29 (a) Doyle, A.G. and Jacobsen, E.N. (2007) *Chemical Reviews*, **107**, 5713–5734; (b) Akiyama, T. (2007) *Chemical Reviews*, **107**, 5744–5758; (c) Adair, G., Mukherjee, S. and List, B. (2008) *Aldrichimica Acta*, **41**, 31–39.

30 (a) Bandini, M., Eichholzer, A., Tragni, M. and Umani-Ronchi, A. (2008) *Angewandte Chemie*, **120**, 3282–3285; (2008) *Angewandte Chemie – International Edition*, **47**, 3238–3241.

31 (a) Corma, A., GarcMa, H., Moussaif, A., Sabater, M.J., Zniber, R. and Redouane, A. (2002) *Chemical Communications*, 1058–1059; (b) Zhang, Y., Zhao, L., Lee, S.S. and Ying, Y.J. (2006) *Advanced Synthesis and Catalysis*, **348**, 2027–2032; (c) Yu, P., He, J. and Guo, C. (2008) *Chemical Communications*, 2355–2357; (d) Chi, Y., Scroggins, S.T. and Fréchet, J.M.J. (2008) *Journal of the American Chemical Society*, **130**, 6322–6323.

32 For recent reviews on metal-catalyzed transformations of unactivated olefins see: Au(I)/(III): (a) Hashmi, A.S.K. and Hutchings, G.J. (2006) *Angewandte Chemie*, **118**, 8064–8105; (2006) *Angewandte Chemie – International Edition*, **45**, 7896–7936; (b) Hashmi, A.S.K. (2007) *Chemical Reviews*, **107**, 3180–3211; (c) Jiméez-Núñez, E. and Echavarren, A.M. (2007) *Chemical Communications*, 333–346; (d) Shen, H.C. (2008) *Tetrahedron*, **64**, 3885–3903; (e) Skouta, R. Li, C.-J. (2008) *Tetrahedron*, **64**, 4917–4938; (f) Widenhoefer, R.A. (2008) *Chemistry – A European Journal*, **14**, 5382–5391; (g) Arcadi, A. (2008) *Chemical Reviews*, **108**, 3266–3325; (h) Li, Z., Brouwer, C. and He, C. (2008) *Chemical Reviews*, **108**, 3239–3265; Pt(II): (i) Chianese, A.R., Lee, S.J. and Gagné, M.R. (2007) *Angewandte Chemie*, **119**, 4118–4136; (2007) *Angewandte Chemie – International Edition*, **46**, 4042–4059; (j) Liu, C., Bender, C.F., Han, X. and Widenhoefer, R.A. (2007) *Chemical Communications*, 3607–3618; Au and Pt π-acids: Fürstner, A. and Davies, P.W. (2007) *Angewandte Chemie*, **119**, 3478–3519; (2007) *Angewandte Chemie – International Edition*, **46**, 3410–3449.

2
Michael Addition

2.1
Chelating α,β-Unsaturated Compounds

Jesús M. García, Mikel Oiarbide, and Claudio Palomo

Summary

Catalytic and enantioselective Friedel–Crafts alkylation reactions between a variety of bidentate Michael acceptors and electron-rich arenes have been developed. These results demonstrated that, in most cases, the substrate capacity for coordinating the catalyst in a bidentate manner is crucial for achieving high selectivity. In many instances, previously well established catalysts, like the amply used Cu(II)-bisoxazoline complexes, suffice, while in other cases new or less developed catalysts have been employed, such as the Cu(II)-trisoxazoline complexes, Sc(III)-PyBOX complexes or even metal-free N-triflyl phosphoramides. In general the chemical and stereochemical efficiencies of the catalytic systems are high for selected combinations of reactants, particularly those involving indoles, N-alkylated indoles, and N-alkyl pyrrole as the arene component. For some substituted indoles, N−H pyrroles, and activated benzenes results are, for most cases, less satisfactory. With regard to the Michael acceptor component, substrates bearing an aryl group at the β-carbon are usually less efficient than their alkyl counterparts, while catalytic and selective methods amenable for β,β-disubstituted acceptors remain an unmet goal. In this context, the development of new catalysts and the finding of improved catalyst–acceptor combinations continue to be of great interest. In prospect, the steadily increasing number of efficient catalytic asymmetric FC alkylation protocols and the structural relevance of products obtained therefrom, allow one to expect important applications of these technologies in addressing practical synthetic problems.

2.1.1
Chelating Compounds

In the current century, recent developments in the field of catalytic enantioselective Friedel–Crafts (FC) type addition reactions of aromatic and heteroaromatic C−H bonds to activated alkenes have been reported [1]. Asymmetric alkylations of electron-rich arenes such as indoles and pyrroles are of great importance for the synthesis of many natural products and pharmaceuticals [2].

Two approaches involving, respectively, chiral secondary amine-catalyzed and chiral Lewis acid-catalyzed reactions are very promising [3]. The latter process usually requires the substrates to be capable of chelating to the metal center of the Lewis acid catalyst complex and most often involves bidentate compounds. Monodentate systems tend to be less selective because lone pair discrimination between the two possible Lewis acid-carbonyl coordination complex geometries is difficult due to steric reasons (Figure 2.1). Although less common, chiral Brønsted acids can also effect catalytic enantioselective FC alkylations by means of an activation mechanism similar to that of Lewis acids.

A series of bidentate compounds (templates) have been studied in the context of the asymmetric catalytic FC reaction. As shown in Figure 2.2, templates of both types of coordination ability A and B have been used, and include: alkylidene malonates (**T1**), and nitroacrylates (**T2**) both capable of forming type A patterns; α,β-unsaturated thioesters (**T3**), and α,β-unsaturated acyl phosphonates (**T4**) which consist of an acyl–heteroatom linkage; β,γ-unsaturated α-ketoesters (**T5**), α,β-unsaturated 2-acyl imidazoles (**T6**), and the related templates **T7**–**T9**, which exhibit an acyl-carbon sp^2 linkage; and, finally, α′-hydroxy enones (**T10**) and α′-phosphonate enones (**T11**) featuring an acyl–carbon sp^3 linkage.

Regarding the pool of chiral metal complexes that have been studied for the catalytic enantioselective addition of electron-rich aromatic and heteroaromatic compounds to bidentate templates, Cu(II) species are by far the most widely studied. Sc(III) salts, and to a lesser extent Pd(II) complexes, have also been applied successfully. For these types of metal centers a number of chiral ligands have been explored, in particular bis(oxazolines) [4], including pincer pyridine-bis(oxazolines) [5], pseudo-C_3-symmetric tris(oxazolines) [6], and BINAP [7]. In the following a

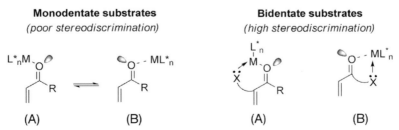

Figure 2.1 Plausible coordination patterns of complexes formed between a chiral Lewis acid (ML_n^*) and either monodentate or bidentate substrates.

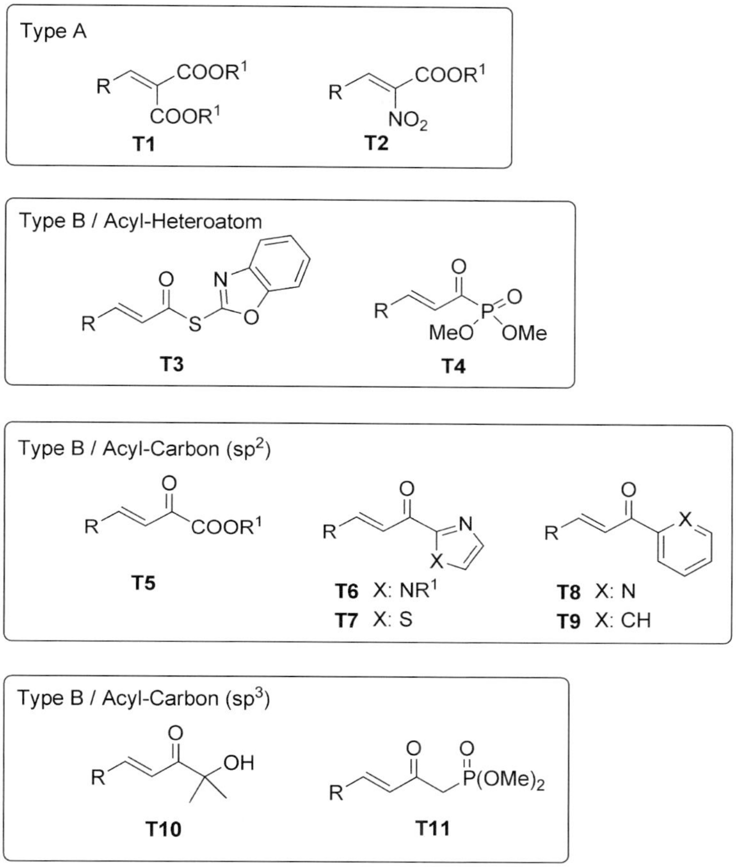

Figure 2.2 Bidentate templates employed for catalytic asymmetric FC alkylation reactions.

succinct description of most remarkable methods is made according to the type of bidentate substrate involved.

2.1.1.1 Alkylidene Malonates

Jørgensen *et al.* documented the first effective catalytic and enantioselective FC alkylation of olefins. The method consists of the conjugate addition of indoles to the aryl-substituted methylidene malonates **1** in the presence of a catalytic amount (10 mol%) of **L1**-Cu(OTf)$_2$ complex as promoter [8]. No reaction took place in MeCN or MeNO$_2$ as solvents, while in Et$_2$O or THF high conversions at 0 °C were achieved with *ee* ranging from 46 to 69% (Table 2.1). Competent alkylidene malonates were restricted to those bearing aryl groups at the Cβ position. The diethyl 2-chlorobenzylidene malonate derivative was less reactive than the other malonates and required 30 °C for the reaction to go to completion (entry 5). Apart from NH-indole, *N*-methylindole, *N*-methylpyrrole, pyrrole, and 2-methylfuran also reacted smoothly giving excellent yields, but the enantioselectivity of the products was comparatively low (12–48% ee).

Table 2.1 Enantioselective FC alkylation reactions of alkylidene malonates with various NH-indoles catalyzed by **L1**-Cu(OTf)$_2$.

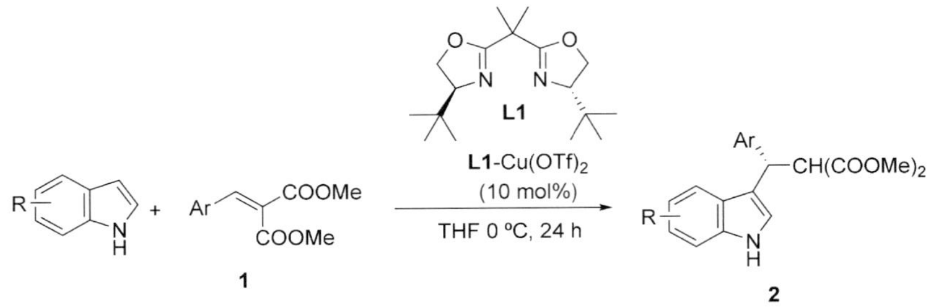

Entry	R	Ar	Yield (%)	ee (%)
1	H	Ph	73	60[a]
2	H	Ph	95	50
3	H	4-NO$_2$Ph	92	56
4	H	4-BrPh	45	50[a]
5	H	2-ClPh	87	69[a,b]
6	5-OMe	Ph	91	60[a,c]
7	5-OMe	4-NO$_2$Ph	99	58
8	4-Cl	4-NO$_2$Ph	62	46[b]

[a] Reaction carried out with the corresponding diethyl arylidene malonate.
[b] Reaction carried out at 30 °C.
[c] Reaction carried out at room temperature.

The sense of asymmetric induction observed in these reactions can be rationalized on the basis of the stereomodel depicted in Figure 2.3, with a copper atom adopting a distorted square planar geometry and the nucleophilic arene approaching the less shielded *Re* face of the olefin β-carbon.

A practical advantage of this FC alkylation reaction is that all products are solid and the optical purity could be greatly enhanced to >90% *ee* by crystallization. The FC adducts **2** can undergo decarboxylation (NaCl, DMSO, 160 °C, 12 h) to give the corresponding mono ester **3** in high yield, showing that the present reaction can formally be considered as a FC alkylation of cinnamate esters (Scheme 2.1) [9].

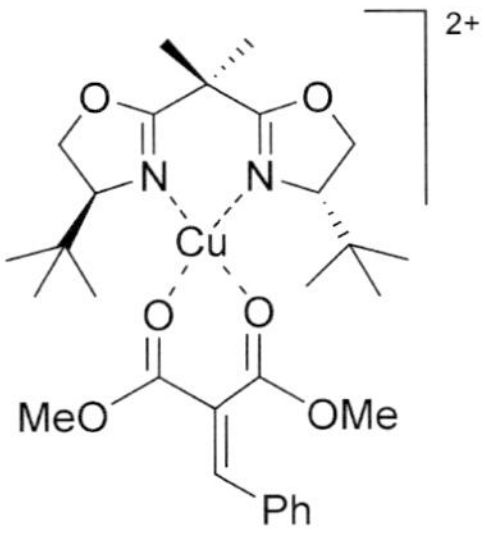

Figure 2.3 Proposed stereomodel for the **1**:**L1**-Cu substrate–catalyst complex [8].

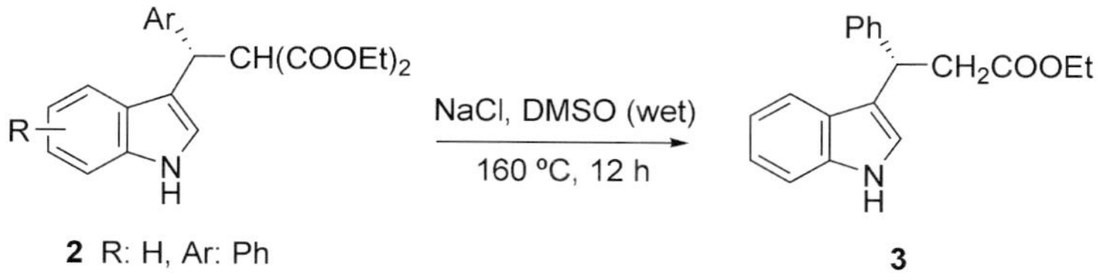

Scheme 2.1

In related work, Yamazaki [10] introduced ethenetricarboxylates and acyl-substituted methylenemalonates **4** as new reaction partners to afford, in the presence of catalytic amounts of copper complex **L1**-Cu(OTf)$_2$ (10 mol%), alkylated products **5** in high yields and with *ee* ranged from 27 to 95% (Table 2.2). The nature of the

Table 2.2 FC reaction of ethenetricarboxylates and acyl methylenemalonates with various indoles catalyzed by **L1**-Cu(OTf)$_2$.

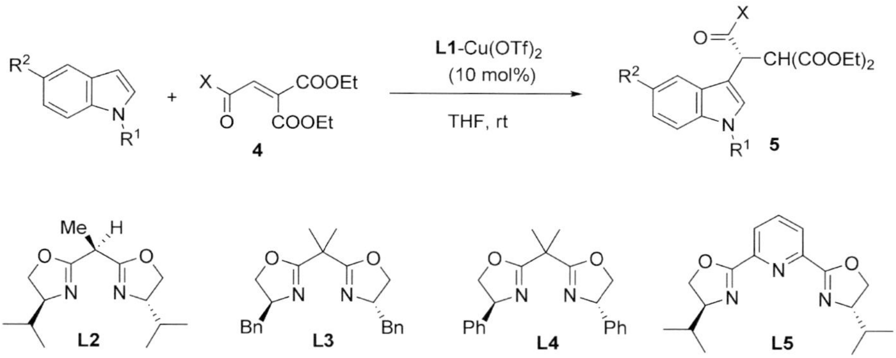

Entry	X	R^1	R^2	Yield (%)	ee (%)
1	OEt	H	H	96	68
2	O^{i}Pr	H	H	71	74
3	O^{t}Bu	H	H	75	73
4	OBn	H	H	81	84
5	Me	H	H	86	38
6	Ph	H	H	92	69
7	O^{t}Bu	Me	H	71	83
8	OBn	Me	H	74	86
9	Me	Me	H	79	27
10	Ph	Me	H	87	95
11	O^{t}Bu	H	OMe	96	86
12	O^{t}Bu	H	Cl	88	74
13	Ph	H	Cl	75	93
14	O^{t}Bu	H	Br	85	80
15	Ph	H	Br	72	83

X group in **4** has an important effect on both yield and selectivity. For instance, *i*Pr, *t*Bu and Bn esters reacted with indole to give the alkylated products in 74–90% yield and 73–84% *ee*. Acetyl and benzoyl derivatives provided the FC adducts in 38 and 69% *ee*, respectively, (entries 5 and 6). *N*-Methylindole and 5-substituted indoles also gave FC alkylated products with variable yields and enantioselectivity (entries 7–15). The use of other bis(oxazoline) ligands such as **L2** and **L3** led to similar *ee* values, while phenyl derivative **L4** was inferior and the pincer ligand **L5** was totally ineffective.

The key importance of the judicious selection of the solvent for these catalytic transformations was clearly shown by Tang and coworkers [11]. Considering the reaction of NH-indoles and alkylidene malonates catalyzed by **L2**-Cu(OTf)$_2$ complex, the enantioselectivity of the reaction varies from high, in alcoholic solvents, to low, in acetone–Et$_2$O mixture, to high but the reverse, in CH$_2$Cl$_2$ or 1,1,2,2-tetrachloroethane (TTCE) (Table 2.3, entries 1–3). The observed reversal of enantioselectivity just by changing between reaction solvents of distinctly different coordinating ability is explained by solvent-induced change of the coordination geometry around the copper center in the transition state.

While the above described results might lead one to conclude that simple bis(oxazoline) ligands can hardly provide selectivities above 90% for the FC reaction of arylidene malonates with indoles, recent work by Reiser and coworkers [12] has

Table 2.3 Enantioselective FC alkylation reactions of alkylidene malonates with various indoles catalyzed by **L2**-Cu(OTf)$_2$.

Entry	R^1	R^2	Solvent	Yield (%)	ee (%)
1	H	Ph	^{i}BuOH	75	93
2	H	Ph	Acetone-Et$_2$O	80	6^a
3	H	Ph	CH$_2$Cl$_2$	90	−78^b
4	H	4-BrPh	^{i}BuOH	68	92
5	H	2-ClPh	^{i}BuOH	56	97
6	H	4-NO$_2$Ph	^{i}BuOH	94	83^b
7	Me	Ph	^{i}BuOH	70	91
8	OMe	Ph	^{i}BuOH	70	94

^{a}Reaction carried out at 15 °C.
^{b}Reactions carried out at 0 °C.

Table 2.4 Enantioselective FC alkylation reactions of benzylidene malonates with indole catalyzed by Cu(OTf)$_2$:azabis(oxazoline) or bis(oxazoline) ligands.

Entry	R^1	R^2	Yield (%)	ee (%)
1	H	Ph	97	99
2	H	Ph	89^a	99
3	H	4-MePh	80	93
4	H	4ClPh	91	98
5	H	2-BrPh	89	85
6	H	4-NO$_2$Ph	94	80
7	5-OMe	Ph	80	90

^{a}Using ligand **L2**.

demonstrated that the ligand-to-copper ratio is crucial for selectivity of the title reaction and that under optimized reaction conditions copper complexes of simple bis(oxazoline) ligands such as **L2**, or aza-bis(oxazolines) **L6**, can afford selectivities of up to 99% ee (Table 2.4). In marked contrast to the usual observations in asymmetric catalysis, an excess of chiral ligand with respect to the metal is detrimental to the enantioselectivity, and a ligand to copper ratio of 1.04 : 1 was found to be optimal. Under optimized conditions, the catalyst also showed good enantioselectivity in the reaction of 5-methoxyindole with diethyl benzylidene malonate in ethanol at room temperature (entry 7).

Tang's group developed a pseudo-C_3-symmetric tris(oxazoline)-Cu complex [13] (**L7-Cu**) for the enantioselective FC type addition reaction. The tris(oxazoline) ligand is cheap, easy to access, air-stable and water-tolerant. Compared with C_2-symmetric bisoxazolines, coordination of three nitrogen atoms of tris(oxazoline) to the metallic center induces a more tunable pseudo-C_3-symmetric chiral space, to shield distant reactive sites. Tridentate ligand is also expected to increase the stability of active intermediates with respect to ordinary bidentate bisoxazolines and the sidearmed oxazoline may be able to modulate the Lewis acidity of the metal center. Thus the metal complex may be tolerant to moisture and it is possible to improve the catalytic activity.

Table 2.5 Enantioselective FC alkylation reactions of alkylidene malonates with various indoles catalyzed by **L7**-Cu(ClO$_4$)$_2$·6H$_2$O.

Entry	R^1	R^2	Yield (%)	ee (%)
1	H	Ph	84	89
2	H	4-NO$_2$Ph	99	91
3	H	3-NO$_2$Ph	99	91
4	H	4-BrPh	95	90
5	H	2-ClPh	99	92
6	H	CH$_3$	84	60^a
7	5-OMe	Ph	73	91
8	5-Me	Ph	92	93
9	4-OMe	Ph	97	91

^{a}Reaction was carried out at −35 °C.

Indeed, the FC reaction of NH-indoles and arylidene malonates using **L7**-Cu(ClO$_4$)$_2$·6H$_2$O complex as the catalyst in acetone–Et$_2$O at −20 °C afforded products with enantioselectivities higher than 90% (Table 2.5). Under these conditions, the reactions become too slow, but the addition of 2 equiv. of hexafluoroisopropanol (HFIP) can greatly improve reactivity without loss of *ee*. Of practical importance, the catalyst system maintained the same catalytic activity even when the reaction was carried out in air.

In subsequent investigations, a dramatic solvent effect was once again found in both reactivity and enantioselectivity [14]. Thus, while the enantioselectivity was improved in alcoholic solvents such as *i*BuOH, weak coordinating solvents such as CCl$_4$, toluene, CH$_2$Cl$_2$ or TTCE gave the product with the opposite configuration. 1,1,2,2-Tetrachloroethane (TTCE) proved to be the best solvent for the production of the opposite enantiomer (*ee* up to 89%). The influence of varying the nature of the copper salt, the ester group on malonates, the ligand-to-metal ratio, the use of additives, and so on, was studied in some detail. In general, the method showed broad scope with respect to the arylidene malonate but was more limited when alkylidene malonates were employed. The method also tolerates several substituted NH-indoles.

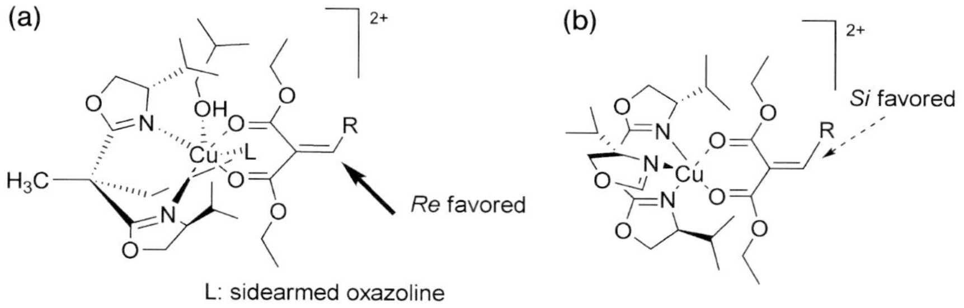

Figure 2.4 Proposed transition state in iBuOH (a) and in TTCE (b).

Finally, an array of homo- and hetero-tris(oxazoline) ligands was easily prepared on a gram scale and their efficiency was tested in the FC reaction of indoles [14b].

Figure 2.4 shows the proposed transition-state models that explain the observed stereochemical reversal of the reaction outcome. In both models the chiral ligand is triple-coordinated to copper. However, while in (a) (iBuOH as the solvent) the alcoholic solvent occupies an additional coordination site with the ligand sidearm back-oriented, in (b) (non-coordinating solvents) the ligand sidearm is on the front position with copper exhibiting a distorted square-pyramidal geometry. Thus in (a) nucleophilic attack of the indole derivative is favored from the Re-face to afford S-enantiomers, while in (b) the Si-face attack is favored.

Adducts from the FC reactions of alkylidene malonates can be subjected to various ulterior transformations owing to the versatility of the malonic framework. For instance, FC alkylation adducts **7** can be easily converted to the corresponding β-substituted tryptophans **9**, of interest in medicinal chemistry, through a route that begins with selective hydrolysis of one ester group to afford **8** (Scheme 2.2) [15].

2.1.1.2 β,γ-Unsaturated α-Ketoesters

In 2001 Jørgensen and coworkers [16] established the potential of β,γ-unsaturated α-ketoesters (**T5**) as alkylating reagents of indoles, furans, and other electron-rich aromatic compounds in Cu-catalyzed asymmetric FC reactions. The ketoester moiety not only exhibits a strong activating effect on the olefinic substrates but also offers a platform for easy elaboration of adducts into the corresponding amino acids or α-hydroxy acids. The FC alkylation reactions are typically run at low temperature ($-78\,°C$ to $0\,°C$) using 10 mol% of **L1**-Cu(OTf)$_2$ complex as a catalyst in CH$_2$Cl$_2$ or (preferably) Et$_2$O as the solvent. However, catalyst loadings as low as 2 mol% proved

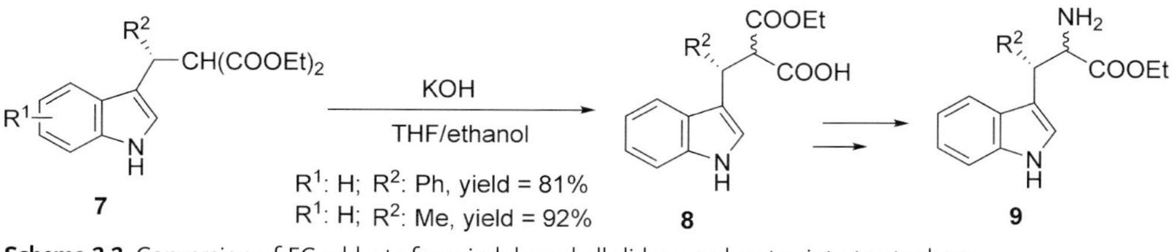

Scheme 2.2 Conversion of FC adducts from indole and alkylidene malonates into tryptophans.

Table 2.6 Catalytic enantioselective FC reactions of indoles with 4-substituted 2-oxo-3-butenoate esters.

Entry	R^1	R^2	R^3	T (°C)	t (h)	Yield (%)	ee (%)
1	H	Ph	Me	−78 → −30	64	77	99.5[a]
2	5-OMe	Ph	Me	−78	1	95	>99.5[b]
3	6-Cl	Ph	Me	−20 → 0	16	69	97[c]
4	6-CO$_2$Me	Ph	Me	−20 → 0	16	82	94[c]
5	H	Me	Et	−78	16	96	95[b]
6	5-OMe	Me	Et	−78	1	95	>99.5[b]
7	6-Cl	CH$_2$OBn	Et	−20 → 0	16	70	80[c]
8	5-OMe	CH$_2$OBn	Et	−78	1	98	95[b]

[a] 2 mol % catalyst was used.
[b] 5 mol % catalyst was used.
[c] 10 mol % catalyst was used.

sufficient in some instances to achieve good yields and high enantioselectivity, although long reaction times were needed (Table 2.6, entry 1). Interestingly, zinc complex **L1-Zn(OTf)$_2$**, which is more air-stable than the corresponding copper complex, was also effective for this reaction, giving the expected product with up to 87% ee. The scope of the enantioselective FC alkylation reaction promoted by **L1-Cu(OTf)$_2$** is shown in Table 2.6. As can be seen, indole itself and various substituted NH-indoles react efficiently with different β,γ-unsaturated α-ketoesters **10**. In most cases, the FC alkylation products were obtained in high yields and with excellent enantioselectivities, even when γ-alkyl β,γ-unsaturated α-ketoesters were used (entries 5–8). Only indoles substituted at the 2-position provided somewhat more limited yields and enantioselectivities (60–70% yield and about 50% ee). In addition, chromatographic purification of the reaction product was not necessary in many instances as most of the products could be isolated by filtration of the chiral catalyst followed by evaporation of the solvent.

The same authors showed that other heteroaromatic compounds such as 2-methylfuran and aromatic compounds such as 1,3-dimethoxybenzene can also be engaged in the FC reaction with β,γ-unsaturated α-ketoesters. For instance, reactions of 1,3-dimethoxybenzene carried out at 0 °C afforded the corresponding products with

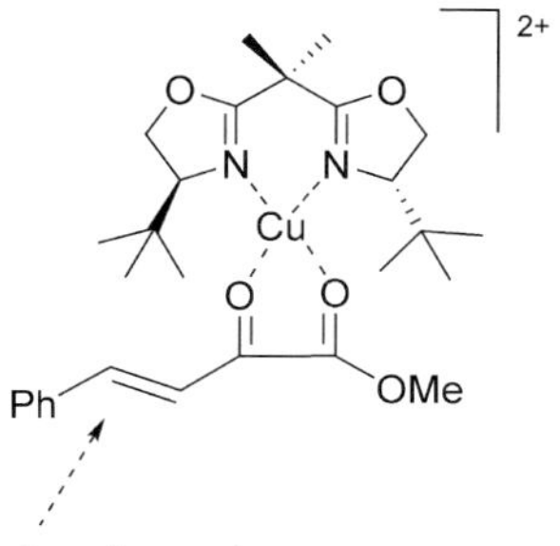

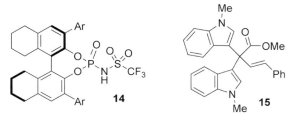

Re-face favored

Figure 2.5 Proposed stereomodel for the **10:L1**-Cu substrate–catalyst complex [16].

yields of about 90% and enantioselectivity in the range of 79–88%, while in the case of 2-methylfuran and using complex **L1**-Cu(SbF$_6$)$_2$ as the catalyst, the corresponding adducts were obtained in 65–68% yield and 60–89% *ee*.

Based on the absolute configuration of the chiral FC product formed, it was postulated that the unsaturated α-ketoesters coordinate in a bidentate fashion to the catalyst metal center which adopts a square-planar geometry (Figure 2.5) [17]. According to this model the chelated intermediate has the *Re*-face of the reacting alkene carbon atom available for approach of the aromatic compound.

Complementing Lewis acids, chiral (protic) Brønsted acids have also been shown to be capable of promoting FC alkylation reactions enantioselectively. In particular, Rueping and coworkers have recently reported a highly enantioselective Brønsted acid catalyzed addition of *N*–Me indoles to β,γ-unsaturated α-ketoesters using a *N*-triflylphosphoramide catalyst **12** derived from binaphthol [18]. The Friedel–Crafts reactions were carried out between various *N*-methylindoles and γ-aryl β,γ-unsaturated α-ketoesters in the presence of 5 mol% of the catalyst in dichloromethane at −75 °C giving rise to the corresponding substituted α-ketoesters in yields from moderate to good (43–88%) and with high enantioselectivities (80–92% *ee*) (Table 2.7).

Fine tuning of electronic and steric properties of the *N*-triflyl phosphoramide catalyst through alteration of the substituents on the BINOL skeleton resulted in determination of the catalytic profile. Indeed, the analog catalyst **14**, under otherwise similar reaction conditions, led to the formation of bisindole **15** as the major, or almost exclusive, product (Figure 2.6).

Figure 2.6 Modified phosphoramide catalyst and major bisindole product obtained [18].

Table 2.7 Enantioselective Brønsted acid catalyzed indole addition to β,γ-unsaturated α-ketoesters.

Entry	R^1	R^2	t (h)	Yield (%)	ee (%)
1	H	Ph	15	62	88
2	5-Br	Ph	24	43	86
3	7-Me	Ph	22	78	84
4	H	4-ClPh	22	65	88
5	H	4-BrPh	24	60	90
6	H	4-MePh	20	69	92
7	5-Br	4-MePh	22	55	80
8	H	4-MeOPh	18	88	86
9	H	2-naphthyl	18	70	90

2.1.1.3 α′-Hydroxy Enones

α′-hydroxy enones **16** have been developed by the group of Palomo as efficient Michael acceptors in a variety of C–C and C–heteroatom bond forming reactions mediated by both Lewis and Brønsted acid-assisted activation and likely taking place through 1,4-metal(proton) chelated reactive intermediates [19]. In the context of the catalytic asymmetric FC reactions, α′-hydroxy enones react smoothly with indoles in the presence of 10 mol% of **L1**-Cu(OTf)$_2$ as catalyst to give the addition products in high yield and very high *ee* [20]. The reactions were carried out in dichloromethane at 0 °C, room temperature and even under refluxing conditions, providing the corresponding FC adducts in good to excellent yields (65–95%) and enantiomeric excesses (83–98%) (Table 2.8). Catalyst loading as low as 2 mol% suffices for achieving satisfactory levels of reactivity and selectivity (entry 2).

Enones bearing branched alkyl chains or aromatic substituents at Cβ showed attenuated reactivity and lower selectivity using catalyst **L1**-Cu(OTf)$_2$, but using catalyst **L8**-Cu(OTf)$_2$, high enantiomeric excesses could be attained (entries 4 and 5). The reactions with differently substituted NH–indoles and even N–Me indole worked nicely (entries 7–9).

On the other hand, α′-hydroxy enones also react with pyrrole and N-methylpyrrole (2 equiv.) to provide, with the same catalyst, the corresponding FC adducts in excellent yields and enantiomeric excess (92–97%) (Table 2.9). Here, aryl-substituted

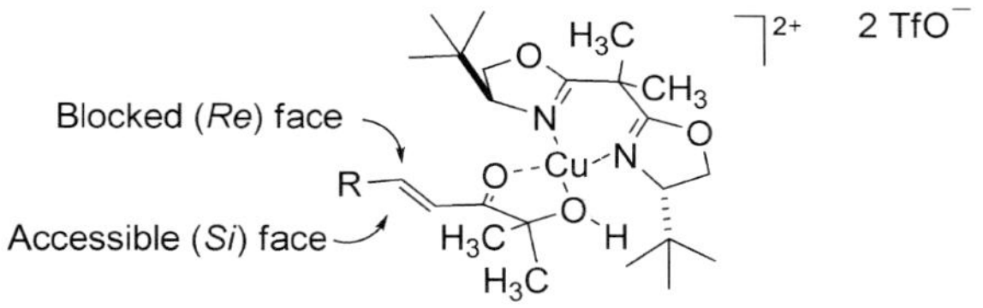

Figure 2.7 Proposed stereomodel for the substrate–catalyst complex [20].

enones represented an exception, providing the corresponding alkylated product less selectively (68% ee, entry 9).

The sense of asymmetric induction observed is consistent with a model which assumes a distorted square planar geometry around copper.

The potential of this catalytic approach is demonstrated by the versatile elaboration of adducts through oxidative cleavage of the ketol moiety. For example, sequential reduction and oxidative cleavage of the alkylated product **17** afforded the corresponding aldehyde **19** in good yield and high enantiomeric purity (Scheme 2.3a). Alternatively, the FC adduct could be transformed directly into carboxylic acid **20** by oxidation with periodic acid in diethyl ether (Scheme 2.3b). A sequential alkyllithium addition to the

Table 2.8 FC alkylation of indoles with various α'-hydroxy enones catalyzed by complex **L1**-Cu(OTf)$_2$.

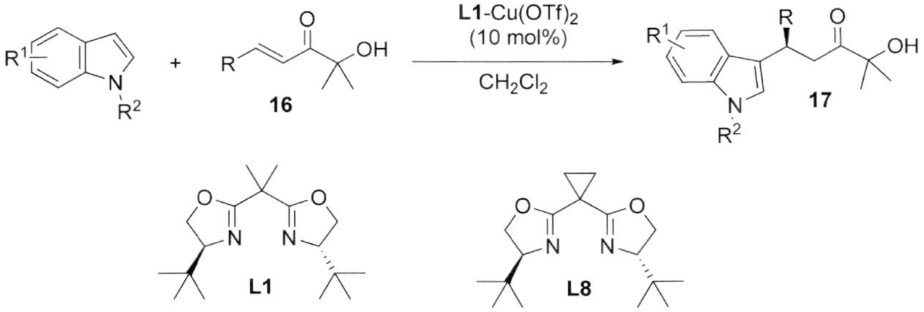

Entry	R^1	R^2	R	T (°C)	t (h)	Yield (%)	ee (%)
1	H	H	PhCH$_2$CH$_2$	reflux	0.5	85	98
2	H	H	CH$_3$(CH$_2$)$_5$	0	12	85	96[a]
3	H	H	(CH$_3$)$_2$CH	reflux	4	81	95
4	H	H	c-C$_6$H$_{11}$	25	24	80	96[b]
5	H	H	4-ClPh	0	48	95	83[c]
6	H	H	Me	0	3	65	98
7	2-Me	H	PhCH$_2$CH$_2$	25	2	89	93
8	5-OMe	H	PhCH$_2$CH$_2$	25	2	96	97
9	H	Me	PhCH$_2$CH$_2$	25	2	86	98

[a] 2 mol% catalyst was used.
[b] Using catalyst **L8**-Cu(OTf)$_2$.
[c] Using 30 mol% catalyst **L8**-Cu(OTf)$_2$.

Table 2.9 FC alkylation of N-methylpyrrole and pyrrole with various α′-hydroxy enones catalyzed by complex **L1**-Cu(OTf)$_2$.

Entry	R	R′	T (°C)	t (h)	Yield (%)	ee (%)
1	PhCH$_2$CH$_2$	Me	25	2	86	92
2	PhCH$_2$CH$_2$	H	−20	2	83	90
3	CH$_3$(CH$_2$)$_5$	Me	−20	6	82	96
4	CH$_3$(CH$_2$)$_5$	H	−20	0.5	87	91[a]
5	(CH$_3$)$_2$CH	Me	0	20	86	95
6	c-C$_6$H$_{11}$	Me	25	4	84	97[b]
7	CH$_3$CH$_2$	Me	−20	18	88	94
8	(CH$_3$)$_2$CHCH$_2$	Me	−20	12	86	94[b]
9	Ph	Me	25	24	95	68[b]

[a] Using catalyst **L8**-Cu(OTf)$_2$.
[b] Using 6 molar equiv of N-methylpyrrole.

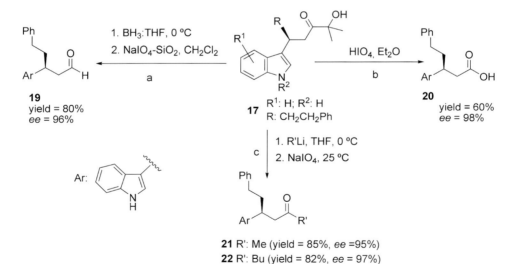

Scheme 2.3 Elaboration of adducts **17** into enantioenriched indole-substituted aldehydes, carboxylic acids and ketones.

carbonyl group of the FC adduct **17**, followed by treatment with NaIO$_4$, constitutes a practical entry to ketone derivatives **21/22** in high yields and excellent enantioselectivities (Scheme 2.3c). Moreover, in these transformations, acetone is the only byproduct formed, an additional aspect of the approach that is of practical interest.

Table 2.10 Copper-catalyzed FC alkylations of indoles with α′-phosphonate enones **23**.

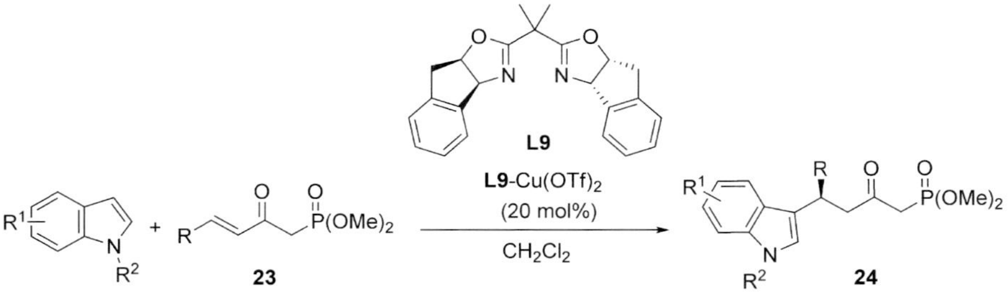

Entry	R¹	R²	R	T (°C)	t (h)	Yield (%)	ee (%)
1	H	H	Me	−78	120	98	95
2	H	H	Et	0	24	98	90
3	H	H	(CH$_3$)$_2$CH	0	120	62	86
4	H	Me	Me	0	60	98	86
5	H	allyl	Me	0	144	39	82
6	H	H	PhCH$_2$CH$_2$	−78	120	99	96
7	H	Me	PhCH$_2$CH$_2$	−40	504	95	94
8	5-OMe	H	Me	−40	48	90	95
9	5-Cl	H	Me	0	48	86	88
10	4-OMe	H	Me	0	48	95	90
11	5-OMe	H	PhCH$_2$CH$_2$	−40	18	95	97(98)[a]

[a]The reaction was carried out in the presence of MS 4 Å.

2.1.1.4 α′-Phosphonate Enones

Kim and coworkers studied the asymmetric FC alkylations of indoles with α′-phosphonate enones **23** using 20 mol% of complex **L9**-Cu(OTf)$_2$ as the chiral catalyst [21]. This reaction exhibited good chemical yields and high enantioselectivities (up to 98% ee) when alkyl substituted enones and either NH- or N-Me indole were employed in toluene or dichloromethane as solvents (Table 2.10). Reactions with N-allyl-substituted indole (entry 5) and branched chain substituted enone (entry 3) were slow and gave the products in low yield.

The catalytic system also worked with N-methylpyrrole affording the corresponding adducts **25** in good yields and selectivities (Scheme 2.4). A chelated transition

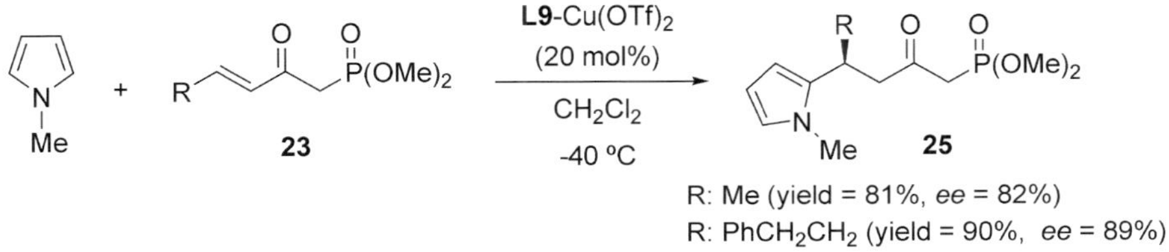

Scheme 2.4 FC reaction of N-methylpyrrole with α′-phosphonate enones catalyzed by **L9**-Cu(OTf)$_2$.

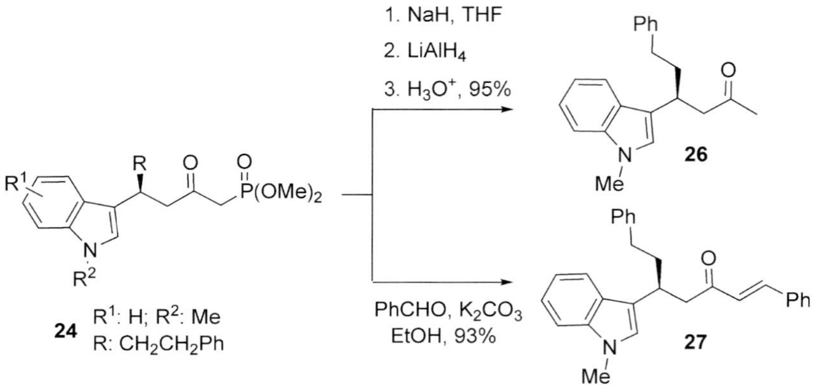

Scheme 2.5 Conversion of β-keto phosphonate adducts **24** into ketone and enone products.

state is proposed with both carbonyl and phosphonyl oxygen atoms of the substrate coordinated to copper.

Finally, the FC adducts bearing the phosphonyl moiety can be elaborated following alternative routes. Thus, treatment with sodium hydride (1.1 equiv.) in THF followed by the addition of lithium aluminium hydride (3 equiv.) and hydrolysis afforded the corresponding methyl ketones **26**, while a Horner–Emmons reaction with benzaldehyde in ethanol at room temperature for 9 h led to α,β-unsaturated ketones **27** (Scheme 2.5). This latter transformation into different ketones represents a formal FC alkylation of simple aliphatic enones, a reaction that bears considerable difficulty in achieving high enantiomeric excesses [22].

2.1.1.5 α,β-Unsaturated Acyl Compounds

2.1.1.5.1 α,β-Unsaturated Acyl Phosphonates

α,β-Unsaturated acyl phosphonates **28** are activated forms of unsaturated esters and their use in catalytic asymmetric FC reactions has been implemented by Evans' group using bis(oxazolinyl)pyridine (pybox)-scandium(III) triflate complexes **L10**-Sc(OTf)$_3$ [23]. The FC reactions of N-substituted indoles with α,β-unsaturated acyl phosphonates in the presence of 10 mol% **L10**-Sc(OTf)$_3$ complex (CH$_2$Cl$_2$ −78 °C) afforded the corresponding adducts in good yields and in excellent enantiomeric excesses (up to >99%) (Table 2.11). Although the parent indole reacted with diminished selectivity (entry 6), N-methyl, N-allyl, or N-benzylindoles behave remarkably (90 to >99% enantioselectivity). Indole derivatives with either electron-withdrawing or electron-donating substituents at C-5 or C-4 were also competent substrates affording the alkylated products with excellent enantioselectivity (entries 9–12) even though lower conversion was obtained in the latter case (entry 12). An increase in the steric requirements at the 2-position of the indole nucleus leads to lower levels of stereoinduction (entries 13 and 14). β-Aryl acyl phosphonates provided lower selectivity than β-alkyl substrates requiring longer reaction times and higher catalyst loadings (entry 5).

Table 2.11 Scandium-catalyzed alkylations of indoles with representative α,β-unsaturated acyl phosphonates.[a]

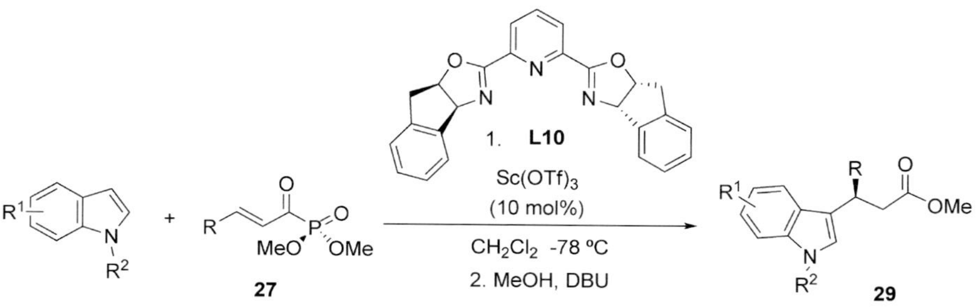

Entry	R¹	R²	R	t (h)	Yield (%)	ee (%)
1	H	Me	Me	4	75	97
2	H	Me	Et	17	65	97
3	H	Me	$(CH_3)_2CH$	20	82	99
4	H	Me	$CH_2OTBDPS$	17	57	94
5	H	Me	Ph	48	85	80[b]
6	H	H	Me	3	83	83
7	H	allyl	Me	5	76	98
8	H	Bn	Me	20	85	99
9	5-Cl	Bn	Me	19	66	>99
10	5-CO_2Me	Bn	Me	16	68	97
11	5-OMe	Bn	Me	19	67	97
12	4-Cl	Bn	Me	48	51	90
13	2-Me	Me	Me	2	94	86
14	2-Ph	Me	Me	20	62	65

[a]Adducts were isolated in the form of their methyl ester derivatives.
[b]Using 20 mol% of catalyst.

In addition to indole derivatives, the method is well suited for other electron-rich arenes. For example, with 3-dimethylaminoanisole the corresponding adduct **30** was obtained in 78% yield and 87% ee (Scheme 2.6).

As shown in Table 2.11 and Scheme 2.6, the resulting adducts can be smoothly transformed to the corresponding methyl esters by *in situ* treatment with DBU/MeOH. Other carboxylic acid derivatives are also possible. For example, the proce-

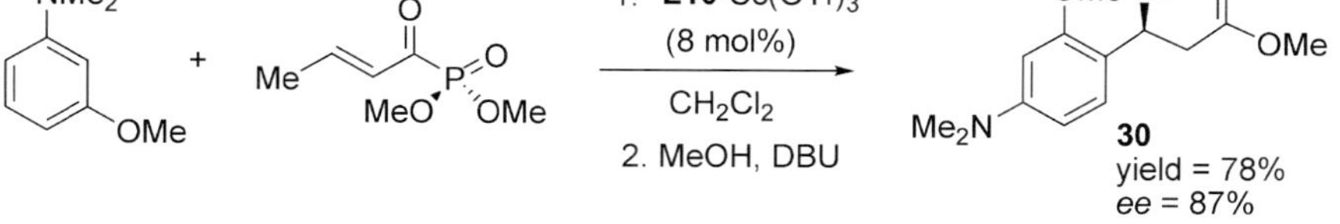

Scheme 2.6

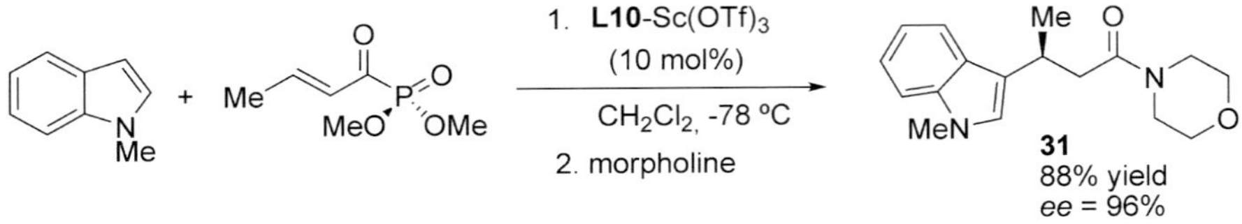

Scheme 2.7

dure may be altered to afford amide products like **31** directly by quenching the mixture with morpholine at room temperature (Scheme 2.7).

2.1.1.5.2 α,β-Unsaturated 2-Acyl Imidazoles Evans and coworkers found that the FC alkylation of electron-rich arenes catalyzed by **L10**-Sc(OTf)$_3$ complex can be extended to α,β-unsaturated 2-acyl imidazoles **32** [23b, 24]. The method represents one of the most general and efficient protocols developed so far. As shown in Table 2.12, a representative selection of indole derivatives react with α,β-unsaturated

Table 2.12 Scandium-catalyzed alkylations of α,β-unsaturated 2-acyl imidazoles **32** with indoles.

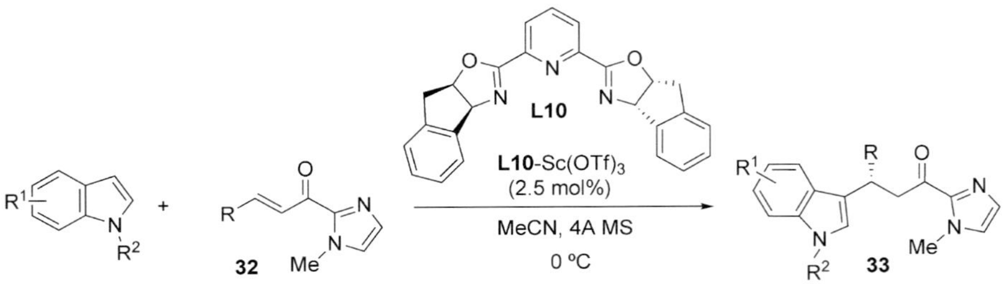

Entry	R^1	R^2	R	t (h)	Yield (%)	ee (%)
1	H	Me	Me	3	93	93 (98)[a]
2	H	Me	Et	12	97	92
3	H	Me	(CH$_3$)$_2$CH	8	78	94
4	H	Me	n-Bu	12	95	93
5	H	Me	CO$_2$Et	12	94	96
6	H	Me	Ph	8	83	91
7	H	H	Me	20	80	65
8	H	allyl	Me	24	80	88
9	H	Bn	Me	8	90	98
10	5-Cl	Bn	Me	20	70	95[b]
11	5-Me	Bn	Me	20	99	97
12	5-OMe	Bn	Me	20	99	97
13	6-OMe	Bn	Me	20	99	95[b]
14	2-Me	Me	Me	2	88	91
15	2-Ph	Me	Me	90	43	66[b]

[a] Using 1 mol% of catalyst at −40 °C.
[b] Using 5 mol% of catalyst.

2-acyl *N*-methylimidazoles in acetonitrile at 0 °C in the presence of 2.5 mol% of the catalyst and 4 Å molecular sieves.

In general, the desired alkylation products from β-alkyl 2-acyl imidazoles were formed in good yields and enantioselectivities, the β-phenyl case also being tolerated (entry 6). Although the study focused on the commercially available *N*-methylimidazole derivatives on the basis of economic considerations, improved enantioselectivities in marginal cases may be achieved by increasing the steric requirements of this moiety (i.e., *N*-isopropylimidazole derivatives).

One of the features of this catalytic system is the broad tolerance toward arene donors. For instance, apart from indoles, 2-methoxyfuran, and pyrrole are also competent nucleophiles for the illustrated reaction (Scheme 2.8). In the case of pyrrole derivatives as nucleophilic reaction partners, it was found that more sterically demanding *N*-substituents on the imidazole moiety produced **35** with increased stereocontrol. Moreover, highly enantiomerically enriched α-substituted tetrahydrocarbazole **38** also proved to be a reliable target compound via intramolecular FC alkylation of **37**.

As shown in Table 2.13, the FC reaction carried out with pyrrole and α,β-unsaturated 2-acyl *N-iso*-propylimidazole **39** in acetonitrile at 0 °C in the presence

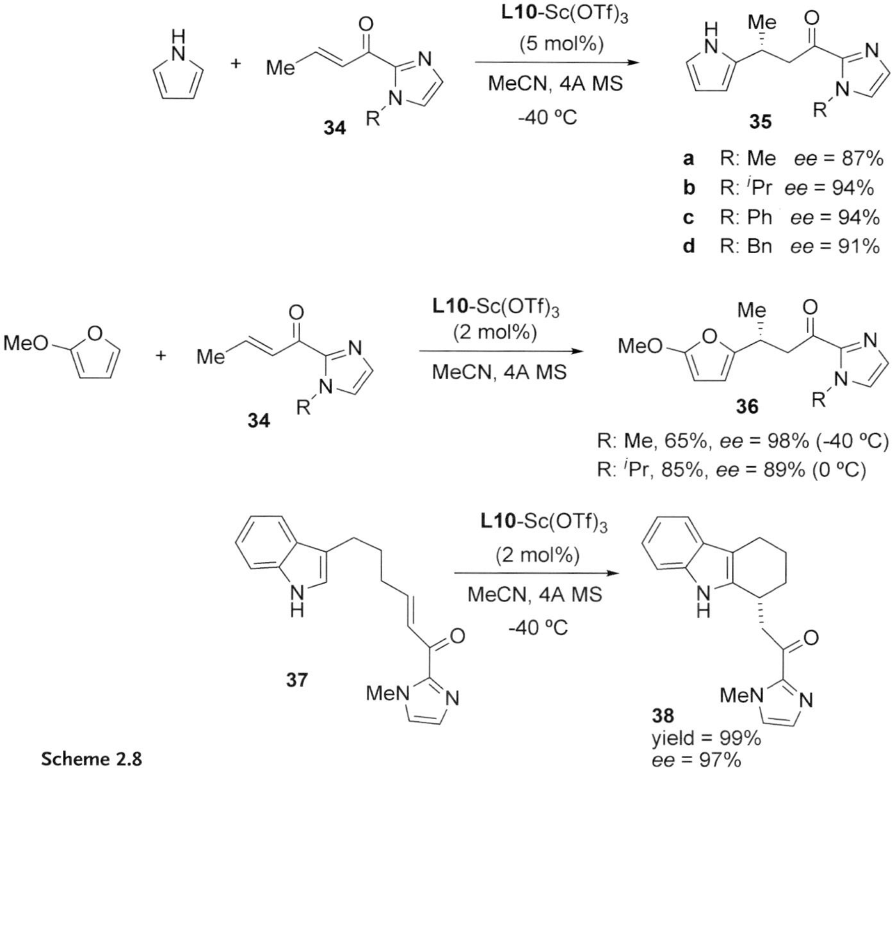

Scheme 2.8

Table 2.13 Scandium-catalyzed alkylations of α,β-unsaturated 2-acyl imidazoles **39** with pyrrole.

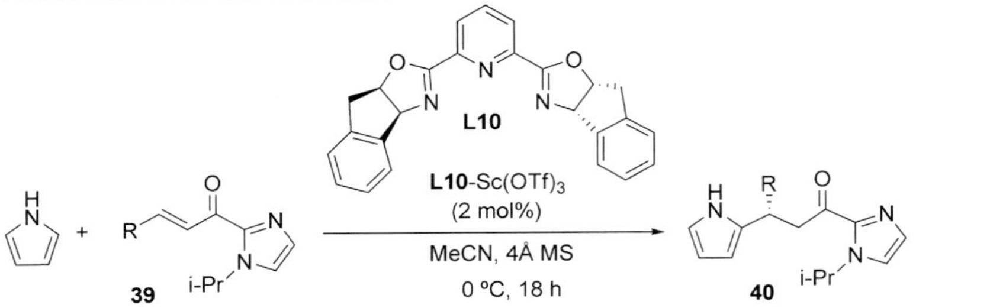

Entry	R	Yield (%)	ee (%)
1	Et	91	86
2	iPr	99	91
3	CO_2Et	99	84
4	Ph	99	96
5	4-OMe	98	92
6	4-MeO_2CPh	99	96
7	2-furyl	95	91

of 4 Å molecular sieves and 2 mol% of **L10**-Sc(OTf)$_3$ provided the corresponding FC adducts in excellent yields (90–99%) and enantiomeric excesses (84–96%). β-Alkyl and β-aryl substitutions were well tolerated in the reaction. It is worth noting that N-substituted and 3-substituted pyrroles provided the corresponding alkylated products with lower enantiomeric excesses (11–78% ee), while 2-substituted pyrroles were competent nucleophiles in this FC reaction (93% ee).

Of particular significance, one may also access the 2-position of the indole nucleus if 4,7-dihydroindole **41** is employed as the nucleophile component and the addition step is immediately followed by an aromatization process (Scheme 2.9) [25].

The 2-acyl imidazole residue may be transformed into a range of carboxylic acid derivatives. The imidazole group can be methylated and the resulting N-methyl imidazolium derivative treated under a variety of nucleophilic conditions to provide esters, amides, and carboxylic acids in good yields in one-pot operations (Table 2.14).

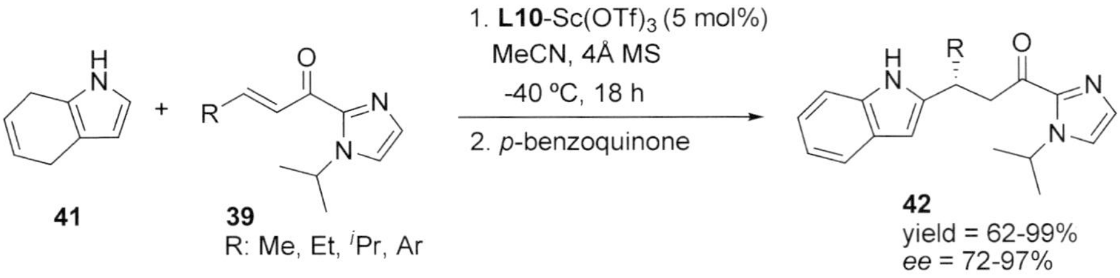

Scheme 2.9 4,7-Dihydroindole alkylations of α,β-unsaturated 2-acyl imidazoles and subsequent one-pot aromatization.

Table 2.14 Conversion of alkylated imidazole to carboxylic acid derivatives **43**.

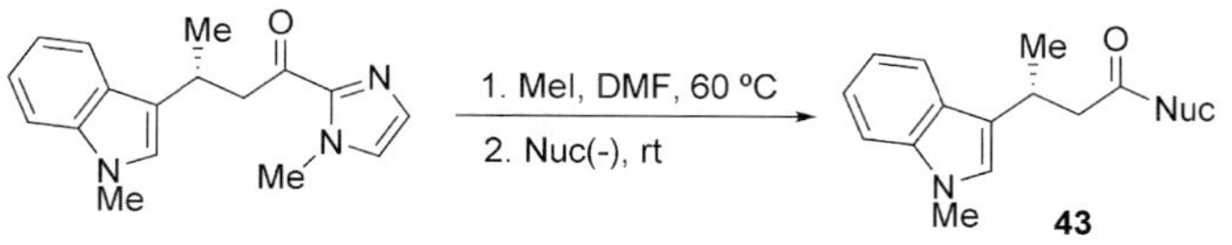

Entry	Nuc(-) conditions	Nuc	Time	Yield (%)
1	MeOH/DBU in CH_2Cl_2	-OMe	30 min	93
2	EtOH/DBU in CH_2Cl_2	-OEt	30 min	86
3	iPrOH/DBU in CH_2Cl_2	$-OCH(CH_3)_2$	30 min	95
4	H_2O/DBU in DMF	-OH	30 min	87
5	$iPrNH_2$ in DMF	$-NHCH(CH_3)_2$	20 min	77
6	morpholine in CH_2Cl_2	morpholine	1 h	88
7	aniline in DMF	-NHPh	12 h	84

On the other hand, the alkylated product can be reduced to the secondary alcohol with sodium borohydride or treated with Grignard reagent to give the tertiary alcohol. The resulting imidazole groups can be methylated and subsequently eliminated under basic conditions to liberate aldehyde **44** or ketone **45** in good yields (Scheme 2.10).

Finally, the alkylated products could be transformed in interesting substrates by means of cleavage of the 2-acyl imidazole. On the one hand, Boc protection [26] of the pyrrole nitrogen, methylation and treatment of the resulting product with MeOH/DBU provided the corresponding methyl ester **46** in 92% yield (Scheme 2.11). On the other hand, if the imidazole cleavage in the pyrrole series was performed in the absence of an external nucleophile, the pyrrole nitrogen was internally acylated to give a 2,3-dihydro-1H-pyrrolizine **47**, precursor of (+)-heliotridane [27].

Mechanistically, the divergent behavior of α,β-unsaturated 2-acyl imidazoles **32/39** and α,β-unsaturated acyl phosphonate **28** substrates in these Sc(III)-catalyzed FC alkylations is quite interesting. For example, the same catalyst system produces the opposite enantiomers with acyl imidazole and with corresponding acyl phosphonates. On the other hand, upon examination of the catalyst loading, an inverse relationship between the amount of catalyst employed and the enantioselectivity of

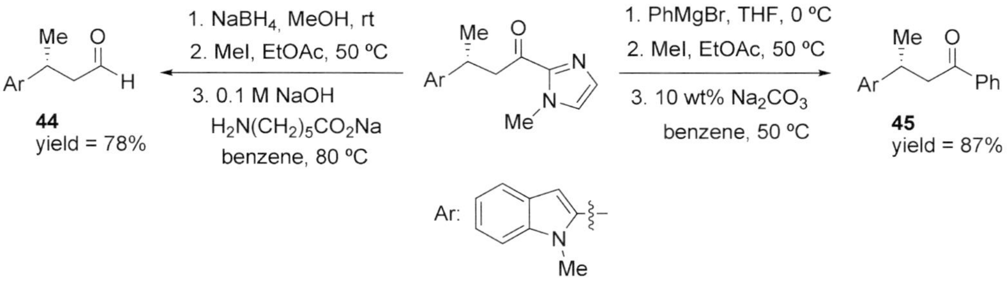

Scheme 2.10 Elaboration of adducts into indole-substituted aldehydes and ketones.

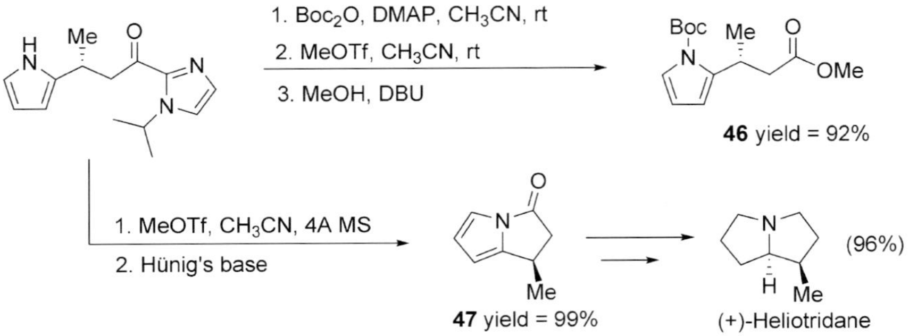

Scheme 2.11 Cleavage of 2-acyl imidazoles and preparation of 2,3-dihydro-1H-pyrrolizine, precursor of (+)-heliotridane.

the process was observed when α,β-unsaturated 2-acyl imidazole was used as the electrophile (catalyst loading/reaction ee: 1 mol%/98%; 10 mol%/90%; 20 mol%/78%; 50 mol%/11%). In contrast, when the reaction is performed with α,β-unsaturated acyl phosphonate substrates, the representation of the reaction's ee versus the catalyst loading exhibits a flat profile. To explain these experimental observations, and based on X-ray crystal structure data of various Sc(III)-pybox complexes, the FC reaction of α,β-unsaturated 2-acyl imidazole was assumed to proceed through a seven-coordinate 1:1:1 product/substrate/catalyst complex that is favored at lower catalyst loading and is more enantioselective than the corresponding 1:1 substrate/catalyst complex that would be favored at higher catalyst loadings.

Finally, the authors evaluated the above catalytic **L10**-Sc(III) system in indole FC reactions involving several other bidentate substrates (Scheme 2.12). While variable

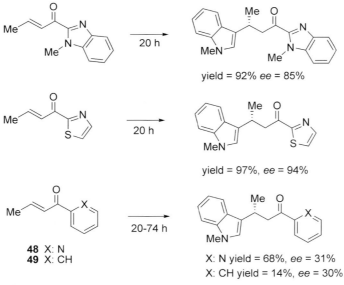

Scheme 2.12

reaction efficiency and selectivity were observed, depending on the substrate employed, the low reactivity and selectivity of monodentate analog **49** demonstrate once again the significance of bidentation in catalytic asymmetric synthesis.

2.1.1.6 α,β-Unsaturated Thioesters

Bandini, Umani-Ronchi, and coworkers described enantioselective FC alkylation of indoles with α,β-unsaturated S-(1,3-benzoxazol-2-yl) thioesters **50** catalyzed by cationic [(Tol-BINAP)Pd(II)] complexes (**L11**-Pd(II)) [28]. The reactions are carried out in acetonitrile at room temperature using 20 mol% of the catalyst which was prepared *in situ* by dissolving commercially available $PdCl_2(MeCN)_2$ (1 equiv.) and (S)-Tol-BINAP (1 equiv.) in toluene, followed by an exchange reaction with $AgSbF_6$ (2 equiv.) in acetonitrile. The alkylated indoles were obtained in moderate to good yields and with enantiomeric excesses up to 86% (Table 2.15).

The absolute configuration of products can be effectively rationalized by considering a two-side binding interaction (six-membered coordination) between an α,β-unsaturated system and the cationic Tol-BINAP-Pd(II) complex (Figure 2.8). In particular, by using (S)-Tol-BINAP as the chiral auxiliary, the shielding effect of the aryl phosphorus substituents calls for the indole to approach the *Re*-face of the electrophile.

Table 2.15 Asymmetric FC alkylation of indoles with α,β-unsaturated thioesters catalyzed by cationic chiral palladium complex **L11**-Pd(II).

Entry	R^1	R^2	t (h)	Yield (%)	ee (%)
1	H	H	18	80 (53)[a]	78 (85)[a]
2	2-Me	H	2	80 (75)[a]	73 (80)[a]
3	5-OMe	H	18	20[b]	77
4	2-Ph	H	24	50	86
5	2-Me	Me	18	70	70
6	5-BnO	H	72	35[c]	86

[a]Reaction carried out at 0 °C.
[b]Presence of by-products.
[c]Unreacted starting material (51%) was recovered after purification.

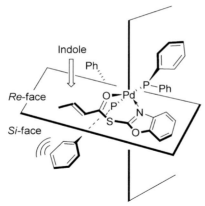

Figure 2.8 Hypothetic trajectory of the incoming indole during the enantiodiscriminating step of the FC alkylation.

The alkylated products could be converted under mild conditions into esters **52/53** and amides **54/55** in high yields and without racemization and the starting auxiliary could be recovered during chromatographic purification (Scheme 2.13).

2.1.1.7 Nitroacrylates

Recently, Liu, Chen and coworkers have described the asymmetric FC alkylation of NH-indoles with nitroacrylates **T2** catalyzed by a chiral metal-bis(oxazoline) complex [29]. Screening of different combinations of Cu(II), Zn(II), and Yb(III) salts with

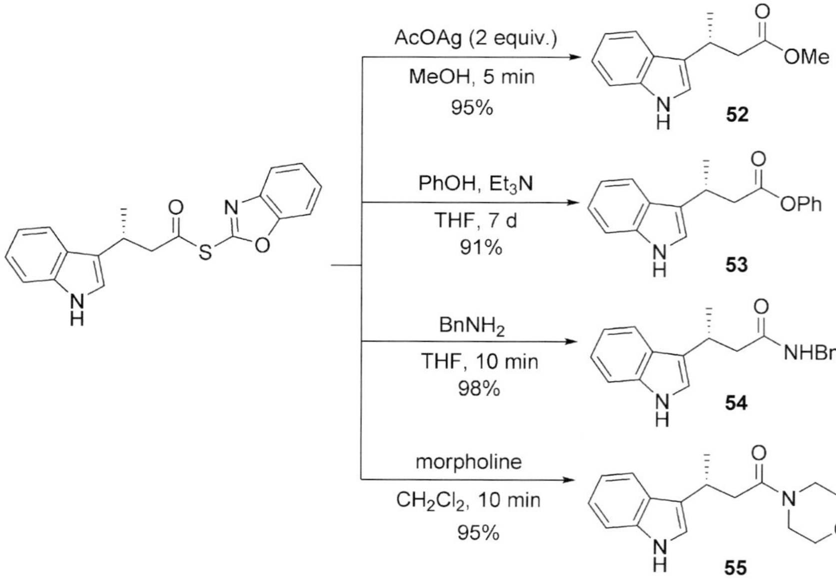

Scheme 2.13 Conversion of S-(1,3-benzoxazol-2-yl) thioester **51** into esters and amides **52–55**.

Table 2.16 FC reaction of nitroacrylates with indoles catalyzed by (4R,5S)-diPh-BOX-Cu(OTf)$_2$ complex.

Entry	R^1	R^2	t (h)	Yield (%)	anti:syn	ee (%) anti/syn
1	H	Me	80	85	72:28	94/58
2	H	OMe	100	90	62:38	92/48
3	H	Cl	70	70	58:42	96/53
4	2-Me	Me	60	60	55:45	69/47
5	5-OBn	Me	100	60	45:55	74/46
6	5-OMe	Me	70	82	62:38	90/55
7	5-OMe	MeO	96	79	55:45	82/47
8	5-Br	Me	80	78	72:28	98/41
9	5-Br	MeO	88	81	71:29	85/54
10	5-Br	Cl	96	78	72:28	99/95
11	4-Br	Cl	100	80	n.d.^{a}	78/89
12	5-MeO$_2$C	Me	100	85	71:29	96/44
13	5-MeO$_2$C	OMe	100	80	67:33	93/40

^{a}Not determined.

several bis(oxazoline) ligands revealed **L12**-Cu(OTf)$_2$ as the best catalyst. The reactions were performed using 10 mol% of the catalyst and when CH$_2$Cl$_2$ was used as the solvent the alkylation between indole and ethyl p-methylbenzylidene nitroacrylate provided a mixture of *anti–syn*-isomers (d.r. = 72 : 28) in 85% yield and with 94% *ee* of *anti*-isomer and 58% *ee* of *syn*-isomer (Table 2.16, entry 1). The reaction in toluene gave high yield and diastereoselectivity (70 : 30), and the enantiomeric excess of the *anti*-isomer was high (96%) while that of the *syn*-isomer was very poor (9%). The yields of the product decreased when the reaction was carried out in THF or *i*PrOH.

Generally, the indoles bearing the substituents at the 5-position reacted with Z-nitroacrylates smoothly to afford the corresponding products in good yields with excellent enantioselectivities of *anti*-isomers (82–99% *ee*), but moderate diastereoselectivities and enantioselectivities of *syn*-isomers. In contrast, 2-substituted indole furnished disappointing results in this FC reaction (entry 4). The diastereomers could be separated by flash chromatography on silica gel and the optical purities of the products could be improved through a co-solvent recrystallization.

The authors invoked that Z-nitroacrylate was partially converted to E-nitroacrylate during the course of the reaction, based on the test experiment shown in Scheme 2.14. This E/Z equilibration might be one cause of diminished selectivity.

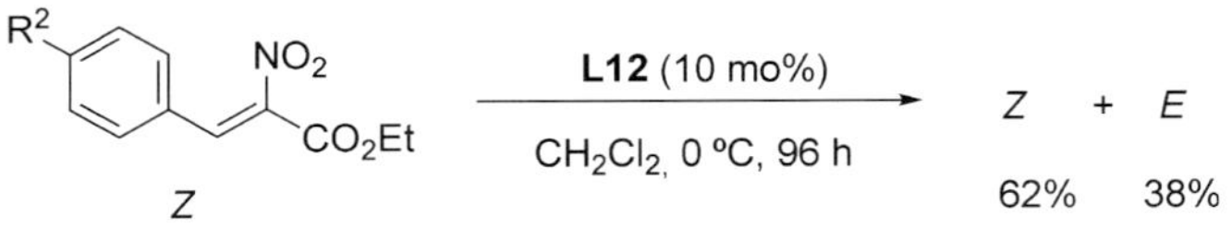

Scheme 2.14

The stereochemical outcome of the reactions can be explained through a four-coordinated Cu(II) complex with a tetrahedral geometry. Thus, when nitroacrylate coordinates with the complex of (4R,5S)-diPh-BOX-Cu(OTf)$_2$, a tetrahedral transition state is formed (Figure 2.9), driven by the stacking interaction of the two phenyl groups of the ligand and nitroacrylate, leading to shielding of the *Re*-face.

The alkylated products could be easily reduced to the corresponding optically active tryptophan analogs, such as **57**, in good yield and without decrease in enantiomeric purity through hydrogenation of the nitro group by Zn/H$^+$ and acylation (Scheme 2.15) [30].

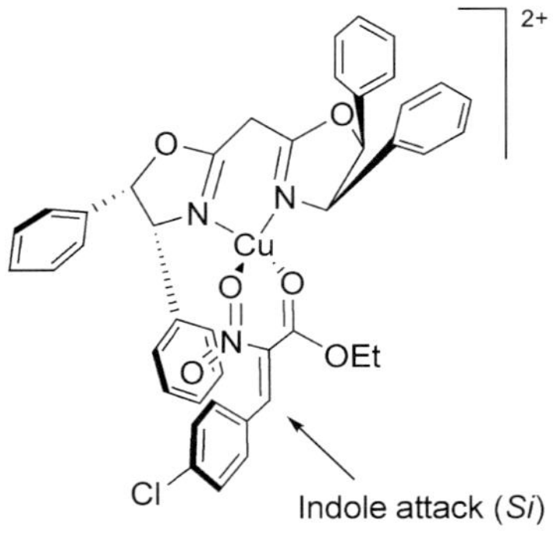

Figure 2.9 Proposed transition state of the reaction.

2.1.1.8 Experimental: Selected Procedures

Typical experimental procedure for the catalytic asymmetric FC reaction of indoles and alkylidene malonates T1 using L2-Cu(OTf)$_2$ as catalyst (Table 2.3) [11]

Under air atmosphere, to a Schlenk tube was added Cu(OTf)$_2$ (0.10 equiv.), followed by (S)–isopropyl–bisoxazoline **L2** (0.10–0.11 equiv.). Then iBuOH was added, and the concentration of Cu^{2+} was maintained at about 0.01 mol mL^{-1}. The resulting purple solution soon turned blue, and was stirred at room temperature (25 °C) for 2 h before arylidene malonate (1.0 equiv.) was added to the mixture. The resulting mixture was kept stirring at room temperature for 15 min, then cooled to −25 °C and stirred for another 15 min before the indole (1.2 equiv.) was added. After the reaction was complete (monitored by TLC), the reaction mixture was concentrated under reduced pressure at room temperature, and the residue was purified by flash column chromatography on silica gel [eluted with CH$_2$Cl$_2$/petroleum ether (1/1, v/v) then pure CH$_2$Cl$_2$] to afford the desired product.

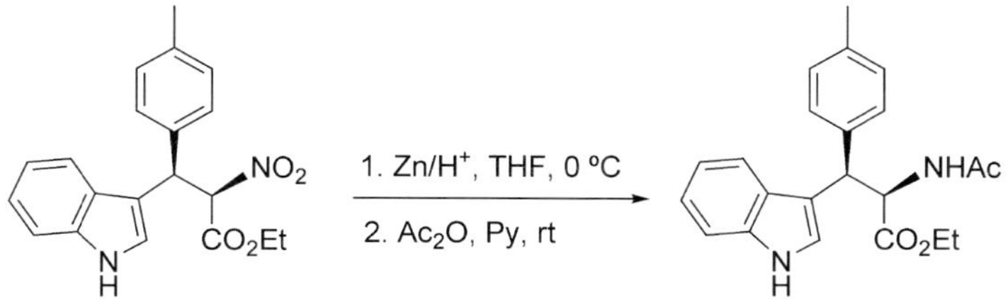

Scheme 2.15 Conversion of alkylated product **56** to tryptophan analog **57**.

Typical experimental procedure for the catalytic asymmetric FC reaction of indoles and alkylidene malonates T1 using L6-Cu(OTf)$_2$ as catalyst (Table 2.4) [12]

To a Schlenk tube ligand **L6** (12.0 mg, 0.05 mmol) and Cu(OTf)$_2$ (18.1 mg, 0.05 mmol) were added under air atmosphere. Ethanol (2 mL) was added and the mixture was stirred for 1 h at room temperature (20–25 °C). To the resulting blue–green solution malonate (1 mmol, 1.0 equiv.) in EtOH (2 mL) was added and stirring was continued for 20 min before the indole (1.2 mmol, 1.2 equiv.) was added. After stirring for 8 h at room temperature, the red colored solution was concentrated under reduced pressure. The crude product was purified by column chromatography (performed with hexanes/CH$_2$Cl$_2$ 1 : 1, followed by CH$_2$Cl$_2$).

Data for (S)-ethyl 2-ethoxycarbonyl-3-(3-indolyl)-3-phenyl propanoate **7** (R^1: H, R^2: Ph): ^{1}H NMR (300 MHz, CDCl$_3$): δ = 0.93–1.06 (m, 6 H), 3.93–4.06 (m, 4 H), 4.30 (d, $J = 11.8$ Hz, 1 H), 5.09 (d, $J = 11.8$ Hz, 1 H), 7.00–7.07 (m, 1 H), 7.09–7.31 (m, 6 H), 7.37 (d, $J = 7.4$ Hz, 2 H), 7.56 (d, $J = 8.0$ Hz, 1 H), 8.07 (brs, 1 H); ^{13}C NMR (75 MHz, CDCl$_3$): δ = 168.1, 167.9, 141.4, 136.2, 128.4, 128.2, 126.8, 126.7, 122.3, 120.9, 119.5, 119.4, 117.0, 111.0, 61.5, 61.4, 58.4, 42.9, 13.8, 13.8; MS (CI): m/z (%) = 383 (MNH$_4^+$, 89), 366 (MH$^+$, 3), 206 (100), 178 (5); mp 174–176 °C; HPLC analysis (Chiralcel OD/OD-H, 10% iPrOH/n-hexane, 0.5 mL min^{-1}, 254 nm; tr (minor) 26.67 min, tr (major) = 31.40 min); >99% ee; $[α]_D^{25} = +65.4°$ (20 mg/2 mL, CH$_2$Cl$_2$).

Typical experimental procedure for the catalytic asymmetric FC reaction of indoles and alkylidene malonates T1 using L7-Cu(OTf)$_2$ as catalyst (Table 2.5) [14]

Under air atmosphere, to a Schlenk tube was added Cu(OTf)$_2$ (0.10 equiv.), followed by (S)-isopropyl-trisoxazoline **L7**. Then iBuOH was added, and the concentration of Cu^{2+} was maintained at about 0.005 mol mL^{-1}. The resulting blue–green solution was stirred at room temperature (10–25 °C) for 2 h before alkylidene malonate (1.0 equiv.) was added to the mixture. The resulting mixture was kept stirring at room temperature for 15 min, then cooled to −25 °C and stirred for another 15 min before the indole (1.2 equiv.) was added. After the reaction was complete (monitored by TLC), the reaction mixture was concentrated under reduced pressure at room temperature, and the residue was purified by flash column chromatography on silica gel [eluted with CH$_2$Cl$_2$/petroleum ether (1/1, v/v) then pure CH$_2$Cl$_2$].

Enantioselective addition of indoles to β,γ-unsaturated α-ketoesters T5 catalyzed by L1-Cu(OTf)$_2$ (Table 2.6) [16]

To a flame-dried Schlenk tube were added Cu(OTf)$_2$ (14.5 mg, 0.04 mmol) and the ligand 2,2'-isopropylidenebis[(4S)-4-*tert*-butyl-2-oxazoline] **L1** (13.0 mg, 0.044 mmol) under a stream of N$_2$. The mixture was dried under vacuum for 1–2 h, Et$_2$O (2.0 mL) was added, and the resulting suspension was stirred vigorously for 1–2 h. To the solution of catalyst was added first the enone (0.8 mmol). The solution was stirred at room temperature for 15 min, then cooled to the desired reaction temperature where the appropriate indole (0.8 mmol) was added. The solution was then stirred at −78 °C or from −20 to 0 °C until the reaction was complete. Pentane (1.0 mL) was added to the reaction mixture. The heterogeneous mixture was filtered through a 40 mm plug of silica gel. The silica was washed with 5–10 mL of 60% pentane in Et$_2$O followed by 5–10 mL of CH$_2$Cl$_2$ and the combined fractions were evaporated. The crude product exists both in its keto and enol form, however, this equilibrium shifts towards the keto form in MeOH at room temperature. The keto form was purified by flash chromatography. In many cases the last purification is not necessary as the reaction is very clean.

Data for (−)-(4R)-4-(1H-indol-3-yl)-2-oxo-4-phenyl-butyric acid methyl ester **11** (R^1: H, R^2: Ph, R^3: Me): ^{1}H NMR (400 MHz, CDCl$_3$): δ = 3.60 (dd, J = 16.8, 7.6 Hz, 1 H), 3.69 (dd, J = 16.8, 7.6 Hz, 1 H), 3.77 (s, 3 H), 4.93 (t, J = 7.6 Hz, 1 H), 7.01–7.44 (m, 10 H), 8.00 (s, br, 1 H); ^{13}C NMR (100 MHz, CDCl$_3$): δ = 37.7, 45.6, 52.9, 111.1, 118.3, 119.4, 119.5, 121.4, 122.3, 126.4, 126.6, 127.7, 128.5, 136.5, 143.1, 161.3, 192.6; mass (TOF ES$^+$): m/z 330 (M + Na)$^+$; HRMS calc. for C$_{19}$H$_{17}$NNaO$_3$ 330.1106, found 330.1108; mp = 98 °C; HPLC analysis (Chiralcel OD-R column, MeOH, 0.5 mL min^{-1}, tr (minor) = 8.9 min, tr (major) = 10.1 min); 99.5% *ee*; $[\alpha]_D^{25} = -23.9°$ (0.0100 g mL^{-1}, CHCl$_3$).

Typical experimental procedure for the catalytic asymmetric FC reaction of indoles and α'-hydroxy enones T10 using L1-Cu(OTf)$_2$ or L8-Cu(OTf)$_2$ as the catalyst (Table 2.8) [20]

2,2'-Isopropylidene bis[(4S)-4-*tert*-butyl-2-oxazoline] (*t*Bu-BOX) **L1** (or 1,1'-bis[2-((4S)-(1,1-dimethylethyl)-1,3-oxazolinyl)]cyclopropane) **L8** (0.05 mmol) was weighed in a flame-dried flask and placed under N$_2$. Cu(OTf)$_2$ (18 mg, 0.05 mmol) was then added by rinsing with dry CH$_2$Cl$_2$ (1.0 mL) from a weighing boat directly into the reaction flask. After stirring at room temperature for 3 h, a solution of the corresponding α'-hydroxy enone (0.5 mmol) in dry CH$_2$Cl$_2$ (0.25 mL) was transferred via cannula into the solution of catalyst, and the resulting mixture was stirred for a further 10 min at room temperature. Then the corresponding indole (1 mmol) was dissolved in 0.25 mL CH$_2$Cl$_2$ and added drop-wise by syringe into the reaction flask, and the mixture was stirred at rt (or as otherwise stated) under N$_2$ until the disappearance of the starting enone (TLC monitoring). The resulting reaction mixture was diluted with NH$_4$Cl (10 mL) and extracted with CH$_2$Cl$_2$ (3 × 10 mL). The CH$_2$Cl$_2$ layers were combined, dried over anhydrous MgSO$_4$ and concentrated. The crude product was purified by column chromatography on SiO$_2$ eluting with 90% hexane and 10% EtOAc to yield pure alkylated compounds.

Data for 2-hydroxy-5-(1H-indol-3-yl)-2-methyl-7-phenyl-heptan-3-one **17** (R^1: H, R^2: H, R: PhCH$_2$CH$_2$): ^{1}H NMR (300 MHz, CDCl$_3$): δ = 8.02 (s, br, 1 H), 7.69 (d,

$J = 7.8$ Hz, 1 H), 7.41 (d, $J = 7.8$ Hz, 1 H), 7.14–7.29 (m, 7.0 Hz), 7.06 (d, $J = 2.1$ Hz, 1 H), 3.65 (m, 1 H) 3.12 (dd, $J = 7.2$ Hz, 1H), 2.92 (dd, $J = 8.1$ Hz 1 H), 2.59 (m, 2 H), 2.13 (m, 2 H), 1.33 (s, 3 H), 1.09 (s, 3 H); ^{13}C NMR (75 MHz, CDCl$_3$): $\delta = 213.5$, 142.2, 136.6, 128.3, 128.3, 126.3, 125.7, 122.0, 121.6, 119.4, 119.3, 118.2, 111.4, 76.3, 42.5, 36.9, 34.1, 32.5, 26.2, 25.8; IR (CH$_2$Cl$_2$) 3428, 1706, 1499, 1456, 742, 702 cm^{-1}; Elemental analysis: Calc. for C$_{22}$H$_{25}$NO$_2$ (335.44): C 78.77, H 7.51, N 4.18; found: C 78.23, H 7.87, N 4.15; HPLC analysis (Chiralcel OD column; hexane: iPrOH 90 : 10; 0.5 mL min^{-1}, 254 nm, tr (minor) = 40.05 min, tr (major) = 54.47 min); 98% ee; $[\alpha]_D^{25} = +27.2°$ (1.0, CH$_2$Cl$_2$).

Typical experimental procedure for the catalytic asymmetric FC reaction of indoles and α′-phosphonate enones T11 using L9-Cu(OTf)$_2$ as catalyst (Table 2.10) [21]

To an oven-dried flask in a dry-box was added an appropriate amount of Cu(II) triflate (20 mol%, 0.022 mmol, 8.1 mg) and (S)-In-Box **L9** (22 mol%, 0.024 mmol, 8.7 mg). The flask was capped with a septum and purged with CH$_2$Cl$_2$ (1 mL). The reaction was allowed to stir at rt for 2 h to prepare the catalyst [{(S)-In-Box}Cu(OTf)$_2$]. A solution of enone (1.0 equiv., 0.11 mmol) in CH$_2$Cl$_2$ (1 mL) was added to the solution of the catalyst. After 0.5 h, the reaction mixture was cooled to 0 °C, and the solution of an indole derivative (1.2 equiv., 0.13 mmol) in CH$_2$Cl$_2$ (1 mL) was added to the reaction mixture. The reaction was maintained at 0 °C until complete consumption of the α′-phosphonate enone determined by TLC. After completion of the reaction, the alkylated compound was purified by flash column chromatography.

Data for dimethyl 4-(1H-indol-3-yl)-2-oxohexylphosphonate **24** (R^1: H, R^2: H, R: CH$_3$CH$_2$): ^{1}H NMR (400 MHz, C$_6$D$_6$): $\delta = 7.72$ (d, 1H), 7.58 (b, 1 H), 7.20–7.14 (m, 4 H), 6.69 (d, 1 H), 3.54 (m, 1 H), 3.29 (dd, 6 H), 3.02 (dd, 1 H), 2.90 (dd, 1 H), 2.66 (m, 2 H), 1.77 (m, 1 H), 0.86 (t, 3 H); ^{13}C NMR (100 MHz, CDCl$_3$): $\delta = 201.72$ (201.66), 136.52, 126.43, 121.74, 121.56, 119.15, 119.02, 117.95, 111.28, 52.91 (52.87), 49.97, 41.99, 40.72, 34.13, 28.35, 11.97; IR (film): 3409, 3276, 2958, 2926, 1715, 1458, 1398, 1339, 1248, 1184, 1034, 932, 894, 745 cm^{-1}; HRMS (EI): Exact mass calc. for C$_{16}$H$_{22}$NO$_4$P [M]$^+$: 323.1268. Found: 323.1286; Elemental analysis calc.: C 59.44, H 6.86, N 4.33; Found C 59.39, H 6.85, N 4.33; HPLC analysis (Chiralcel AD-H (IA), 20% i-PrOH/hexane, 0.2 ml min^{-1}, 254 nm, tr (major) = 45.4 min, tr (minor) = 49.0 min; $[\alpha]_D^{25} = +6.31°$ (0.4, CH$_2$Cl$_2$).

Typical experimental procedure for the catalytic asymmetric FC reaction of indoles and α,β-unsaturated acyl phosphonates T4 using L10-c(OTf)$_3$ as catalyst. General procedure for the preparation of scandium(III) triflate complex (L10-Sc(OTf)$_3$) (Table 2.11) [23a]

A 2-dram oven dried vial was charged with a stir bar, Sc(OTf)$_3$ (16.2 mg, 0.033 mmol, 0.1 equiv.), and pybox ligand **L10** (0.040 mmol, 0.12 equiv.) in a dry box. The vial was capped with a septum and removed from the dry box. Dichloromethane (1.0 mL) was added to the vial under an atmosphere of dry N$_2$. The resulting mixture was stirred vigorously at room temperature for 2 h until the reaction became homogeneous.

General catalytic procedure: The resulting catalyst solution was cooled to −78 °C, and the acyl phosphonate (1.0 equiv., 0.33 mmol) was added. After 1 min, a 1.0 M solution of the nucleophile (1.2 equiv., 0.40 mmol) was added slowly to the reaction along the side of the vial. The reaction was stirred at the same temperature until consumption of the acyl phosphonate as monitored by TLC. Methanol (0.50 mL), followed by DBU (0.10 mL), were added directly to the reaction mixture. The cold bath was removed and the reaction was allowed to stir for 30 min at room temperature. The reaction was diluted with EtOAc (20 mL), then washed with sat. NH_4Cl and brine. The organic layer was dried over anhydrous Na_2SO_4, filtered, and concentrated *in vacuo*. The remaining material was purified by flash column chromatography (SiO_2) using a mixture of EtOAc and hexanes as eluent to afford γ-indolyl carbonyl compounds.

Data for (S)-3-(1-methyl-1H-indol-3-yl)-butyric acid methyl ester **29** (R^1: H, R^2: Me, R: Me): ^{1}H NMR (500 MHz, $CDCl_3$): δ = 7.63 (m, 1 H), 7.30–7.20 (m, 2 H), 7.12–7.09 (m, 1 H), 6.85 (s, 1 H), 3.74 (s, 3 H), 3.65 (s, 3 H), 3.60 (m, 1 H), 2.81 (dd, J = 15.1, 5.9 Hz, 1 H), 2.56 (dd, J = 14.7, 8.3–Hz, 1 H), 1.40 (d, J = 7.3 Hz, 3 H); ^{13}C NMR (100 MHz, $CDCl_3$): δ 173.3, 137.1, 126.7, 124.8, 121.5, 119.3, 119.2, 118.6, 109.3, 51.5, 42.4, 32.6, 27.9, 21.2; IR (film): 3053, 2953, 2825, 1732, 1483, 1134, 1012 cm^{-1}; HRMS (ES): Exact mass calc. for $C_{14}H_{18}O_2N$ $[M + H]^+$: 232.1337. Found: 232.1328; HPLC analysis (Chiralcel OD-H, 3% *i*PrOH/2% EtOH/hexanes, 0.8 mL min^{-1}, 254 nm, tr (minor) = 12.9 min, tr (major) = 16.3 min; $[\alpha]_D^{25}$ = +8.1° (2.0, CH_2Cl_2).

Typical experimental procedure for the catalytic asymmetric FC reaction of indoles and α,β-unsaturated 2-Acyl imidazoles T6 using L10-Sc(OTf)$_3$ as catalyst. General procedure for the preparation of Sc(III) triflate complex (L10-Sc(OTf)$_3$) (Table 2.12) [23b]

To an oven-dried 2-dram vial in a dry-box was added an appropriate amount of scandium (III) triflate, 4 Å MS (15 mg/0.13 mmol of substrate), and 1.2 equiv. of (S)-Indapybox ligand **L10**. The vial was capped with a septum and purged with 1 mL of dichloromethane. The catalyst was allowed to age for 2 h at rt. The dichloromethane was removed by a steady stream of N_2 to yield a white solid.

General catalytic procedure: The resulting solid was dissolved in 1 mL of acetonitrile and cooled to 0 °C before the corresponding 2-acyl imidazole (1.0 equiv.) was added to the flask, either neat or as a 0.74–1.56 M solution in acetonitrile. After 1 min, the nucleophile (1.2 equiv.) was added either neat or as a 1.0 M solution in acetonitrile. The reaction was maintained at 0 °C until consumption of the 2-acyl imidazole as monitored by TLC. The alkylated compounds were isolated directly from the reaction solution by silica chromatography.

Data for (R)-1-(1-methyl-1H-imidazol-2-yl)-3-(1-methyl-1Hindol-3-yl)butan-1-one **33** (R^1: H, R^2: Me, R: Me): ^{1}H NMR (500 MHz, $CDCl_3$): δ = 7.67 (d, J = 8.1 Hz, 1 H), 7.27 (m, 1 H), 7.21 (t, J = 7.0 Hz, 1 H), 7.15 (s, 1 H), 7.09 (d, J = 7.0 Hz, 1 H), 7.00 (s, 1 H), 6.94 (s, 1 H), 3.94 (s, 3 H), 3.89–3.83 (m, 1 H), 3.73 (s, 3 H), 3.56 (dd, J = 16.1 and 6.2 Hz, 1 H), 3.46 (dd, J = 15.7 and 8.1 Hz, 1 H), 1.43 (d, J = 7.0 Hz, 3 H); ^{13}C NMR (100 MHz, $CDCl_3$): δ = 192.3, 143.4, 137.0, 128.9, 126.9, 126.8, 125.0, 121.4, 119.9, 119.4, 118.5, 109.1, 46.8, 36.1, 32.6, 27.1, 21.9; IR (film): 3107.6, 3055.2, 2959.9, 1670.7, 1467.7, 1406.1, 1372.4, 1328.1, 1270.8, 1239.5, 1155.0, 1132.1, 1106.0, 1003.6, 980.0, 915.0, 740.6, 695.8, 642.8 cm^{-1}; HRMS (CI): Exact mass calc. for $C_{17}H_{19}N_3O$ $[M + H]^+$: 282.1606. Found: 282.1599; HPLC analysis (Chiralcel

OD-H, 10% iPrOH/hexanes, 0.8 mL min^{-1}, 310 nm, tr (major) = 17.6 min, tr (minor) = 19.3 min; $[\alpha]_D^{25} = -4.4°$ (1.0, CH$_2$Cl$_2$).

Typical experimental procedure for the catalytic asymmetric FC reaction of indoles and α,β-unsaturated thioesters T3 using cationic [Pd(II)(Tol-Binap)] complex (L11-Pd(II)) as catalyst (Table 2.15) [28]

In a dried, two-neck flask, PdCl$_2$(MeCN)$_2$ (5.2 mg, 0.02 mmol) and Tol-Binap **L11** (13.6 mg, 0.02 mmol) were suspended in anhydrous toluene (0.5 mL). The orange mixture was stirred for about 30 min, until a bright-yellow precipitate formed. The solvent was removed under reduced pressure, and anhydrous MeCN was added (2.0 mL). To the resulting yellow solution was added AgSbF$_6$ (13.8 mg, 0.04 mmol), and the precipitation of AgCl was observed. After 15 min of stirring, α,β-unsaturated thioester (0.1 mmol) and the indole (0.2 mmol) were added, and the mixture was stirred for 18–72 h. The reaction was quenched with saturated aqueous NaHCO$_3$ solution and extracted with AcOEt. The organic layers were collected, dried (Na$_2$SO$_4$), and evaporated under reduced pressure.

Data for S-(1,3-benzoxazol-2-yl) (R)-3-(1H-Indol-3-yl)butanethioate **51** (R^1: H, R^2: H): ^{1}H-NMR (200 MHz, CDCl$_3$): δ = 8.16 (s, br, 1 H), 8.03–7.49 (m, 1 H), 7.78–7.64 (m, 2 H); 7.49–6.98 (m, 5 H), 6.62–6.55 (m, 1 H), 4.08–3.85 (m, 3 H); 1.55 (d, J = 6.6 Hz, 3 H); ^{13}C-NMR (50 MHz, CDCl$_3$): δ = 178.1, 173.3, 145.9, 135.8, 131.0, 129.9, 126.7, 125.8, 125.2, 120.6, 119.1, 118.8, 115.7, 113.9, 110.3, 109.5, 45.5, 27.6, 21.1; Anal. calc. for C$_{19}$H$_{16}$N$_2$O$_2$S (336.4): C 67.84, H 4.79, N 8.33; found: C 67.08, H 4.76, N 8.32; IR (nujol): 3416, 2952, 2919, 2853, 1732, 1474, 1454, 1348, 1248, 1012; $[\alpha]_D^{25} = +30.4$ (0.56, CHCl$_3$).

Typical experimental procedure for the catalytic asymmetric FC reaction of indoles and nitroacrylates T2 (Table 2.16) [29]

Cu(OTf)$_2$ (5.4 mg, 0.015 mmol) and (4R,5S)-diPh-BOX **L12** (7.6 mg, 0.0165 mmol) were charged in a dried 5 mL tube under argon, followed by addition of CH$_2$Cl$_2$ (0.9 mL). The solution was stirred at room temperature for 30 min and the corresponding nitroacrylate (0.15 mmol) was added. The mixture was stirred for 10 min at room temperature, then for 15 min at 0 °C. Indole (0.18 mmol) was added. After stirring for 60–130 h, water (6 mL) was added, followed by extraction with CH$_2$Cl$_2$. The combined organic phases were dried with Na$_2$SO$_4$ and the solvent was removed under reduced pressure. The residue was purified and the diastereomers were separated by flash chromatography on silica gel (eluent: petroleum ether/ethyl acetate 4 : 1) to afford the alkylated product.

Data for ethyl 3-(1H-indol-3-yl)-2-nitro-3-p-tolylpropanoate **56** (R^1: H, R^2: Me): *anti*-isomer: ^{1}H NMR (300 MHz, CDCl$_3$): δ = 8.05 (s, br, 1 H), 7.48 (d, J = 7.9 Hz, 1 H), 7.32 (d, J = 9.2 Hz, 1 H), 7.28–7.22 (m, 2 H), 7.18–7.13 (m, 2 H), 7.09–7.03 (m, 3 H), 5.89 (d, J = 11.4 Hz, 1 H), 5.34 (d, J = 11.4 Hz, 1 H), 4.07–4.01 (m, 2 H), 2.27 (s, 3 H), 1.02 (t, J = 7.1 Hz, 3 H); ^{13}C NMR (75 MHz, CDCl$_3$): δ = 163.5, 137.5, 136.3, 134.4, 129.5, 128.4, 126.3, 122.7, 120.5, 119.9, 118.9, 114.0, 111.3, 91.8, 62.9, 44.2, 21.1, 13.6; FTIR (KBr): 3420, 2924, 1745, 1561, 1358, 1097, 1013, 742 cm^{-1}; HRMS (FAB) m/z calc. for C$_{20}$H$_{20}$N$_2$O$_4$ (M$^+$): 352.1417, found: 352.1414; mp: 146–148 °C; HPLC analysis (Chiralpak AD-H, hexane/iPrOH 90 : 10, 0.5 mL min^{-1}, tr (major) = 58.1 min, tr (minor) = 77.2 min; ee 99% after recrystallization; $[\alpha]_D^{25}$ = −58.2 (0.55, CHCl$_3$). syn-Isomer: ^{1}H NMR (300 MHz, CDCl$_3$): δ = 8.05 (s, br, 1 H), 7.59 (d, J = 7.9 Hz, 1 H), 7.33–7.26 (m, 4 H), 7.18–7.08 (m, 4 H), 5.89 (d, J = 11.6 Hz, 1 H), 5.32 (d, J = 11.6 Hz, 1 H), 4.03–3.94 (m, 2 H), 2.27 (s, 3 H), 0.91 (t, J = 7.1 Hz, 3 H); ^{13}C NMR (75 MHz, CDCl$_3$): δ = 163.4, 137.3, 136.0, 135.6, 129.5, 127.5, 126.2, 122.7, 121.7, 120.0, 119.0, 113.2, 111.6, 91.9, 62.9, 43.5, 21.0, 13.4; FTIR (KBr): 3419, 2984, 1744, 1561, 1457, 1182, 1031, 744 cm^{-1}; HRMS (EI) m/z calc. for C$_{20}$H$_{20}$N$_2$O$_4$ (M$^+$): 352.1423, found: 352.1425; mp: 145–146 °C; HPLC analysis (Chiralpak AD-H, hexane/iPrOH 90 : 10, 0.5 mL min^{-1}, tr (major) = 46.3 min, tr (minor) = 77.7 min; ee 58%.

Abbreviations

Ac	acetyl
Ar	aryl
BINAP	2,2′-bis(diphenylphosphino)-1,1′-binaphthalene
Bn	benzyl
Boc	tert-butoxycarbonyl
BOX	bisoxazoline
Bu	butyl
DBU	1,8-diazabicyclo[5.4.0]undec-7-ene
DMF	N,N-dimethylformamide
DMPA	4-(dimethylamino)pyridine
DMSO	dimethyl sulfoxide
ee	enantiomeric excess
eq or equiv.	equivalent
Et	ethyl
FC	Friedel-Crafts
HFIP	hexafluoro-iso-propanol
L.A.	Lewis acid
Me	methyl
MeO	methoxy
MS	molecular sieves
Nuc	(−)nucleophile
OTf	triflate
Ph	phenyl
Pr	propyl
Py	pyridine
Pybox	bis(oxazolinyl)pyridine
rt	room temperature
TBDPS	tert-butyldiphenylsilyl
TLC	thin layer chromatography
Tol	p-tolyl
Tol-BINAP	2,2′-bis(di-p-tolylphosphino)-1,1′-binaphthyl
TTCE	1,1,2,2-tetrachloroethane

2.2
Simple α,β-Unsaturated Substrates

Giuseppe Bartoli and Paolo Melchiorre

Summary

The catalytic asymmetric Michael-type addition of aromatic and hetero-aromatic substrates to α,β-unsaturated carbonyl compounds represents a powerful FC alkylation strategy for stereoselective C–C bond formation. Despite its synthetic value, such transformations have represented a highly challenging task for chemists involved in the development of new asymmetric FC approaches. This chapter describes the particularly intensive investigations that, over the past few years, have allowed identification of novel and efficient solutions for the catalytic, highly enantioselective Michael additions of aromatic substrates to simple unsaturated carbonyl compounds. The advent of the LUMO-lowering activation strategy, an organocatalytic approach for the activation of α,β-unsaturated aldehydes via the reversible formation of iminium ions with chiral secondary amines, was essential for performing the asymmetric FC alkylation of unsaturated aldehydes, whereas the first efficient Michael addition to non-chelating enones exploited a metal-based chiral catalyst. In a broader sense, these studies represent a wonderful validation of the complementary relationship between organo- and metal-catalysis: the use of purely organic molecules as chiral catalysts complements the traditional organometallic and biological approaches to asymmetric catalysis, enabling synthetic chemists to move closer to being able to build any chiral scaffold in an efficient, rapid, and stereoselective manner.

2.2.1
Introduction

The catalytic asymmetric Michael-type addition of aromatic and hetero-aromatic substrates to both α,β-unsaturated aldehydes and ketones represents a powerful FC alkylation strategy for stereoselective C–C bond formation [1a, b, 31]. Despite its synthetic value, such transformations have represented a highly challenging task for chemists involved in the development of new asymmetric FC approaches. The stereocontrolled FC alkylation of α,β-unsaturated aldehydes is complicated by the high tendency of electron-rich aromatics to undergo acid-catalyzed 1,2-carbonyl addition instead of the desired conjugate addition manifold [32]. On the other hand, in the metal-catalyzed asymmetric Michael addition to α,β-unsaturated ketones the steric similarity of the two carbonyl substituents does not generally permit high levels of lone pair discrimination in the metal association step, which is an essential requirement for high stereocontrol to be achieved. These aspects have limited for a long time the asymmetric FC reactions to bidentate chelating carbonyls (see Section 2.1), a necessary architectural prerequisite to ensure high levels of stereoselectivity [33].

Over the past few years, the particularly intensive investigations on this topic have allowed chemists to identify novel and efficient solutions for the catalytic, highly enantioselective Michael additions of aromatic substrates to simple unsaturated carbonyl compounds. Noteworthy is the advent of the LUMO-lowering activation strategy, an organocatalytic approach for the activation of α,β-unsaturated aldehydes via the reversible formation of iminium ions with chiral secondary amines [3d, 34], which was essential for performing the asymmetric FC alkylation of unsaturated aldehydes [3a], whereas the first efficient Michael addition to non-chelating enones exploited a metal-based chiral catalyst [22, 35]. In a broader sense, these studies represent a wonderful validation of the complementary relationship between organo- and metal-catalysis: the use of purely organic molecules as chiral catalysts complements the traditional organometallic and biological approaches to asymmetric catalysis, enabling synthetic chemists to move closer to being able to build any chiral scaffold in an efficient, rapid, and stereoselective manner.

2.2.1.1 α,β-Unsaturated Aldehydes

The asymmetric Michael addition of aromatics to α,β-unsaturated aldehydes has represented a benchmark reaction for the development of new organocatalytic approaches. The importance of the reaction and the compatibility of the conjugate addition manifold with chiral amines, that minimize the known tendency toward the 1,2-addition under acid-catalysis, have stimulated a great deal of effort toward the optimization of new FC strategies based on the design of novel organocatalysts and new activation modes.

2.2.1.1.1 Organocatalysis

At the turn of the millennium, there has been explosive growth of interest in an area of asymmetric synthesis now known as organocatalysis, catalysis mediated solely by small organic molecules [36]. In particular, the use of chiral secondary amine catalysis (asymmetric aminocatalysis) was established as a powerful synthetic tool for the chemo- and enantioselective functionalization of carbonyl compounds [37]. In this area, MacMillan and colleagues introduced a novel catalytic activation concept – termed iminium catalysis – that provided the platform for the development of a large range of asymmetric transformations involving α,β-unsaturated aldehydes [3d, 34, 38]. This organocatalytic activation mode exploits the reversible condensation of a chiral amine such as **1** with an unsaturated aldehyde to form an iminium ion intermediate (Figure 2.10). This kind of carbogenic system, which exists as a rapid equilibrium between an electron-deficient and an electron-rich state, effectively lowers the energetic potential of the LUMO π-system, enhancing susceptibility toward nucleophilic attack.

Iminium Catalysis

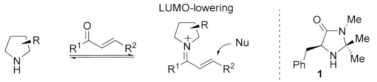

Figure 2.10 LUMO-lowering activation of unsaturated carbonyls via iminium ion formation with chiral secondary amines.

Origin of the Stereoselectivity

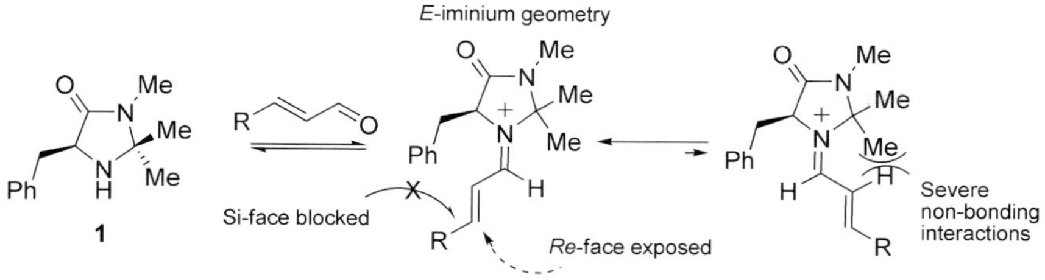

Figure 2.11 Iminium geometry control and π-facial shielding by imidazolidinone catalyst **1**.

Central to the success of imidazolidinone **1** as a stereoselective iminium activator is its ability to form, effectively and reversibly, a reactive iminium ion intermediate with high levels of both geometry control and π-facial discrimination. (Figure 2.11) The catalyst-activated iminium intermediate predominantly exists in the (*E*)-conformation to avoid severe non-bonding interactions between the substrate double bond and *gem*-dimethyl groups on the catalyst scaffold. Selective π-facial coverage by the benzyl group of the imidazolidinone framework leaves the iminium ion *Re*-face exposed for the nucleophilic attack, resulting in highly enantioselective bond formation.

The high efficiency of the easily available chiral imidazolidinone **1** was exploited in different highly enantioselective cycloaddition reactions involving enals [38]. To further demonstrate the value of iminium ion catalysis, MacMillan and coworkers sought to extend the LUMO-lowering activation to promote mechanistically distinct transformations of α,β-unsaturated aldehydes in a highly enantioselective fashion. In particular, they developed an asymmetric FC alkylation of pyrroles [3a] with α,β-unsaturated aldehydes to generate β-pyrrolyl carbonyls, useful synthons for the construction of a variety of biomedical agents (Table 2.17). More importantly, this study represented the first example of an asymmetric, catalytic Michael addition of electron-rich aromatics to simple α,β-unsaturated compounds.

Central to the implementation of this FC asymmetric reaction was the observation that the unsaturated iminium ions arising from condensation of **1** with enals were inert toward 1,2-addition, due to the steric constraints imposed by the catalyst framework. Iminium activation delivered unique reactivity, selectively partitioning pyrrole nucleophiles toward a non-conventional 1,4-addition manifold while enforcing high levels of enantiocontrol in the carbon–carbon bond-forming event. The method proved to be general with respect to pyrrole architecture as well as tolerant to variation in the steric contribution of the olefin substituent, affording products in high optical purity (enantiomeric excess ranging from 87 to 97%) [3a].

Driven by the ability of iminium catalysis to mediate enantioselective coupling of pyrroles and enals, MacMillan's group sought to extend this organocatalytic Friedel–Crafts strategy to other heteroaromatics. In particular, they focused on the FC alkylation of indole, a privileged structural motif of established value in medicinal chemistry and complex target synthesis [39]. Indeed, a poor reaction rate was observed in the conjugate addition of indole to unsaturated aldehydes under catalysis by imidazolidinone **1**. This is consistent with the notion that the pyrrole π-system,

Table 2.17 The first aminocatalytic and asymmetric FC alkylation process.

Pyrrole Friedel–Crafts Alkylation

R: Alkyl, Aryl

yield = 68–90%
ee = 87–97%

Entry	R	R'	R''	Time (h)	Temp. (°C)	Yield (%)	ee (%)
1	Me	Me	H	72	−60	83	91
2	i-Pr	Me	H	72	−50	80	91
3	Ph	Me	H	42	−30	87	93
4	CH_2OBn	Me	H	72	−60	90	87
5	CO_2Me	Me	H	104	−50	72	90
6	Ph	allyl	H	72	−30	83	91
7	CO_2Me	H	H	42	−60	74	90
8	Ph	Me	2-Bu	120	−60	87	90
9	Ph	Me	3-Pr	120	−60	68	97

despite the structural similarities with the indole framework, is significantly more activated toward electrophilic substitutions. The poor results in terms of both reactivity and enantioselectivity obtained in the reaction catalyzed by **1** highlighted the need for a new, more reactive and versatile amine catalyst, which would allow the enantioselective catalytic addition of less reactive nucleophiles, thus providing new reaction manifolds that were not previously possible. It was theorized that replacement of the *trans*-methyl group (with respect to the benzyl moiety) with a hydrogen substituent would lessen the steric hindrance on the participating nitrogen lone pair, increasing its nucleophilic tendency to rapidly engage in iminium ion formation, with a beneficial effect on the overall reaction rate (Figure 2.12). At the same time, replacement of the *cis*-methyl group with a larger substituent, such as a *t*-butyl moiety, provided increased iminium geometry control and better coverage of the blocked *Si*-enantioface. In addition, the lack of a methyl group in catalyst **2** allowed the nucleophile to approach the *Re* face of the formed chiral iminium ion without steric obstruction [3b, 40].

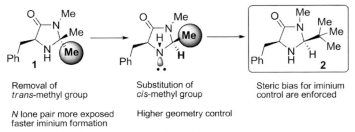

Removal of *trans*-methyl group

N lone pair more exposed faster iminium formation

Substitution of *cis*-methyl group

Higher geometry control

Steric bias for iminium control are enforced

Figure 2.12 Imidazolidinone catalysts evolution.

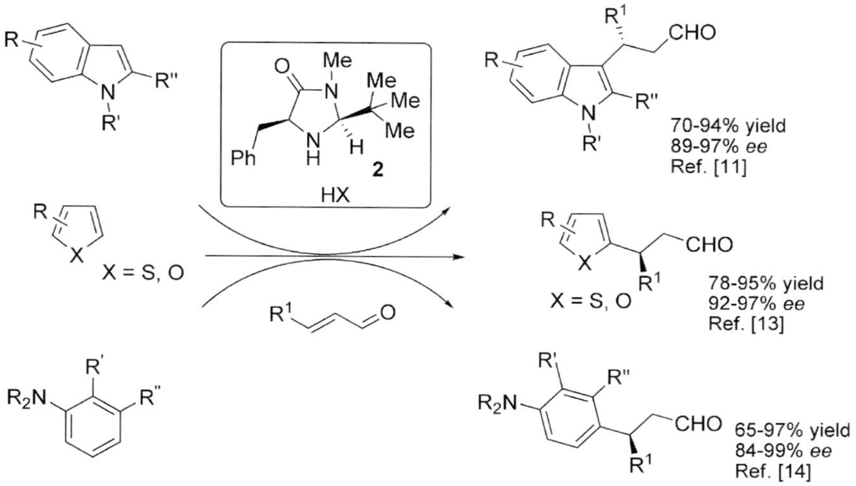

Scheme 2.16 Asymmetric FC alkylation of π-rich aromatics by iminium catalysis.

Since its introduction in 2002, the new imidazolidinone catalyst **2** has been successfully applied to effectively catalyze a wide range of highly enantioselective FC reactions of unsaturated aldehydes with different π-nucleophiles such as indoles [3b], furans [3b], thiophenes [41] and electron-rich benzenes [3c] (Scheme 2.16).

Due to its efficiency and synthetic utility, the amine-catalyzed asymmetric FC alkylation of unsaturated aldehydes has found a wide application for the preparation of valuable enantioenriched biological active compounds. MacMillan and coworkers exploited the organocatalytic alkylation of the 5-methoxy-2-methyl indole **3** with crotonaldehyde, promoted by imidazolidinone **2**, followed by oxidation of the formyl moiety for the direct synthesis of the indolobutyric acid **4**, a COX-2 inhibitor [3b] (Scheme 2.17a).

The organocatalytic FC alkylation of indoles was also the key synthetic step for accessing the highly potent and selective serotonin reuptake inhibitor **7** (BMS-594726, Scheme 2.17b) [42]. In this study, the second generation MacMillan imidazolidinone catalyst **6** was employed to promote the asymmetric Michael addition of substituted indoles to α-branched α,β-unsaturated aldehydes such as **5**, a particularly challenging class of substrate for iminium ion activation. The high reactivity and enantioselectivity observed in the key FC alkylation allowed the elegant and efficient synthesis of the desired compound **7** [42a].

Aminocatalytic Michael addition to α,β-unsaturated aldehydes ensures highly rigid regiocontrol with respect to ring functionalization [43]: for example, the nucleophilic attack takes place at the 3-position of the indole nucleus whereas the pyrrole 2-position is selectively functionalized. One of the ultimate goals of chemical research is the engineering and development of new approaches that enable previously unknown transformations. A wonderful validation of this pattern in the field of organocatalytic FC alkylation was recently proposed by MacMillan's group, who expanded the value of iminium catalysis by describing the asymmetric Michael addition of π-deficient

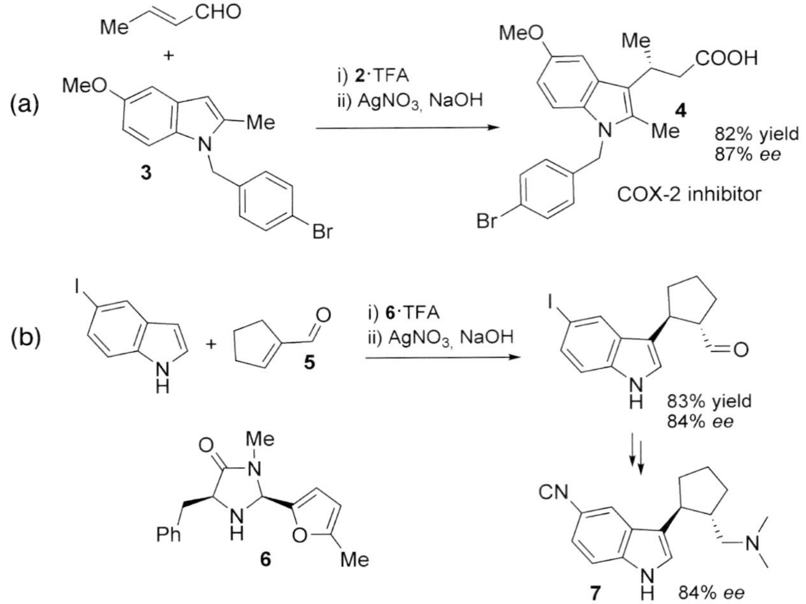

Scheme 2.17 Synthetic utility of the iminium catalyzed asymmetric FC of enals.

aromatics to enals [44]. The novel approach allows the site-specific alkylation of aromatic nucleophiles outside the constraints of FC regioselectivity (Figure 2.13).

Here, heteroaryl trifluoro borate salts are viable substrates for amine-catalyzed Michael addition to unsaturated aldehydes. In particular, upon incorporation of a BF_3K moiety, electron-deficient heteroaromatics that are traditionally inert to iminium catalysis, such as 2-formyl furans, benzofurans and N-Boc indoles, become excellent π-nucleophiles (Scheme 2.18). Most importantly, these salts can provide non-traditional regiocontrol as part of a FC pathway: for example, alkylation of an electron-deficient indole at the 2-position was accomplished in high yield and high enantioselectivity (yield = 79%, ee = 91%) when the imidazolidinone **8** was used as the catalyst [17]. Interestingly, the use of HF plays a central role in the catalytic cycle, as it is necessary for sequestration of the reaction by-product, boron trifluoride, by forming a BF_4K precipitate.

The discovery that BF_3K salts can direct the regioselectivity of the FC alkylation of enals will most likely have a profound impact in the field of aromatic electrophilic

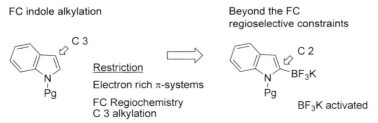

Figure 2.13 Beyond the Friedel-Crafts regioselectivity constraints.

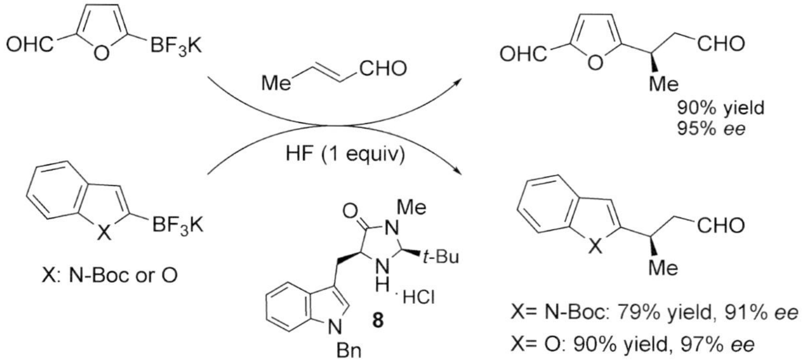

Scheme 2.18 Organocatalytic FC alkylations with trifluoroborate salts.

substitutions, opening new routes toward the synthesis of biologically relevant compounds.

2.2.1.1.2 **Organocatalytic Domino Reactions** Recently, enantioselective cascade catalysis has been recognized as a new synthetic solution to the stereoselective construction of molecular complexity [45]. This bio-inspired strategy is based upon the combination of multiple asymmetric transformations in a cascade sequence, providing rapid access to complex molecules containing multiple stereocenters from simple precursors and in a single operation. Once again, the asymmetric FC alkylation of α,β-unsaturated aldehydes has been used as a benchmark reaction to demonstrate the ability of simple organic molecules to efficiently catalyze complexity- and diversity-generating processes.

The first example of an organocatalytic domino reaction involving FC alkylation was presented in 2004 by MacMillan's group. The addition-cyclization of tryptamines **9** with enals in the presence of imidazolidinone catalyst **1** provides the asymmetric construction of pyrroloindoline architecture **10**, an important scaffold found in many natural compounds, in one single step and in high yield and excellent enantioselectivities (Scheme 2.19) [46].

First, the nucleophilic indole **9** attacks the chiral iminium ion formed by the imidazolidinone **1** with the α,β-unsaturated aldehydes (Scheme 2.19). This highly enantioselective formation of a quaternary all-carbon stereocenter is then followed by the trapping of the indolinium ion (intermediate **A**) by the pendant alcohol or aminoprotected carbamate moieties; central to the implementation of this strategy is the fact that the quaternary carbon-bearing indolium cannot undergo rearomatization by means of proton loss, in contrast to the analogous 3-H indole addition pathway. In one single and simple operation, the pyrroloindolines **10** are thus synthesized with high diastereomeric and enantiomeric ratio. Many analogs of naturally occurring compounds could be accessed as a result of this approach. (−)-Flustramine **11** was synthesized in just five steps starting from the product of the organocatalytic reaction.

The next step forward toward the application of chiral secondary amine catalysis in asymmetric organocascade involving FC processes was advanced in 2005 by MacMillan [47]. In this study, the ability of chiral secondary amines to integrate orthogonal activation modes of carbonyl compounds (enamine and iminium ion

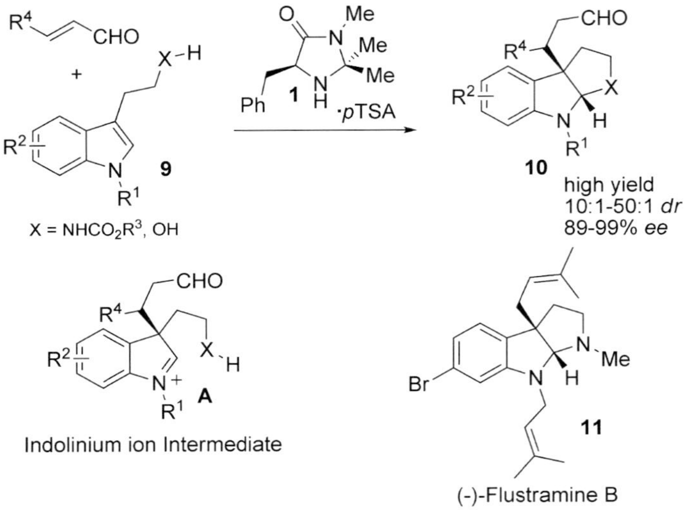

Scheme 2.19 Domino conjugate addition – cyclization reaction.

catalysis) into more elaborate reaction sequences was successfully exploited, providing powerful, new complexity-generating transforms [48].

The group of MacMillan applied a variation of their chiral imidazolidinones (catalyst **8**) to combine the enantioselective conjugate additions of a large number of diverse aromatic π-nucleophiles with the α-chlorination step, a new multicomponent reaction in which the two stereoselective steps are intermolecular reactions (Scheme 2.20) [47]. An extremely appealing feature of these domino sequences is that the interaction between the chiral catalyst and the chiral intermediate, resulting from the first conjugate addition, induces a remarkable enantio-enrichment in the final enamine step, affording rapid access to products with enantiomeric excess generally over 99% for the *syn*-diastereoisomer.

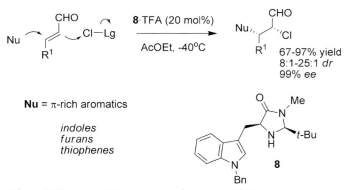

Scheme 2.20 Asymmetric organocascade: aminocatalytic conjugate addition-halogenation sequence.

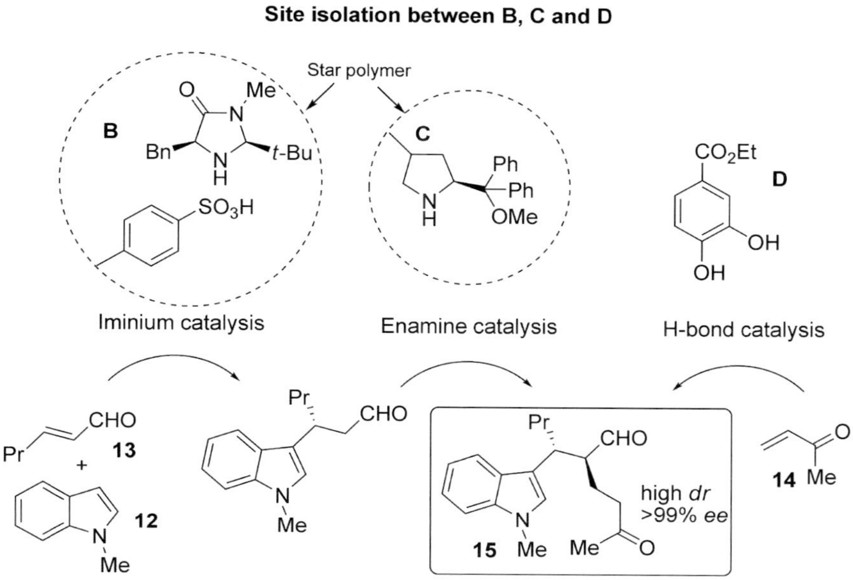

Scheme 2.21 One-pot multi-catalysts cascade reactions.

Recently, this FC organocascade strategy was further ameliorated by using non-interpenetrating star polymer catalysts to combine iminium, enamine and hydrogen-bond catalysis [49]. The proper site isolation with star polymers enables the combination of otherwise incompatible catalysts for sophisticated asymmetric cascade reactions generating complex molecules with multiple chiral centers. This strategy was tested in the sequential double-Michael addition of *N*-Me indole **12** to 2-hexenal **13** and subsequent aldehyde addition to methyl vinyl ketone **14** (Scheme 2.21). Moreover, it was demonstrated that individual access to all four possible stereoisomers of the cascade product **15** can easily be achieved via judicious selection of the amine enantiomer involved in each catalytic cycle.

2.2.1.1.3 Experimental: Selected Procedures

General procedure for the organocatalytic Friedel–Crafts alkylation of indoles (Scheme 2.16) [41]

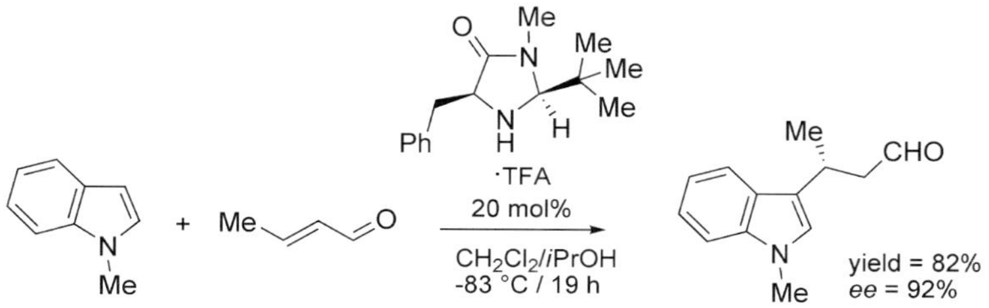

In an ordinary 2-dram vial equipped with a Teflon-coated stir bar, the catalyst (2*S*,5*S*)-5-benzyl-2-*tert*-butyl-3-methyl-imidazolidin-4-one (24.6 mg, 0.1 mmol, 20 mol%) and trifluoroacetic acid (7.7 μl, 0.1 mmol, 20 mol%) were dissolved in a mixture of CH_2Cl_2 (0.85 mL) and isopropanol (0.15 mL), and then placed in a bath at −83 °C. After 5 min stirring, crotonaldehyde (125 μL, 1.50 mmol) was added, and stirring continued for an additional 10 min. Then 1-methyl-1*H*-indole (64 μL, 0.50 mmol) was added in one portion, the vial was capped with a rubber stopper and stirring was continued at constant temperature for 19 h. The crude reaction mixture was diluted with Et_2O and flushed through a short plug of silica gel. Solvent was removed *in vacuo*, and the residue was purified by silica gel chromatography (benzene) to yield the title compound as a colorless oil (83 mg, yield = 82%, *ee* = 92%). IR (film) 3054, 2960, 2824, 2722, 1720, 1616, 1550, 1474, 1374, 1329, 1241, 740 cm^{-1}; ^{1}H NMR (300 MHz, $CDCl_3$) δ 9.75 (dd, *J* = 2.1, 2.1 Hz, 1H, CHO), 7.63 (d, *J* = 7.8 Hz, 1H, ArH), 7.32–7.21 (m, 2H, ArH), 7.12 (ddd, *J* = 1.5, 7.4, 8.1 Hz, 1H, ArH), 6.84 (s, 1H, NCH), 3.75 (s, 3H, NCH_3), 3.68 (dt, *J* = 6.9, 13.8 Hz, 1H, ArCH), 2.88 (ddd, *J* = 2.7, 6.9, 16.2 Hz, 1H, CH_2CO); 2.71 (ddd, *J* = 2.7, 6.9, 16.2 Hz, 1H, CH_2CO); 1.44 (d, *J* = 7.2 Hz, 3H, $CHCH_3$); ^{13}C NMR (75 MHz, $CDCl_3$) δ 202.8, 137.2, 126.6, 125.2, 121.8, 119.1, 118.9, 118.8, 109.5, 51.2, 32.8, 26.0, 21.9; HRMS (CI) exact mass calc. for ($C_{13}H_{15}NO$) requires *m/z* 201.1154, found *m/z* 201.1152. $[\alpha]^D = -4.2$ (*c* = 1.0, $CHCl_3$). The enantiomeric ratio was determined by chiral HPLC analysis of the alcohol, obtained by $NaBH_4$ reduction of the aldehyde, using a Chiracel AD and AD guard column (2 : 98 ethanol/hexanes, 1 mL min^{-1}); S isomer t_r = 25.2 min and R isomer t_r = 27.8 min.

Organocatalytic Friedel–Crafts alkylation with trifluoroborate salts (Figure 2.13) [47]

To a plastic vial equipped with a magnetic stir bar and charged with the catalyst (2*S*,5*S*)-2-*tert*-butyl-5-((1-benzyl-1*H*-indol-3-yl)methyl)-3-methylimidazolidin-4-one (12.4 mg, 0.033 mmol, 20 mol%) and HCl (4 N in dioxane, 8.3 μL, 0.033 mmol, 20 mol%), is added HF (48% wt, 3.5 mg, 0.083 mmol, 1.0 equiv.) followed by 1,2-dimethoxyethane (250 μL, 1 M relative to aldehyde). After 5 min stirring at −20 °C, the reaction is started with the addition of crotonaldehyde (21 μL, 0.25 mmol) immediately followed by the addition of the of potassium 2-(*tert*-butyl 1*H*-indole-1-carboxylate)trifluoroborate (26.9 mg, 0.083 mmol). The reaction is stirred at −20 °C for 24 h and quenched with 1 M HCl (1.0 mL) and stirred with chloroform (1.5 mL) for 30 min. The organic layer is extracted with chloroform (2 × 2.0 mL), dried over Na_2SO_4, filtered through celite (ether wash) and concentrated *in vacuo*. Purification by chromatography (silica gel, 10% ether in pentanes) yields the title compound as a light yellow oil (19.0 mg, yield = 79%, ee = 91%). IR (film) 1728, 1455, 1370, 1327, 1157, 747.4 cm^{-1}; ^{1}H NMR (300 MHz, CDCl$_3$) δ 9.77 (t, 1H, J = 1.8 Hz, CHO), 8.03 (dt, 1H, J = 0.6, 7.8 Hz, aryl H), 7.45 (m, 1H, aryl H), 7.26–7.15 (m, 2H, aryl H), 6.40 (t, 1H, J = 0.9 Hz, aryl H), 4.24 (m, 1H, CHCH$_3$), 2.57 (dd, 1H, CH$_2$), 2.89 (1H, dd, J = 1.8, 5.4 Hz, CH$_2$), 1.37 (t, 3H, J = 6.9 Hz, CH$_3$); ^{13}C NMR (75 MHz, CDCl$_3$) δ 201.95, 150.69, 145.80, 129.22, 123.96, 123.00, 120.28, 115.94, 106.42, 84.47, 50.80, 28.45, 28.02, 21.06; HRMS (EI+) exact mass calculated for [M$^+$] (C$_{17}$H$_{21}$NO$_3$) requires m/z 287.1521, found m/z 287.1533; [α]^D = −6.1 (c = 0.6, CHCl$_3$). The enantiomeric excess was determined by SFC analysis using a Chiralcel OD-H column (5% to 50% MeCN, linear gradient, 100 bar, 35 °C oven, flow = 4.0 mL min^{-1}); (S)-isomer t_r = 2.51 min, (R)-isomer t_r = 2.97 min.

Organocatalytic pyrroloindoline construction (Scheme 2.19) [49]

In an ordinary test tube equipped with a magnetic stirring bar, N-10-BOC-1-prenyl-6-bromotryptamine (258 mg, 0.64 mmol, 1 equiv.), the catalyst (2S,5S)-5-benzyl-2-*tert*-butyl-3-methyl-imidazolidin-4-one (31 mg, 0.13 mmol, 20 mol%) and trifluoroacetic acid (9.8 μl, 0.13 mmol, 20 mol%) were dissolved in CH$_2$Cl$_2$ (4.2 mL, 0.15 M). After 5 min stirring at −84 °C, 4 equiv. of acrolein (0.17 ml, 2.56 mmol), was added, and the mixture was stirred for 72 h until complete consumption of the starting material. To the reaction mixture was added pH 7 buffer, and the whole was extracted with Et$_2$O. Solvent was removed *in vacuo*, and the residue was purified by silica gel chromatography in 10% EtOAc/hexanes to provide the title compound as a colorless oil (231 mg, yield = 78%, ee = 80%). IR (thin film) 2971, 2929, 2717, 1723, 1695, 1601, 1490, 1447, 1394, 1366, 1250, 1219, 1158 cm^{-1}; ^{1}H NMR (300 MHz, CDCl$_3$) δ 9.63 (s, 1H), 6.75 (d, J = 7.8 Hz, 1H), 6.69 (d, J = 7.8 Hz, 1H), 6.38 (s, 1H), 5.25 (d, J = 41.1 Hz, 1H), 5.08 (br s, 1H), 3.65–4.09 (m, 3H), 2.94 (br s, 1H), 1.89–2.43 (m, 6H), 1.76 (s, 3H), 1.71 (s, 3H), 1.43 (s, 9H); ^{13}C NMR (75 MHz, CDCl3) δ 201.2, 154.7, 153.8, 151.8, 134.7, 134.4, 130.3, 123.9, 122.7, 121.5, 120.8, 120.0, 119.7, 109.0, 84.7, 84.1, 80.9, 80.2, 56.5 55.2, 45.7, 45.2, 43.8, 40.3, 39.0, 38.3, 31.2, 28.7, 26.1, 18.5; HRMS (CI) exact mass calcd for (C$_{23}$H$_{31}$BrN$_2$O$_3$ + Na +

CH_3OH^+) requires m/z 517.1678, found m/z 517.1674; $[\alpha]^D\ 20 = -218.9$ ($c = 1.0$, $CHCl_3$).

2.2.1.2 α,β-Unsaturated Ketones

The stereoselective Michael addition to α,β-unsaturated ketones represents a challenging objective in asymmetric catalysis. In metal-catalyzed asymmetric processes, the steric and electronic similarity of the two carbonyl substituents does not generally permit high levels of lone pair discrimination in the metal association step, thus rendering the stereodifferentiation of the two enantiotopic faces of the enones a difficult task. The iminium activation approach, overcoming the necessity of a specific lone pair coordination, can, in, principle constitute a suitable and general platform for accomplishing highly stereoselective transformations of enones. However, the inherent problems of forming congested iminium ions from ketones, along with the issue associated with a more difficult control of the iminium ion geometry, have complicated the development of an efficient chiral organocatalyst for the FC alkylation of enones.

The first solution to the problem of the identification of a suitable catalytic system for the stereoselective FC alkylation of α,β-unsaturated ketones comes from the realm of metal-based catalysis. In 2003, the group of Bandini and Umani-Ronchi described the first asymmetric conjugate addition of indoles with up to 89% *ee* by exploiting the catalytic combination of the chiral [Al(Salen)Cl] complex and 2,6-lutidine [35]. Only four years later an effective organocatalytic system was discovered, based on the iminium activation of enones by chiral primary amines [50], leading to the FC adducts with high stereoselectivity.

2.2.1.2.1 Organometallic Catalysis
The metal-catalyzed asymmetric FC alkylations of non-chelating unsaturated ketones represented a considerable synthetic challenge until 2003. In that year, the first effective enantioselective addition of indoles to (*E*)-arylcrotyl ketones **16** catalyzed by a chiral aluminum complex was presented. In particular, the synergistic cooperation of the chiral [SalenAlCl] complex **17** and a catalytic amount of an achiral base additive such as 2,6-lutidine **18** provided a direct and stereoselective route to β-indolyl ketones **19** in high yield and with interesting enantioselectivity (Table 2.18) [35]. Both theoretical and spectroscopic evidence suggests an interaction between the additive and the aluminum center, leading to the formation of a new, highly stereoselective catalytic entity [22].

Despite the moderate to good selectivity achieved and the restrictions in substrate scope, this report represented an important breakthrough in the field of asymmetric FC reactions as well as in metal catalysis, providing support for the [SalenAl(III)] complexes as highly efficient activators of electrophiles having a simple one-point binding [51].

For years the use of the chiral aluminium complex **17** has constituted the only access to β-indolyl ketones. Recently, Pedro and coworkers have identified a complex of BINOL-based ligand with $Zr(O^tBu)_4$ **20** as a suitable metal catalyst to address the issue of the FC alkylation of simple, non-chelating α,β-unsaturated ketones [52]. The catalytic methodology allows the conjugate addition of both indoles and pyrroles to a

Table 2.18 The first catalytic asymmetric FC indole alkylation with simple enones.

Entry	R	R'	Ar	Time (h)	Temp. (°C)	Yield (%)	ee (%)
1	Me	H	Ph	72	0	48	84
2	H	H	Ph	48	RT	35	64
3	H	OMe	Ph	48	RT	41	65
4	Me	H	pMe-C$_6$H$_4$	48	RT	80	73
5	Me	H	pCl-C$_6$H$_4$	96	−15	68	89
6	Me	H	pBr-C$_6$H$_4$	72	−20	78	86
7	Me	H	C$_6$F$_5$	48	RT	90	88
8	H	H	C$_6$F$_5$	48	RT	67	80

large number of β-substituted α,β-enones in good yields and with enantioselectivity above 95% *ee* in most of the reported examples (Scheme 2.22). Besides the high level of stereocontrol achieved, the value of this report lies in the considerable extension of the scope of the catalytic FC reaction: different types of both π-rich aromatics and enones were employed by using a simple catalytic procedure at room temperature and commercially available chiral ligands.

2.2.1.2.2 **Organocatalysis** Currently, asymmetric aminocatalysis is recognized by the synthetic community as a powerful and reliable strategy for the enantioselective transformation of aldehydes. In comparison, little progress has been achieved in the corresponding asymmetric functionalization of ketones, probably due to the inherent difficulties in generating congested covalent intermediates from ketones and chiral secondary amines. Recently, low molecular weight chiral primary

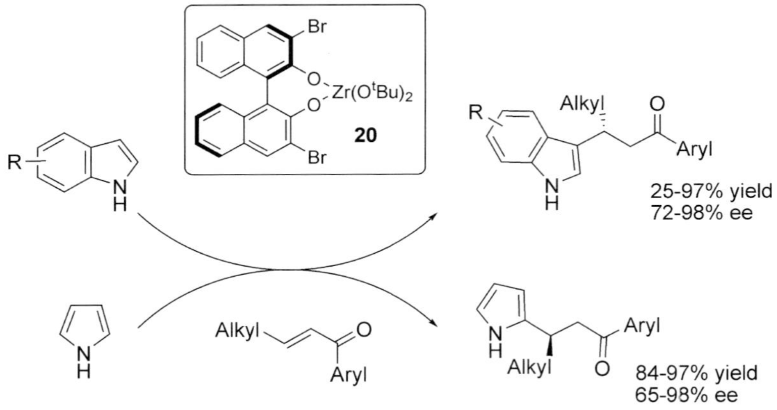

Scheme 2.22 Metal-based catalysis in the FC alkylation of pyrroles and indoles with α,β-unsaturated ketones.

amines have been added to the arsenal as effective amiocatalysts for enamine and iminium ion involving transformations [53]. In particular, owing to reduced steric constraints, chiral primary amine derivatives offer the unique possibility of participating in processes between sterically-demanding partners such as α,β-unsaturated ketones, thus overcoming the inherent difficulties of chiral secondary amine catalysis [53a–e].

These new organocatalytic tools were also exploited in the FC indole alkylation of simple enones, a transformation in which the use of a secondary amine such as the MacMillan imidazolidinone **6** afforded very modest reactivity and enantioselectivity [54]. In 2007, Chen and coworkers illustrated that the CF_3SO_3H salt of the chiral primary amine **18**, easily derived from cinchonine, is able to catalyze the chemo- and stereo-selective addition of indoles to both aliphatic and aromatic ketones (*ee*s ranging from 47 to 89%, Table 2.19, Scheme a) [50a]. Soon after this publication, a similar catalytic system was independently developed by Melchiorre and colleagues. They used the catalytic salt **19**, made by combining two chiral entities such as 9-amino(9-deoxy)*epi*-hydroquinine, derived from hydroquinine [55], and D-*N*-Boc phenylglycine, to promote the asymmetric FC alkylation of different indoles with enones (Table 2.19, Scheme b) [50b]. Both studies illustrated the influence of the steric contribution of the R^2 ketone substituents on the stereoselectivity: more encumbered substituents engender higher selectivity, albeit with slightly lower reactivity. The catalytic salt **19** showed high generality, since a variety of unsaturated ketones can be efficiently activated: both linear compounds, including chalcone, a particularly challenging class of substrates for iminium catalysis, and cyclic enones afforded the expected FC adducts in good yield and high optical purity.

Recently, a novel organocatalytic approach has been introduced for the asymmetric catalytic FC alkylation of indoles with α,β-enones, based on the catalytic behavior of chiral Brønsted acids. At the end of 2006, Xu, Xia and colleagues demonstrated that the combination of D-camphorsulfonic acid **20** and the

Table 2.19 Asymmetric FC alkylation of simple enones catalyzed by chiral primary amine salts.

(a) catalytic salt **18** 30 mol%
$(CF_3SO_3^-)_2$
DCM/iPrOH
3–6 days
−20 / 0 °C
R^1 = alkyl, aryl
R^2 = alkyl
R^3 = H, OMe, Br
R^4 = H, Me
35–99% yield
47–89% ee

(b) catalytic salt **19** 20 mol%
$\left(\text{Boc-HN}\overset{\text{Ph}}{\underset{\text{COO}^-}{\big|}}\right)_2$
Toluene
2–4 days
RT / 70 °C
R^1 = alkyl, aryl
R^2 = alkyl, aryl
R^3 = H, OMe, Cl
R^4 = H, Me
56–99% yield
70–96% ee

Entry	R¹	R²	R³	R⁴	Catalyst 18		Catalyst 19	
					Yield (%)	ee (%)	Yield (%)	ee (%)
1	Ph	Me	H	H	72	65	90	88
2	Ph	Et	H	H	47	81	56	95
3	2-thienyl	Me	H	H	83	50	92	84
4	Alkyl chain[a]	Me	H	H	70	75	91	93
5	Alkyl chain[a]	Me	MeO	H	74	78	76	93
6	Alkyl chain[a]	Me	H	Me	99	65	84	94
7	Alkyl chain[a]	(CH₂)₃	H	H	82	56	65	78
8	CO₂Et	Me	H	H	—	—	99	95
9	Ph	Ph	H	H	—	—	78	82

[a]Alkyl chain: n-propyl for catalyst **18**, n-pentyl for catalyst **19**.

ionic liquid 1-butyl-3-methyl-1H-imidazolium bromide (BmimBr) was an effective catalyst for the asymmetric addition of different indoles to chalcones (Scheme 2.23a) [56]. Despite the modest stereocontrol achieved (ees ranging from 19 to 58%), this study opened up new opportunities for the development of new and efficient catalytic systems. Along this line, it was recently demonstrated that the chiral BINOL-based phosphoric acid **21** is a highly active catalyst for the FC alkylation of indoles with chalcones (Scheme 2.23b) [57]. Once again the enantioselectivity imparted by the chiral acid catalyst was moderate, but the main feature of this study was the low catalyst loading required (as low as 2 mol%) to obtain the products in high yield, thus demonstrating the potential of this novel organocatalytic approach.

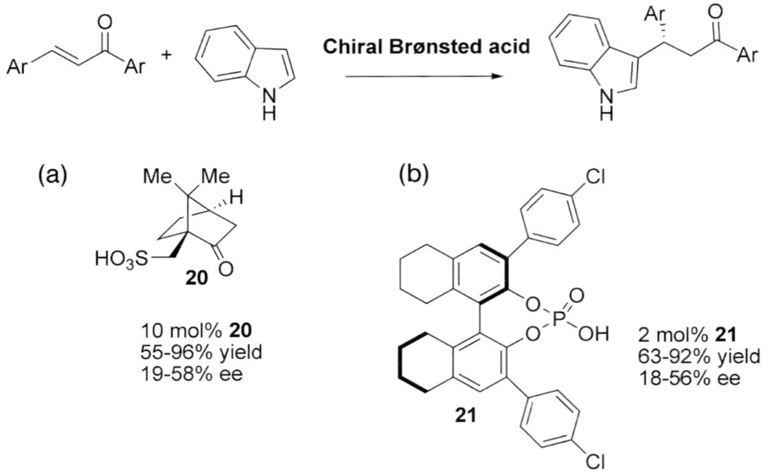

Scheme 2.23 Enantioselective FC alkylation of chalcones catalyzed by chiral Brønsted acids.

2.2.1.2.3 Experimental: Selected Procedures

Metal-catalyzed asymmetric Friedel–Crafts alkylation of indoles with simple α,β-unsaturated ketones (Table 2.18) [36]

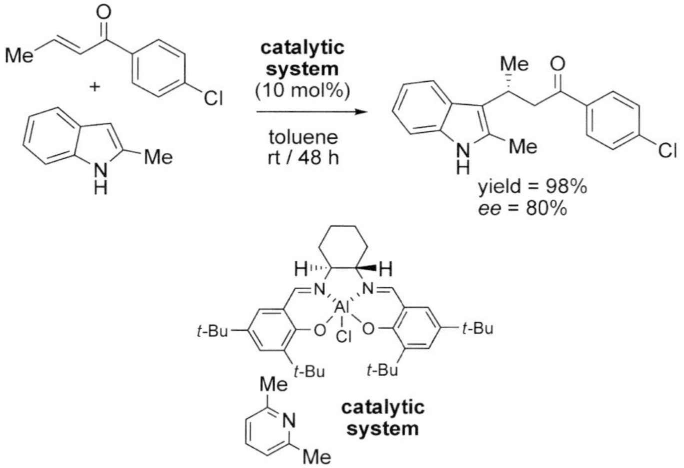

[SalenAlCl] (18 mg, 0.03 mmol, 10 mol%) and 2,6-lutidine (3.5 µl, 0.03 mmol, 10 mol %) were added to anhydrous toluene (1 mL) and the mixture was stirred for 5 min at room temperature. The reaction is started with the addition of the enone (54 mg, 0.3 mmol) immediately followed by the addition of the 2-methyl indole (59 mg, 0.45 mmol). After 48 h stirring, the reaction was quenched with a saturated solution of NaHCO$_3$, and the resulting mixture extracted with AcOEt (3 × 10 mL). The combined organics were washed with brine (5 mL), dried over MgSO$_4$, filtered and concentrated *in vacuo*, and the residue was purified by silica gel chromatography in 15% Et$_2$O/hexanes to provide the title compound as a pale yellow oil (92 mg, yield = 98%, ee = 80%). Chiral analysis was carried out by HPLC (Chiralcel OD iPrOH/hexane (20 : 80), flow rate 0.7 mL min^{-1}, 225 nm; t_r-(S) = 11.44 min, t_r-(R) = 15.11 min); $[\alpha]^D = -54$ (c 0.98 in CHCl$_3$). IR (Nujol) 3398, 3055, 2963, 1695, 1587, 1456, 1091 cm^{-1}; MS (70 eV): m/z (%): 311 (20) [M$^+$], 281 (18), 253 (10), 207 (82), 191 (15), 158 (100), 139 (22), 130 (18), 111 (12), 75 (9); ^{1}H NMR (200 MHz, CDCl$_3$, 25 °C, TMS): δ = 8.4 (br, 1H), 7.77–7.81 (m, 2H), 7.65–7.69 (m, 1H), 7.23–7.36 (m, 4H), 7.06–7.14 (m, 2H), 3.73 (q, J = 7.0 Hz, 1H), 3.43–3.55 (m, 1H), 3.30 (dd, J = 7.0 Hz, J = 16.2 Hz, 1H), 2.38 (s, 3H), 1.50 (d, J = 7.0 Hz, 1H); ^{13}C NMR (75 MHz, CDCl$_3$, 25 °C, TMS) = 198.86, 160.19, 139.10, 135.47, 130.36, 129.37, 128.63, 120.63, 118.91, 118.87, 115.10, 110.53, 45.50, 27.43, 21.02, 11.88.

Organocatalytic asymmetric Friedel–Crafts alkylation of indoles with simple α,β-unsaturated ketones (Table 2.19) [53b]

In an ordinary test tube equipped with a magnetic stirring bar, 9-amino(9-deoxy)*epi*-hydroquinine (0.04 mmol, 13.0 mg, 10 mol%) was dissolved in 1 mL of toluene. After addition of 0.08 mmol (20 mg, 40 mol%) of D-*N*-Boc-phenylglycine, the solution was stirred for 5 min at room temperature. After addition of *trans*-4-phenyl-3-buten-2-one (29.2 mg, 0.2 mmol), the mixture was stirred at room temperature for 10 min. Then indole (28.3 mg, 0.24 mmol) was added in one portion, the tube was closed with a rubber stopper and stirring was continued for 24 h at 70 °C. Then the crude reaction mixture was diluted with Et$_2$O (2 mL) and flushed through a plug of silica, using hexane/Et$_2$O 1/1 as the eluent. Solvent was removed *in vacuo*, and the residue was purified by flash chromatography (hexane/AcOEt = 85/15) to yield the desired product as a white foam (47.5 mg, 90% yield and 88% ee). The ee was determined by HPLC analysis using a Chiralpak AD-H column (80/20 hexane/*i*-PrOH; flow rate 0.75 mL min^{-1}; λ = 214, 254 nm; t_{minor} = 10.2 min; t_{major} = 10.8 min). $[\alpha]^{rt}_D = +20.3$ (c = 0.95, CHCl$_3$, ee = 88%). ^{1}H NMR (400 MHz, CDCl$_3$): δ = 2.07 (s, 3H), 3.16 (dd, J = 7.6, 16.0 Hz, 1H), 3.25 (dd, J = 7.6, 16.0 Hz, 1H), 4.84 (t, J = 7.6 Hz, 1H), 6.96–6.99 (m, 1H), 7.00–7.05 (m, 1H), 7.12–7.20 (m, 2H), 7.23–7.33 (m, 5H), 7.42 (d, J = 8 Hz, 1H), 8.02 (br s, 1H); ^{13}C NMR (150 MHz, CDCl$_3$): δ = 30.3 (CH$_3$), 38.4 (CH), 50.3 (CH$_2$), 111.1 (CH), 118.8 (C), 119.4 (CH), 119.4 (CH), 121.3 (CH), 122.1 (CH), 126.4 (CH), 126.5 (C), 127.7 (CH), 128.5 (CH), 136.6 (C), 143.9 (C), 207.6 (C).

2.2.2
Intramolecular Approach

By exploiting the potential of both metal- and organo-catalysis, the asymmetric FC alkylation of electron-rich aromatics with simple, unsaturated carbonyl compounds has gained a high standard of efficiency. Currently, this catalytic asymmetric FC

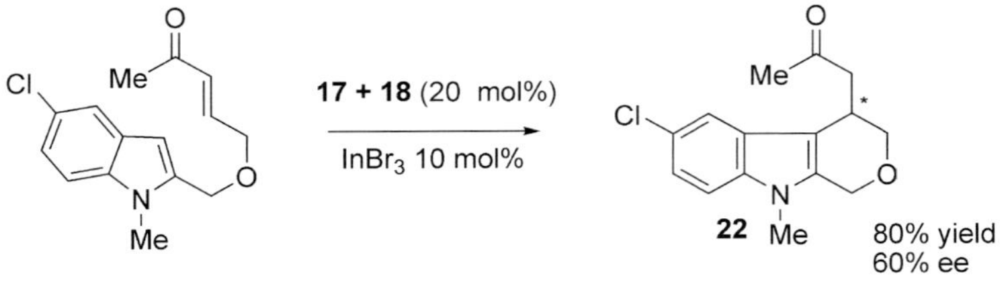

Scheme 2.24 Metal-catalyzed intramolecular FC alkylation of indoles with enones.

approach is recognized as a powerful synthetic tool for the synthesis of important chiral molecules. Recently, a novel and ambitious synthetic target has been identified, that is the development of intramolecular ring-closing FC alkylation of indolyl α,β-unsaturated carbonyls. This intramolecular variant represents an important synthetic step forward since it allows the easy and direct construction of tetrahydro-β-carbolines (THBCs) and tetrahydropyrano indoles (THPIs) that are core structural elements in natural and synthetic organic compounds possessing a wide diversity of important biological activities [58]. Once again, the complementarity between metal- and organo-catalysis has allowed the implementation of such FC type cyclization strategies with both unsaturated ketones and aldehydes, respectively.

The first example came from the laboratories of Bandini and Umani-Ronchi [59]. Exploiting the knowledge accumulated on the (salen)aluminium(III) complex catalyzed FC alkylation of enones [35], they developed a practical catalytic approach to the synthesis of 4-substituted 1,2,3,4-tetrahydro-β-carbolines and 1,2,3,9-tetrahydropyrano[3,4-b] indoles **22** (Scheme 2.24). Despite the moderate level of stereoselectivity achieved (up to 60% *ee*), this study represented an important advance in the field of asymmetric FC alkylation, indicating a novel strategy for the preparation of valuable cyclic molecules. Interestingly, the chiral [SalenAlCl] complex **17** in combination with a catalytic amount of 2,6-lutidine **18**, that afforded high stereocontrol in the intermolecular approach, did not furnish the desired cyclic compounds. The best results, in terms of reactivity as well as selectivity, were achieved by using a heterobimetallic system, made by combining 20 mol% of the chiral aluminum-complex **17** with 10 mol% of InBr$_3$. Spectroscopic and experimental evidence supports that a type of catalysis under a cooperative regime is likely active.

The intramolecular ring-closing FC alkylation strategy was later extended to indolyl α,β-unsaturated aldehydes by exploiting an aminocatalytic approach [60]. The use of

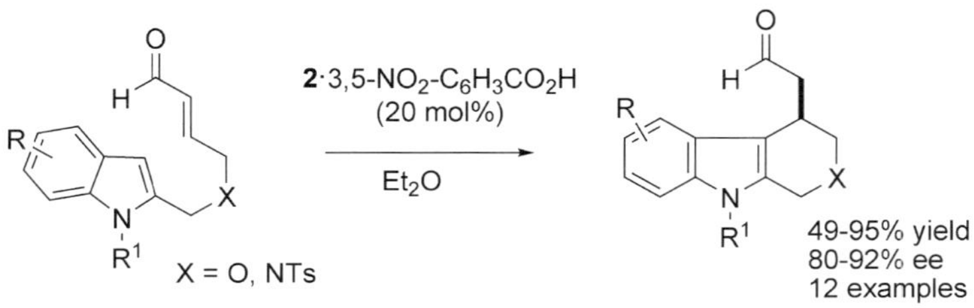

Scheme 2.25 Asymmetric organocatalytic intramolecular FC alkylation of indoles with α,β-unsaturated aldehydes.

the MacMillan imidazolidinone catalyst **2** allowed the synthesis of both THPIs and THBCs with high stereocontrol (up to 93 *ee*, Scheme 2.25).

Acknowledgments

The authors acknowledge the financial support provided by MUR (Ministery of University and Research) in the frame of "*Stereoselection in Asymmetric Synthesis: Methodologies and Applications*" – Research projects of national interest and by the University of Bologna: Fundamental Oriented Research (RFO).

Abbreviations

AcOEt	ethyl acetate
Bn	benzyl
Boc	*tert*-butyloxycarbonyl
DME	dimethoxy ethane
FC	Friedel–Crafts
Lg	leaving group
LUMO	Lowest Unoccupied Molecular Orbital
Pg	protecting group
*p*TSA	*p*-toluensulfonic acid
TFA	trifluoro acetic acid
THF	tetrahydrofuran
Ts	tosyl

2.3
Nitroalkenes

Luca Bernardi and Alfredo Ricci

Summary

The literature up to July 2008 dealing with catalytic enantioselective FC alkylation of aromatic compounds with nitroalkenes has been reviewed. The known possibility of activating nitroalkenes for the FC addition of (hetero)aromatic compounds by Brønsted or Lewis acids has inspired the development of several catalytic enantioselective protocols based on H-bond donor organocatalysts and chiral Lewis acids. The use of bifunctional catalysts, able to activate both reaction partners, seems to be, in most cases, a necessary requirement for obtaining satisfactory enantioselectivities. Although indoles were the only (hetero)aromatic reaction partners studied initially in this transformation, more recent protocols have also been developed to encompass pyrroles and electron-rich furans as donors. The synthetic versatility of the nitro group make these adducts useful intermediates for obtaining optically active biologically interesting compounds, such as tryptamine or 1,2,3,4-β-tetrahydrocarboline alkaloids.

2.3.1
Introduction

This chapter deals with catalytic enantioselective Friedel–Crafts alkylations of (hetero) aromatic compounds with nitroalkenes, using different types of chiral catalysts. These transformations can also be considered as Michael-type additions, and nitroalkenes present indeed a high reactivity for the conjugate addition of nucleophilic species [61]. The strong electron-withdrawing properties of the nitro group and its excellent conjugation with the double bond efficiently stabilize negative charges developing during nucleophilic additions, rendering these compounds even more reactive than the corresponding α,β-unsaturated carbonyl compounds. The outstanding synthetic versatility of the nitro group, which can be transformed into amines, carbonyl compounds or removed through standard manipulations, besides giving possibilities of functionalizations at the α-carbon [62], renders Michael-type additions to these electron-poor alkenes extremely useful synthetic tools in organic chemistry.

It has long been recognized that electron-rich heterocyclic compounds can undergo FC type alkylation with nitroalkenes [63]. However, this transformation was not developed in a catalytic asymmetric fashion until 2005. This is in sharp contrast to asymmetric FC alkylations with other acceptors, which have seen flourishing progress from the end of the 1990s [1, 64]. As the activation of the nitro functionality by a chiral Lewis or Brønsted acid is the most obvious option for the realization of an asymmetric FC addition to nitroalkenes, this surprising lack of enantioselective protocols might be ascribed to the poorly defined interactions between the nitro group and the catalyst or promoter. Both oxygen atoms can, in principle, interact with the Lewis or Brønsted acid, even at the same time, giving rise to multiple activated complexes, each having its proper reactivity and stereoselectivity. This unpredictable geometry thus renders extremely challenging the design and development of a chiral species able to provide a suitable asymmetric environment around the alkene.

In the first part of this chapter will be briefly described how the understanding of the geometry of the interactions between nitro compounds and Brønsted acids, combined with the tremendous recent development of chiral Brønsted acid catalysis, have paved the way for the first asymmetric versions of FC alkylation with nitroalkenes. In the second part, chiral Lewis acid catalyzed asymmetric reactions will be considered, showing the development of fruitful combinations of metals and chiral ligands leading to successful and synthetically useful protocols.

2.3.2
Organocatalytic Enantioselective Reactions

2.3.2.1 Friedel–Crafts Alkylation of Indoles
The indole nucleus is one of the most common structures in naturally occurring alkaloids and related compounds, often showing significant biological activity. Indoles are electron-rich heteroaromatic compounds with a remarkable nucleophilic reactivity

at the 3-position, tunable by changing the electronic properties of the substituent at the nitrogen atom. These features make them ideal candidates for the development of asymmetric FC alkylations and, consequently, they were the first aromatic compounds investigated for catalytic enantioselective FC additions to nitroalkenes.

2.3.2.1.1 Thiourea Catalysts The hydrogen bond interaction between the basic oxygen atoms of nitroalkenes and different acidic promoters has been widely exploited for activation of these electron-poor alkenes towards FC alkylations. However, due to the aforementioned unpredictable geometry of the resulting complexes, it has always been difficult to realize a suitable chiral Brønsted acid able to shield preferentially one of the two prochiral faces of the alkene. Two major discoveries prepared the ground for the development of the first examples of catalytic asymmetric FC reactions of nitroalkenes: the striking utility and efficacy of chiral urea and thiourea catalysts developed by Jacobsen and Takemoto for several asymmetric transformations [65], and the observed recognition of the nitro group by N,N'-diarylureas in the solid state, giving crystals showing a well-defined double hydrogen bond interaction between the two oxygen atoms of the nitro group and the acidic protons of the urea moiety (Figure 2.14) [66].

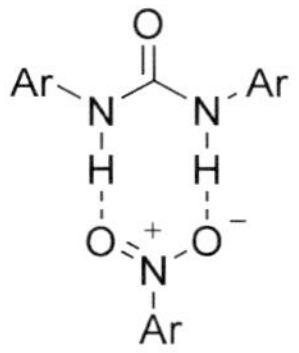

Figure 2.14 Double hydrogen bond interaction between ureas and nitroarenes observed in the solid state.

This well-defined interaction was first exploited by Herrera and Ricci for the acceleration of the FC alkylation of electron-rich indoles, pyrroles and aniline derivatives with nitroalkenes (Scheme 2.26) [67] using the 1,3-bis(3,5-bis(trifluoromethyl)phenyl)thiourea catalyst **1**, previously introduced by Schreiner for carbonyl group activation [68]. A double hydrogen bond interaction of the catalyst with the nitroalkene, according to the geometry observed by X-ray (Figure 2.14), was believed to decrease the electron density at the β-carbon, thus promoting the addition.

This work was the basis for the development of an asymmetric version of this transformation [69]. A screening of thiourea catalysts bearing the 3,5-bis(trifluoromethyl)phenyl moiety on one side and a chiral alcohol on the other, identified catalyst **2** derived from (1R,2S)-cis-1-amino-2-indanol as the most promising structure. Using catalyst **2** the FC alkylation of a series of indoles with nitroalkenes could be performed with generally good yields and moderate to good enantioselectivities (Table 2.20).

Additional experiments were carried out to give some insight into a possible reaction intermediate: N-methyl indole furnished the product in nearly racemic form, and catalysts structurally similar to **2** but lacking the free alcoholic moiety were much less efficient in terms of yield and enantioselectivity (Figure 2.15a).

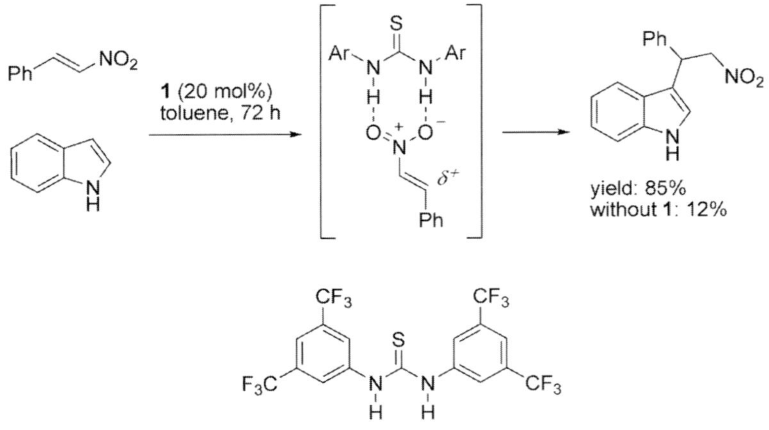

Scheme 2.26 Thiourea **1** catalyzed FC alkylation of indole with *trans*-β-nitrostyrene.

A bifunctional mode of action of catalyst **2**, remembering Takemoto's proposal for the addition of malonates to nitroalkenes [70], was therefore suggested by the authors (Figure 2.15b), with simultaneous activation of the nitroalkene by the thiourea, and of the indole by the alcohol through coordination with the N–H. The result of this unexpected coordination at the indole N–H can be rationalized considering an increase in the electron density of the indole, thus augmenting its nucleophilicity at

Table 2.20 Asymmetric FC alkylation of indoles with nitroalkenes catalyzed by **2**: representative results.

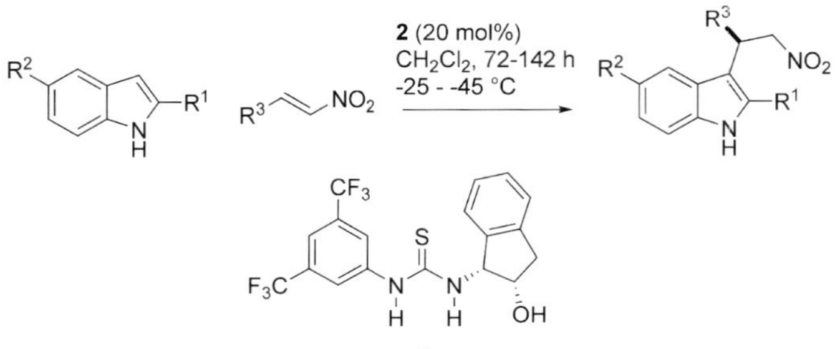

R^1, R^2, R^3	Yield (%)	ee (%)	R^1, R^2, R^3	Yield (%)	ee (%)
H,H,Ph	78	85	H,H,2-furyl	88	73
Me,H,Ph	82	74	H,H,2-thienyl	70	73
H,OMe,Ph	86	89	H,H,*n*-pentyl	76	83
H,Cl,Ph	35	71	H,H,*i*Pr	37	81

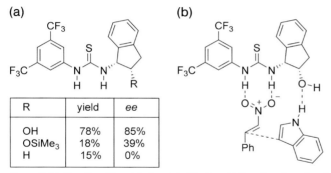

Figure 2.15 Results with catalysts related to 2 and bifunctional mode of action.

C_3 and promoting the addition. However, an alternative explanation for the role of this group is the stabilization of positive charges developing at the indole nitrogen during the addition, or assistance in proton-transfer processes occurring in reaction intermediates, similar to the "proton slide" mechanism operative in many enzymatic reactions [71].

A second catalytic system for FC additions to nitroalkenes, based on the thiourea motif, was proposed by Connon. Reasoning that bis-N,N'-aryl thiourea catalysts are generally more active than the corresponding N-aryl N'-alkyl derivatives, Connon synthesized a series of axially chiral C_2-symmetric bis-urea and bis-thiourea structures derived from BINAM, and tested them in the FC alkylation of N-methyl indole with trans-β-nitrostyrene. The most efficient catalyst 3, bearing two 3,5-(bistrifluoromethyl)phenyl thiourea moieties, was then used in the reaction of N-methyl indole with a series of aromatic and aliphatic nitroalkenes, with good results in terms of yields but generally poor enantioselectivities (Scheme 2.27) [72].

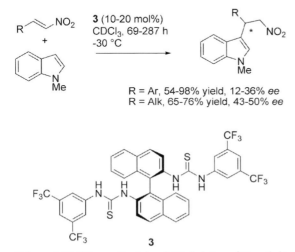

Scheme 2.27 Thiourea 3 catalyzed FC alkylation of N-methyl indole with nitroalkenes.

Nitroalkenes bearing aliphatic substituents gave better enantioselectivities, and in this case N–H indole performed poorly in the reaction. An X-ray structure determination of catalyst 3 showed, surprisingly, an *s-trans,cis* conformation at the two thiourea moieties. In each thiourea the two protons are thus pointing in opposite directions. Considering the considerable distance between the two thioureas (3.53 Å compared with a distance of about 2.15 Å between the two oxygens of nitrostyrene), the authors suggested the intriguing possibility that the catalyst, in this case, might act as a single hydrogen bond donor.

2.3.2.1.2 **Bissulfonamide Catalysts** Bissulfonamides derived from C_2-symmetric chiral amines were employed by Corey as ligands for boron and aluminum salts in Lewis acid catalyzed enantioselective Diels–Alder reactions [73]. Envisioning a possible activation of basic substrates by the very acidic protons of these compounds, Jørgensen investigated their behavior in the FC alkylation of *N*-alkyl indole derivatives with nitroalkenes [74]. The acidity of the catalytic species was found to be crucial for the activation of the nitroalkenes, with only the very acidic trifluoromethylsulfonyl derivatives being able to promote the reaction. Using catalyst 4 derived from (1*R*,2*R*)-1,2-diphenylethane-1,2-diamine (2 mol%), several *N*-alkyl indoles were reacted with a series of nitroalkenes, providing the products in good yields but only moderate enantioselectivities (13–63%), which however could be consistently improved by crystallization (Scheme 2.28).

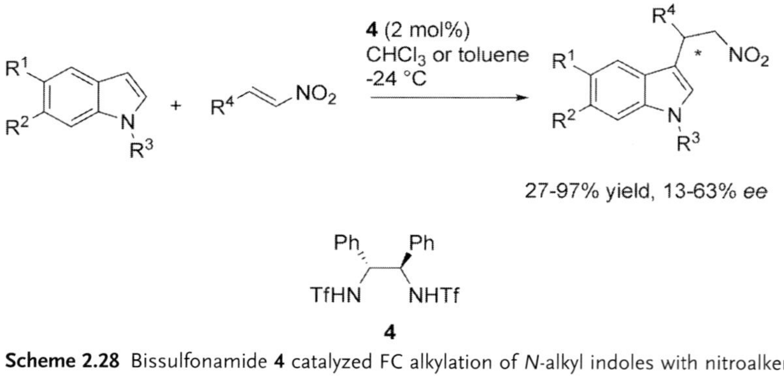

Scheme 2.28 Bissulfonamide 4 catalyzed FC alkylation of *N*-alkyl indoles with nitroalkenes.

On the basis of the X-ray structures of catalyst 4 and other 1,2-diamine-derived bissulfonamides, a model involving a single point binding of the nitroalkene to the catalyst through hydrogen bond was invoked (Figure 2.16a). The enantioselectivity of the transformation depends on the substituents at the carbon backbone of the catalyst, which from the X-ray structures were found to influence profoundly the N–C–C–N dihedral angle; for example, a catalyst with *tert*-butyl groups shows a small dihedral angle of 19.4° (X-ray) and gives the product in nearly racemic form, whereas catalyst 4 bearing phenyl groups has a larger dihedral angle of 64.2° (X-ray) and gives up to 64% *ee* (Figure 2.16b).

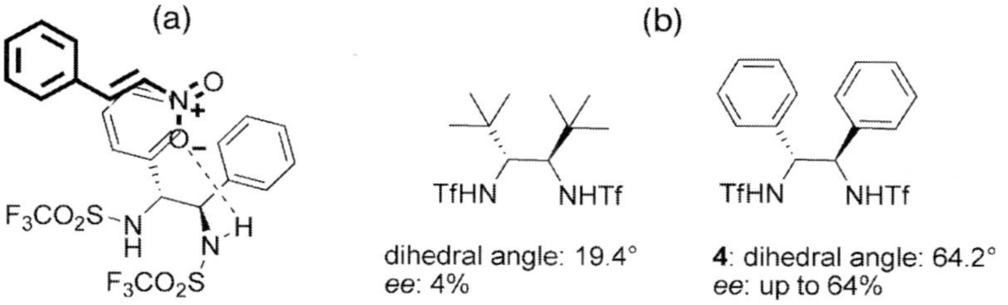

Figure 2.16 One point binding of catalyst **4** and influence of the dihedral angle on the enantioselectivity.

(a) F$_3$CO$_2$S-NH, F$_3$CO$_2$S-NH

(b) TfHN, NHTf — dihedral angle: 19.4°, ee: 4%

4: TfHN, NHTf — dihedral angle: 64.2°, ee: up to 64%

2.3.2.1.3 **Phosphoric Acid Catalysts** Chiral phosphoric acids derived from BINOL have been shown to be extremely powerful catalysts for a variety of asymmetric transformations, mostly involving imines which can be activated through protonation by the rather strong chiral acid [75]. However, Akiyama showed for the first time that these catalysts are also able to activate the less basic nitroalkenes, presumably through a single point binding, specifically for the asymmetric FC alkylation of indoles [76]. Under the optimized conditions, a range of indoles were alkylated with good yields and excellent enantioselectivities by reaction with several nitroalkenes used in excess (2–5 equiv.) (Table 2.21), in the presence of the 3,3′-bis-triphenylsilyl BINOL-derived phosphoric acid catalyst **5** [77]. Molecular sieves (3 Å) were found to influence dramatically the enantioselectivity and the yield of the reaction.

Table 2.21 Asymmetric FC alkylation of indoles with nitroalkenes catalyzed by **3**: representative results.

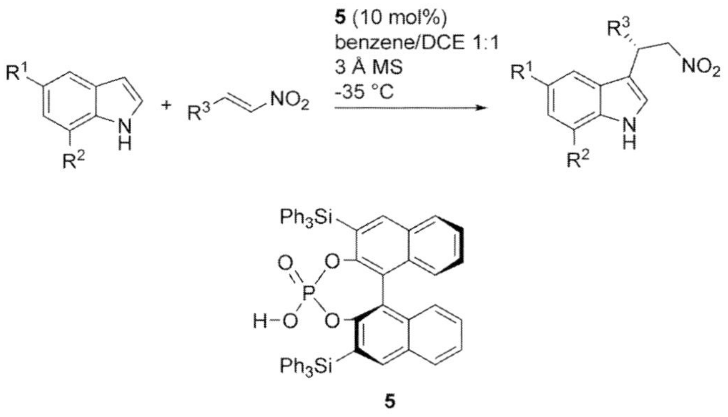

Conditions: **5** (10 mol%), benzene/DCE 1:1, 3 Å MS, −35 °C

Catalyst **5**: 3,3′-bis(Ph$_3$Si)-BINOL phosphoric acid

R^1, R^2, R^3	Yield (%)	ee (%)	R^1, R^2, R^3	Yield (%)	ee (%)
H,H,Ph	76	91	H,H,Ph(CH$_2$)$_2$	57	88
H,H,4-MeC$_6$H$_4$	64	90	H,H,CH$_3$(CH$_2$)$_2$	70	90
H,H,4-MeOC$_6$H$_4$	74	91	Cl,H,Ph	63	90
H,H,2-thienyl	71	90	H,Me,Ph	70	94

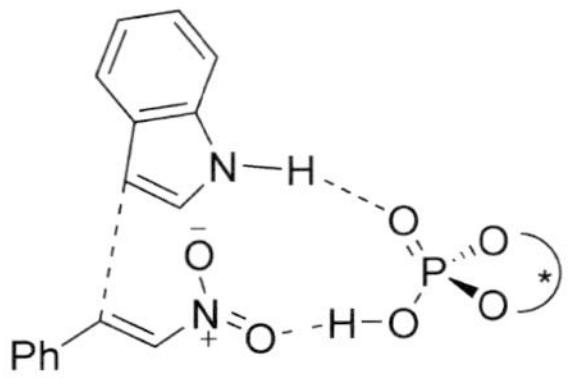

Figure 2.17 Cyclic transition state proposed by Akiyama.

At first sight a single point binding of the nitroalkene to the phosphoric acid catalyst could be considered as solely responsible for the promotion of the reaction. However, on the basis of the poor results obtained with N-methyl indole, Akiyama assumed a bifunctional behavior of catalyst **5**, with the acidic proton coordinating the nitro group and the phosphoryl oxygen the indole N—H (Figure 2.17), in agreement with the cyclic transition state previously proposed by the same author for other types of transformations and with the bifunctional mode of action of the thiourea **2** (Figure 2.15) proposed by Ricci.

2.3.2.1.4 **Synthetic Applications** Tryptamine is a mono-amine alkaloid found in plants and mammals. Biosynthetically derived from the α-amino acid tryptophan, tryptamine and its derivatives (Figure 2.18) often play key roles as neurotransmitters or neuromodulators. The optically active β-nitro-indol-3-yl products deriving from the FC alkylation of indoles with nitroalkenes are obvious precursors of tryptamine derivatives bearing an unusual chiral center at the benzylic position.

Ricci [69], Jørgensen [74] and Akiyama [76] demonstrated the possibility of reducing the nitro group of some of the products deriving from the catalytic reactions giving the corresponding tryptamines. The obtained amine could be isolated in good yield and with no reduction in enantioselectivity after derivatization as exemplified in Scheme 2.29. Alternatively, the obtained tryptamine could be condensed with benzaldehyde by means of a Pictet–Spengler cyclization (Scheme 2.29). This transformation furnishes 1,2,3,4-tetrahydro-β-carboline derivatives, ubiquitous structures in natural

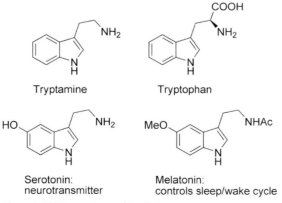

Figure 2.18 Tryptamine and its derivatives.

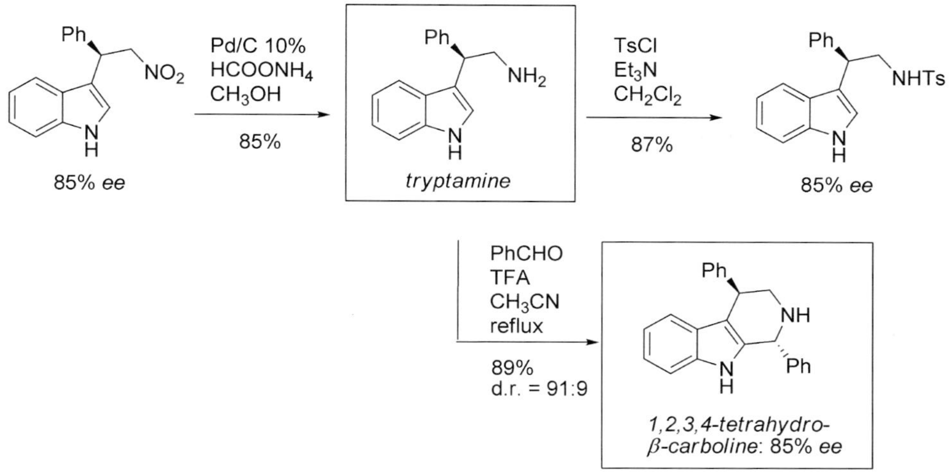

Scheme 2.29 Tryptamines and 1,2,3,4-tetrahydro-β-carbolines from the FC products.

and biologically active alkaloids. Remarkably, many naturally occurring β-carbolines derive from tryptamines via a similar biosynthetic cyclization process.

2.3.3
Friedel–Crafts Reactions of Naphthols

Although, historically, the FC alkylations of phenols and naphthols were amongst the first FC reactions to be developed in a catalytic asymmetric fashion [67a, b], it was not until 2007 that the first example of an enantioselective alkylation of naphthols with nitroalkenes appeared in the literature [78].

2.3.3.1 Thiourea Catalysts
Considering the possibility of a simultaneous activation of 2-naphthol by a Brønsted base and of a nitroalkene by a weak Brønsted acid, Chen explored the capability of different bifunctional organic catalysts for promotion of the reaction between 2-naphthol and *trans*-β-nitrostyrene (Figure 2.19).

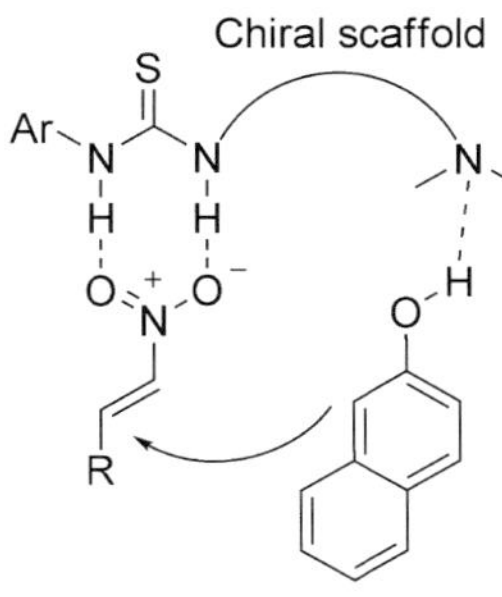

Figure 2.19 Chen's working hypothesis.

Table 2.22 Asymmetric FC alkylation of 2-naphthols: representative results.

R¹, R², R³	Yield (%)	ee (%)	R¹, R², R³	Yield (%)	ee (%)
H,H,Ph	80	93	H,H,3-MeC$_6$H$_4$	72	91
H,H,4-MeC$_6$H$_4$	69	85	H,H,CH$_3$(CH$_2$)$_2$	69	94
H,H,4-MeOC$_6$H$_4$	74	85	MeO,H,Ph	81	91
H,H,2-thienyl	79	94	H,Br,Ph	72	90

The thiourea catalyst **6** derived from cinchonine, previously developed by the same authors [79], was found to give the corresponding FC adduct with excellent levels of enantioselectivity at low temperature, and was employed in the reaction with different nitroalkenes and 2-naphthols (Table 2.22). The FC products were obtained with excellent results, using both aromatic and aliphatic nitroalkenes.

A single example dealing with 1-naphthol, catalyzed by a quinine-derived catalyst structurally related to **6**, was also reported, although the reaction conditions were slightly modified and the results in terms of enantioselectivity (80% *ee*), were not as good as in the case of 2-naphthol. Serendipitously, it was also found that by increasing the reaction time to 144 h, instead of 96 h, a dimeric compound resulting from the condensation of two molecules of FC adduct could be obtained in moderate yield (Scheme 2.30).

The higher enantiomeric excess observed in this dimeric product compared to the parent FC adduct, together with other experimental evidence, was taken as a proof of the involvement of the catalyst **6** in a kinetic resolution step leading to the dimer. On this basis, a reaction pathway involving the attack of a nitronate on the nitro group of a second molecule, followed by loss of water, intramolecular attack of the phenolic oxygen atoms and finally loss of HNO$_2$ was proposed by the authors (Scheme 2.30).

2.3.4
Addition of 1,2,3-Triazoles

1,2,3-Triazoles and derivatives are known to add to electron-deficient alkenes with one of their nitrogen atoms forming a C−N bond, followed by a proton transfer in the

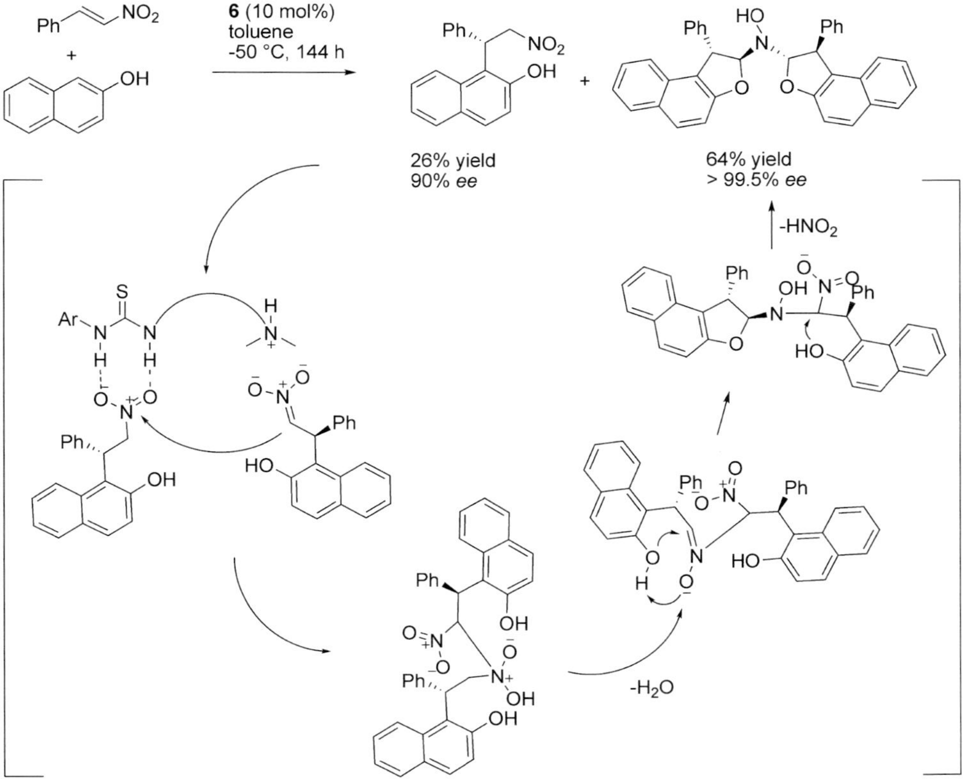

Scheme 2.30 Dimeric product obtained upon prolonged reaction time.

intermediate adducts restoring the heteroaromatic nucleus. Considering the analogy of this process with more classical FC reactions involving the formation of a C–C bond, it is not surprising that the development of the asymmetric addition of these important heteroaromatic compounds to nitroalkenes followed a similar approach to other FC asymmetric reactions.

2.3.4.1 Cinchona Alkaloids Catalysts

The feasibility of the enantioselective addition of benzotriazole to *trans*-β-nitrostyrene was explored by Wang using a series of bifunctional organic catalysts, mostly derived from *Cinchona* alkaloids. Having recognized the crucial role of hydroxy groups in enantioselectivity, catalyst **7** derived from quinidine through demethylation [80] was used in the asymmetric addition of benzotriazole to a series of aromatic, heteroaromatic and aliphatic nitroalkenes with moderate to good results (Scheme 2.31) [81]. The N1 addition product was formed exclusively in all cases and with all catalysts tested, in contrast to the chiral Lewis acid catalyzed addition of this heterocyclic compound to α,β-unsaturated imides and ketones proposed by Jacobsen, which invariably gave a mixture of N1 and N2 adducts [82].

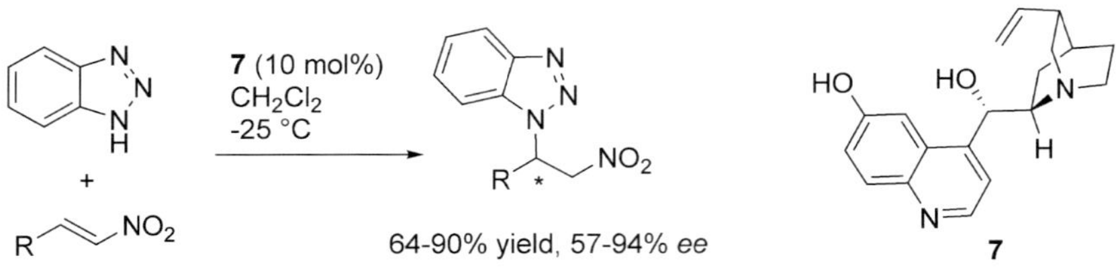

Scheme 2.31 Asymmetric addition of benzotriazole to nitroalkenes.

The procedure was also applied successfully to the addition of other N-heterocycles, such as 1,2,3-triazoles and the important pharmacophores, tetrazoles, although only a single example for each heterocycle was provided. Purines, however, were found to be unreactive.

2.3.5
Conclusion and Outlook

The few examples of organocatalytic asymmetric FC alkylation of indoles with nitroalkenes seem to indicate the requirement of a highly organized transition state in order to obtain good enantioselectivities. The use of thioureas, capable of a double hydrogen-bond interaction between the nitro group and their acidic protons, was considered at the beginning to be a prequisite for a well-defined transition state. However, at present, it seems that a mandatory feature in the achievement of good enantioselectivities is instead the presence of a Brønsted basic functionality in the catalyst structure, able to coordinate the indole N−H. This interaction, capable of activating the indole for the addition, stabilizing developing positive charges in the transition state, and eventually assisting in proton transfer processes, is of fundamental importance for the definition of the geometry of the enantioselectivity determining step. Single point binding catalysts for the activation of the nitroalkene through a hydrogen bond, which in the beginning were employed with only limited success due to the rather unpredictable complexation geometry, could thus also be used with great success, provided that the acidic moiety in the catalyst is flanked by a second Brønsted base functionality, as shown by Akiyama. Similar observations hold for 2-naphthol and 1,2,3-triazoles, although the available data on these systems are extremely scarce. In these cases the need for activation of the FC donor through coordination of a basic catalyst at the acidic phenolic or triazole N1 proton was perceived already in the first stages of the investigations, probably on the basis of an intuitively higher acidity of these protons and literature precedents employing basic catalysts for the activation of 2-naphthols [83].

In conclusion, although these recent advances indicate the feasibility of organocatalytic enantioselective FC additions with nitroalkenes, the reported procedures are limited to a very small number of FC donors. Furthermore, although the reported protocols show without doubt that these transformations are possible, they are still far from satisfactory in terms of their real applicability to medium to large scale synthesis. The reason is the rather high catalyst loading, the cost and the difficult preparation of some of the catalysts used, the long reaction times and the low temperatures and, finally, the difficult separation of the products from the catalysts, at least in the case of some of the thiourea compounds.

This field is thus very far from mature and it is possible to predict that the future will witness tremendous advances directed towards not only to the use of other classes of aromatic and heteroaromatic compounds for the alkylation with nitroalkenes, but also to the development of new and more user-friendly procedures, easily applicable on a large scale.

2.3.6
Experimental: Selected Procedures

Representative procedure for the catalytic asymmetric FC alkylation of indoles with nitroalkenes catalyzed by 2 (Table 2.20) [72]

In a test tube, to a solution of *trans*-β-nitrostyrene (14.9 mg, 0.10 mmol) and catalyst 2 (8.3 mg, 0.02 mmol) in CH_2Cl_2 (previously filtered through basic Al_2O_3, 100 μL), cooled to $-24\,°C$, indole (17.6 mg, 0.15 mmol) was added in one portion. The test tube was then placed in a freezer at $-24\,°C$ for 72 h, then the product 3-(2-nitro-1-phenyl-ethyl)-1*H*-indole was obtained by chromatography on silica gel (*n*-hexane-EtOAc mixtures) as a colorless oil in 78% yield (20.7 mg). The *ee* of the product was determined by HPLC using a Daicel Chiralpak AD-H column (*n*-hexane/iPrOH 90:10, flow rate 0.75 mL min^{-1}, $t_{maj} = 33.0$ min; $t_{min} = 36.4$ min, 85% *ee*). $[\alpha]_D^{23} = -8$ ($c = 0.65$, $CHCl_3$); ^1H NMR (400 MHz, $CDCl_3$) δ 8.08 (br s, 1H), 7.58–6.95 (m, 10H), 5.19 (t, $J = 8.2$ Hz, 1H), 5.07 (dd, $J = 12.4, 7.2$ Hz, 1H), 4.95 (dd, $J = 12.4, 8.2$ Hz, 1H); ^{13}C NMR (100 MHz, $CDCl_3$) δ 139.2, 137.9, 128.9, 127.8, 127.6, 124.7, 122.7, 121.6, 120.0, 118.9, 114.5, 111.4, 79.5, 41.6; ESIMS *m/z* 267 $[M^+ + H]$.

Representative procedure for the catalytic asymmetric FC alkylation of indole with nitroalkenes catalyzed by 5 (Table 2.21) [79]

To a cooled ($-35\,°C$) suspension of activated powder MS 3 Å (40 mg), (*R*)-3,3′-bis (triphenylsilyl)-1,1′-binaphthyl phosphate 5 (17.3 mg, 0.02 mmol) and (*E*)-1-methyl-4-(2-nitrovinyl)benzene (163.1 mg, 1.00 mmol) in a benzene/dichloroethane mixture (1:1, 1.0 mL), indole (26.3 mg, 0.20 mmol) was added. After being stirred at the same temperature for 116 h, the reaction mixture was purified directly by column chromatography on silica gel (*n*-hexane/EtOAc mixtures) giving (*S*)-3-(1-(4-methylphenyl)-2-nitroethyl)-1*H*-indole in 64% yield (35.8 mg). The *ee* of the product was determined by HPLC using a Daicel Chiralcel OD-H column (*n*-hexane/iPrOH = 70:30, flow rate 1.0 mL min^{-1}, $t_{maj} = 18.1$ min; $t_{min} = 28.7$ min, 90% *ee*). $[\alpha]_D^{23} = +13.2$ ($c = 1.0$, CH_2Cl_2); ^1H NMR (400 MHz, $CDCl_3$) δ 8.03 (br s, 1H), 7.43 (d, $J = 7.9$ Hz, 1H), 7.31 (d, $J = 8.1$ Hz, 1H), 7.23–7.16 (m, 3H), 7.11–7.04 (m, 3H), 6.97 (d, $J = 1.8$ Hz, 1H,), 5.13 (t, $J = 8.0$ Hz, 1H), 5.02 (dd, $J = 12.4, 7.6$ Hz, 1H), 4.89 (dd, $J = 12.4, 8.3$ Hz, 1H), 2.29 (s, 3H); ^{13}C NMR (100 MHz, $CDCl_3$) δ 137.2, 136.4, 136.1, 129.5, 127.6, 126.0, 122.6, 121.5, 119.8, 118.9, 114.5, 111.3, 79.6, 41.1, 21.0.

Representative procedure for the catalytic asymmetric FC alkylation of 2-naphthols with nitroalkenes catalyzed by 6 (Table 2.22) [81]

Catalyst **6** (5.6 mg, 0.01 mmol, 10 mol%), 2-naphthol (14.4 mg, 0.10 mmol), and 4 Å MS (20 mg) were stirred in dry toluene (0.80 mL) and cooled to −50 °C under Ar. Then *trans*-β-nitrostyrene (22.3 mg, 0.15 mmol) in dry toluene (0.2 mL) was added. After 96 h, the product was directly purified by flash chromatography on silica gel (previously saturated with cold petroleum ether) to give (*S*)-1-(2-nitro-1-phenylethyl) naphthalen-2-ol in 80% yield (23.4 mg) as a yellow oil. The *ee* of the product was determined by HPLC using a Daicel Chiralcel AS column (*n*-hexane/*i*PrOH = 90 : 10, flow rate 1.0 mL min^{-1}, t_S = 13.9 min; t_R = 16.1 min, 93% *ee*). $[\alpha]_D^{20} = -27.3$ (*c* = 0.26, CHCl$_3$); ^{1}H NMR (300 MHz, CDCl$_3$) δ 8.12 (d, *J* = 8.7 Hz, 1H), 7.79 (d, *J* = 8.1 Hz, 1H), 7.70 (d, *J* = 8.8 Hz, 1H), 7.55–7.50 (m, 1H), 7.39–7.34 (m, 3H), 7.31–7.22 (m, 3H), 6.97 (d, *J* = 8.8 Hz, 1H), 5.87 (t, *J* = 7.5 Hz, 1H), 5.47 (dd, *J* = 7.8, 13.2 Hz, 1H), 5.33 (dd, *J* = 7.1, 13.2 Hz, 1H), 5.28 (s, 1H); ^{13}C NMR (75 MHz, CDCl$_3$) δ 151.4, 139.5, 133.0, 130.1, 129.7, 128.9, 128.8, 127.4, 127.2, 123.6, 122.4, 118.4, 117.6, 78.2, 41.0; ESI-HRMS: calc. for C$_{18}$H$_{15}$NO$_3$ + Na 316.0944, found 316.0941.

2.3.7
Lewis Acid Catalyzed Enantioselective Reactions

The Michael addition of electron-rich heterocycles to nitroalkenes using Lewis acids has been well documented in the literature. However, in numerous cases clearly exemplified by indole and pyrrole, careful control over the acidity is necessary to avoid undesirable side reactions such as dimerization and polymerization, and milder reaction conditions are required to implement a useful FC protocol for these substrates. For this purpose, catalytic amounts of Lewis acids such as lanthanide triflates [84], indium bromide [85], bismuth triflate [86], iodine [87], and solid acids [88] have been proven to be useful for developing a non-asymmetric FC alkylation reaction of arenes with nitroalkenes. On the other hand the asymmetric version of this important reaction, previously only poorly investigated, has been recently disclosed by using chiral ligand–metal complexes as asymmetric catalysts that allow the FC alkylation reaction to occur under mild conditions and achieving excellent enantio-, regio-selectivities and reactivities.

In asymmetric catalysis, ligands induce asymmetry in a reaction, not only through steric factors but also by generating electronic asymmetry on the metal center through the presence of different donor atoms [89]. The most important and widely used of these heterodentate ligands are those which bear P and/or N as their donor atoms. Among the numerous N-containing chiral ligands, differently structured bi- or tridentate C_2-symmetric bisoxazolines (**8–9**), a ligand derived from axially chiral BINAM (**10**), an imidazoline-aminophenol structure (**11**), a Salen-type (**12**) and also a semi-crown designed phenoxide-based molecule (**13**) (Figure 2.20) have been employed in the metal-catalyzed FC alkylation of arenes and heteroarenes with nitroalkenes, to generate, in combination with the metals of choice, mostly zinc, but also aluminum and copper, efficient catalytic species.

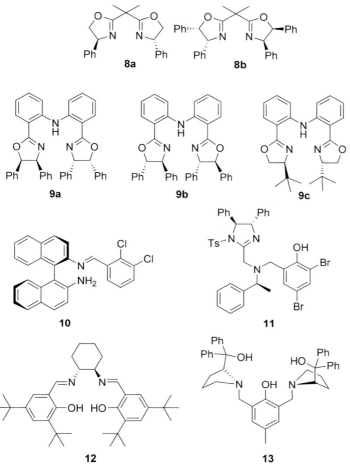

Figure 2.20 Chiral ligands used in asymmetric FC alkylations with nitroalkenes catalyzed by Lewis acids.

2.3.7.1 Friedel–Crafts Reactions of Indoles

2.3.7.1.1 Zinc Catalysts From the beginning the metal-catalyzed FC alkylation of arenes and heteroarenes with nitroalkenes has been mainly addressed to the indole, a privileged structure in the field of medicinal chemistry. A wide range of aromatic, heteroaromatic and also aliphatic nitroalkenes and of substituted indoles have been used in this reaction under different conditions. Selection of the Lewis acid plays a relevant role in the formation of a complex with the chiral ligand capable of displaying the best performance in terms of both chemistry and selectivity. The key role played by the metal species is exemplified for the alkylation of indole in Table 2.23. Zhou and Du showed that a variety of chiral Lewis acid catalysts, generated *in situ* from metal salts and bisoxazolines **8a** [90] and **9a** [91], respectively, led to significantly different

Table 2.23 Asymmetric FC alkylation of indole using different metal salts with ligands **8a** and **9a**.

	Ligand 8a				Ligand 9a		
MX$_n$	t (h)	Yield (%)	ee (%)	MX$_n$	t (h)	Yield (%)	ee (%)
Fe(ClO$_4$)$_2$	60	—	—	Yb(OTf)$_3$	42	51	4
Mg(OTf)$_2$	60	34	0	Cu(OAc)$_2$	122	53	1
Cu(OTf)$_2$	60	67	80	NiCl$_2$	116	87	53
Zn(OTf)$_2$	11	97	79	Zn(OTf)$_2$	8	99	83

reaction outcomes in terms of both yields and enantioselectivities, the best results being obtained by using Zn(OTf)$_2$ as the metal salt.

Moreover with the catalysis by **8a**-Zn(OTf)$_2$ and **9a**-Zn(OTf)$_2$ complexes, a remarkable solvent effect can be observed with toluene, giving better results than polar solvents, while coordinative solvents inhibit the activity of the catalyst (Table 2.24).

On the basis of these results, both zinc complexes could then be used in the asymmetric FC alkylation of a few indole derivatives with several nitroalkenes with good yields and enantioselectivities (Scheme 2.32). Remarkably, these optimized conditions regarding the Lewis acid metal and the reaction medium have found application in the indole alkylation, even in the presence of structurally different ligands such as **10** [92].

Introduction of a 1-Me into the indole ring caused, in the two reactions reported in Scheme 2.32 remarkably different effects, which can be accounted for by the different mechanistic pathways involved. Whereas with **8a**-Zn(OTf)$_2$ catalysis the enantioselectivity was dramatically lowered from 84% *ee* to 31% *ee*, with the **9a**-Zn(OTf)$_2$ complex the reaction gave the expected product with high yields and selectivity (93% yield, 90% *ee*). Regarding the origin of stereoselectivity the binding models of the catalysts to the substrate appear closely related to the ligand structure. A plausible

Table 2.24 Solvents and temperature effects with **8a**-Zn(OTf)$_2$ and **9a**-Zn(OTf)$_2$ complexes.

8a-Zn(OTf)$_2$ catalyst					9a-Zn(OTf)$_2$ catalyst			
Solvent	T (°C)	t (h)	Yield (%)	ee (%)	Solvent	T (°C)	Yield (%)	ee (%)
Toluene	25	5	93	74	Toluene	10	99	90
Toluene	0	15	97	84	Toluene	−20	99	94
Toluene	−10	36	88	83	CH$_2$Cl$_2$	10	91	67
Et$_2$O	15	11	84	70	THF	10	67	16
CH$_2$Cl$_2$	15	11	84	14	hexane	15	94	30
THF	15	11	88	4				

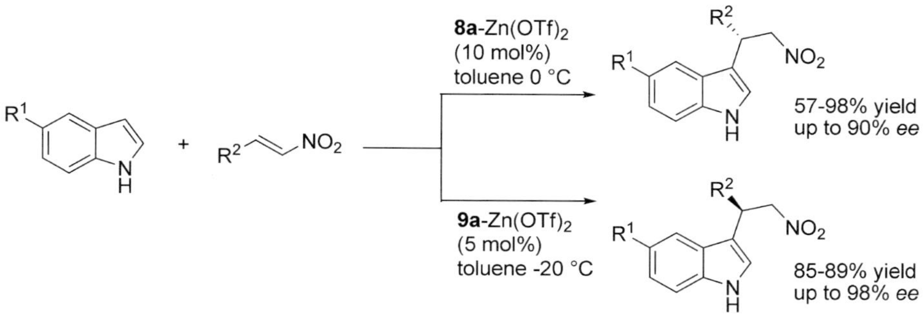

Scheme 2.32 Asymmetric FC alkylation of indole derivatives with nitroalkenes using zinc catalysts.

mechanism for the **8a**-Zn(OTf)$_2$ catalyzed FC reaction between indoles and nitroalkenes implies activation of the nitroalkene by Zn(II) chelation to form a four-membered intermediate (step a) which undergoes nucleophilic addition of the indole to the *Re* face to provide the FC adduct (step b). Subsequently, H-transfer (step c), followed by dissociation (step d), affords the product with the *S* configuration and regenerates the Zn(II)-bisoxazoline catalyst (Scheme 2.33).

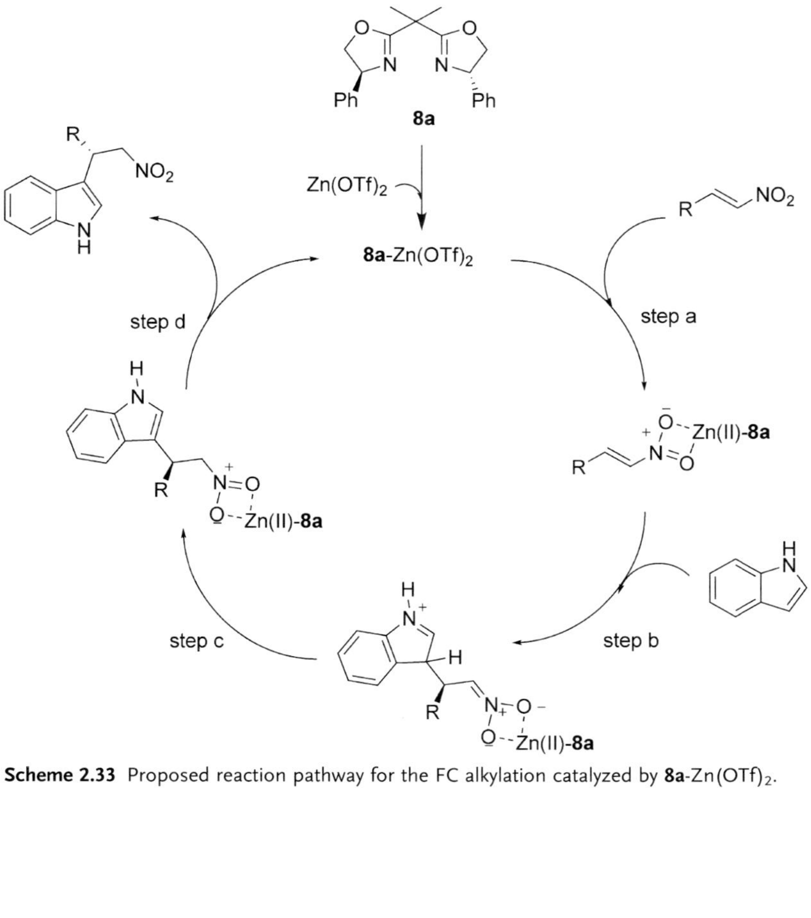

Scheme 2.33 Proposed reaction pathway for the FC alkylation catalyzed by **8a**-Zn(OTf)$_2$.

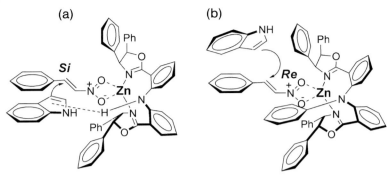

Figure 2.21 Bifunctional mode of action of ligand **9a** and different interactions with its N-phenyl analog.

Analogously, mixing the tridentate ligand **9a** with Zn(OTf)$_2$ in toluene, a 1:1 mononuclear Zn(II) complex is formed. However, in some contrast with the previous complex, this acts in a bifunctional fashion, activating the nitroalkene through the coordination of the nitro group to the Lewis acid center, while the NH group between the two phenyl groups in the ligand acts as a H-bond donor giving a π interaction with the electron-rich cloud of the indole [93]. This second interaction directs the indole attack preferentially to the *Si*-face of the nitroalkene, as depicted in Figure 2.21a. Such a bifunctional mode of interaction in the transition state accounts for the obtainment of the products with *R* configuration. To confirm the role of this π interaction in the origin of the enantioselectivity, a ligand with a large substituent in place of the H atom of the NH fragment (e.g., an *N*-phenyl) was employed, giving the product with an opposite *S* configuration and with very low enantioselectivity (31% *ee*). This result was rationalized by considering that in this case the attack from the back side, instead of being promoted by the aforementioned π-interaction, may be shielded by the bulky phenyl group (Figure 2.21b), thus reversing the absolute configuration of the products.

2.3.7.1.2 Copper and Aluminum Catalysts

Singh demonstrated that copper salts also play an important role in the catalytic asymmetric FC alkylation of indoles [94]. Among a wide series of BOX ligands studied, the C_2-symmetric malonate derived **8b** was the most efficient, in combination with Cu(OTf)$_2$ for the alkylation of indole with nitrostyrene, in CHCl$_3$ as the solvent of choice. The reaction can furnish a variety of nitroalkylated indoles in good to excellent yields (up to 95%) with moderate enantioselectivities (up to 86%). Using the same catalytic system, an interesting variant of this FC alkylation reaction has been recently proposed by Liu and Chen employing nitroacrylates in a catalytic asymmetric FC alkylation of indoles that affords nitro-precursors of chiral tryptophans (Scheme 2.34) [29]. The indoles bearing either electron-withdrawing or electron-donating substituents at the 5-position react with (*Z*)-nitroacrylates smoothly to afford the corresponding products in good yields with excellent enantioselectivities of the *anti*-isomers (82–99%) but moderate diastereoselectivities and enantioselectivities of *syn*-isomers in general.

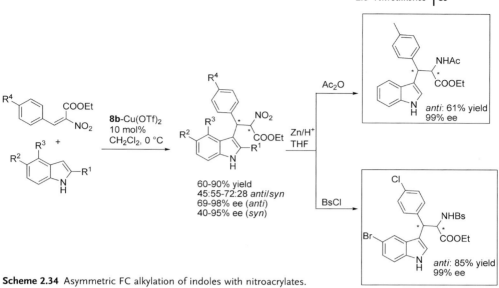

Scheme 2.34 Asymmetric FC alkylation of indoles with nitroacrylates.

The *anti*-alkylation products (99% *ee* after recrystallization) are transformed to tryptophan analogs through hydrogenation of the nitro group by Zn/H$^+$ and acylation or sulfonylation of the primary amino function, without decrease of the optical purity (Scheme 2.34).

The stereochemical outcome has been explained assuming that when nitroacrylate coordinates with the complex **8b**-Cu(OTf)$_2$, a tetrahedral complex is formed, leading to attack at the *Si*-face of the double bond of the nitroacrylate. Although the transition states would adopt two coordination modes (Figure 2.22) which correspond to (*Z*)- and (*E*)-nitroacrylates respectively, both modes give the same configuration (*R*) at the 3-position of the FC adduct.

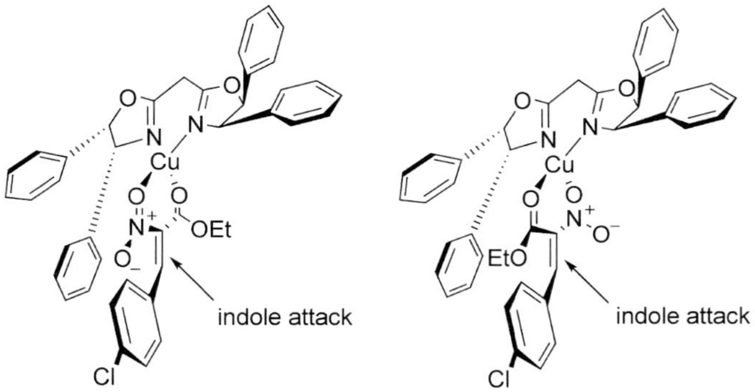

Figure 2.22 (*Z*)-Nitroacrylate (a) and (*E*)-nitroacrylate (b) complexes with the catalyst **8b**-Cu(OTf)$_2$.

The chiral imidazoline-aminophenol ligand **11** was developed by Arai using a combinatorial HTS method [95]. The solution mixtures of several Henry reactions, catalyzed by a combination of copper salts and imidazoline ligands supported on solid phase, were analyzed by CD spectroscopy. The heterogeneous nature of the supported chiral ligand guaranteed no interference in the CD analysis of the reaction solutions. Based on these results, **11** was then synthesized by solution-phase synthesis and was used with very good results in combination with Cu(OAc)$_2$ for a catalytic enantioselective Henry reaction. The same ligand **11**, in combination with a different copper salt ((CuOTf)$_2$–C$_6$H$_6$) was also found to be useful for the enantioselective FC alkylation of indoles with nitroalkenes [95], inspiring a three-component tandem reaction of indole, nitroalkenes, and aldehydes to construct acyclic products with three contiguous stereocenters [96]. This ligand, in combination with (CuOTf)$_2$–C$_6$H$_6$ was highly efficient in promoting a FC and a Henry reaction in two subsequent steps, based on the fact that the reaction of indole with aldehydes is relatively slow compared to the FC reaction of indole with nitroalkenes. With a well organized **11**-CuOTf catalyst satisfying the requirements of the activation of the reaction intermediate (the FC adduct) at the appropriate time in the catalytic cycle, the tandem FC-H product was formed together with variable amounts of the FC adduct, the best results being obtained, as shown in Table 2.25, in the presence of 2 equiv. of HFIP and a 2/1/2 ratio between indole, nitroalkene and aldehyde. By using these optimized conditions, the authors were able to prove the generality of this FC-H reaction in the case of variously substituted aromatic and aliphatic aldehydes and of aromatic and aliphatic nitroalkenes (Table 2.25). Although with electron-deficient aldehydes poor yields were observed at room temperature due to the prevailing formation of a bisindole as the side product, performing the reaction at 0 °C led to the expected FC-H product in good yield and enantioselectivities. The diastereomeric ratio under the optimized conditions appears quite variable, depending upon the reagent structure, but in almost all cases two diastereoisomers were formed with high enantioselectivities, as reported in Table 2.25. *N*-Methylindole can also be used as a substrate with this catalytic system without reduction in the enantiomeric excess and/or chemical yields.

A plausible reaction pathway has been disclosed for this reaction which reasonably starts with the activation of the nitroalkene by the **11**-CuOTf catalyst (Figure 2.23a) giving the FC reaction. The diastereoselective Henry reaction is then promoted by the Cu-nitronate (Figure 2.23b) functionality of the intermediate that results from the FC addition. Although a similar catalyst had been previously used in a nitroaldol reaction

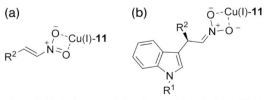

Figure 2.23 Activation of nitroalkenes (a) and of the FC-H intermediate (b) by the **11**-CuOTf complex.

Table 2.25 Asymmetric FC-H alkylation of indole using **11**-CuOTf as the catalyst.[a]

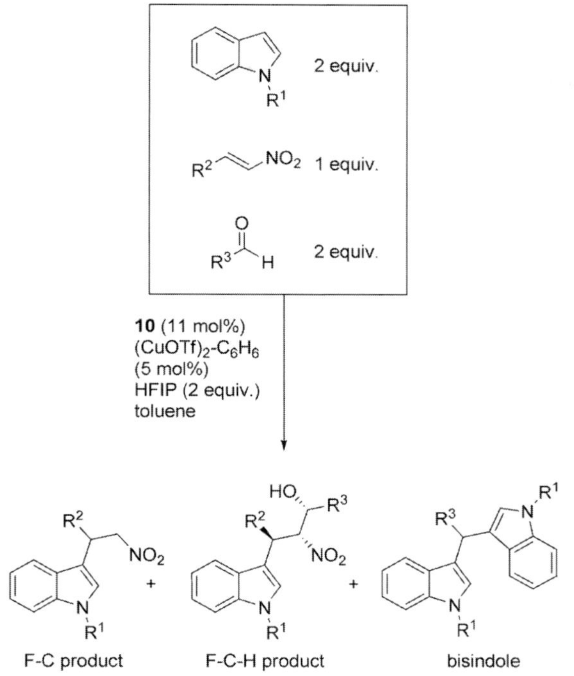

Reaction scheme:
- Indole (2 equiv.) with N-R¹
- R²–CH=CH–NO₂ (1 equiv.)
- R³–CHO (2 equiv.)
- **10** (11 mol%), (CuOTf)₂·C₆H₆ (5 mol%), HFIP (2 equiv.), toluene

Products: F-C product (R² on indole with CH₂NO₂), F-C-H product (R², R³, OH, NO₂ on indole), bisindole.

R¹,R²,R³	t (h)	d.r.	Yield min (%)	ee min (%)	Yield max (%)	ee max (%)
H,Ph,Ph	60	1:19:0	17	70	79	99
H,Ph,4-BrC₆H₄	3	1:9:0	2	7	29	87
H,Ph,4-BrC₆H₄[b]	14	1:16:0	trace	—	84	90
H,Ph,4-ClC₆H₄[b]	17	1:10:0	2	—	82	90
H,Ph,4-NO₂C₆H₄[b]	15	1:10:0	10	18	90	90
H,Ph,n-pentyl	14	1:3:0	14	77	82	99
H,Ph,i-pentyl	21	1:7:0	21	70	79	99
H,n-pentyl,Ph	15	1:2:0	4	—	76	98
H,n-pentyl,4-BrC₆H₄[b]	17	1:1.3:0	—	—	66	99
H,PhC₂H₄,Ph	15	1:1.5:0	12	—	83	90
Me,Ph,Ph[b]	13	1:1.5:0	—	—	72	99

[a]All the reactions were carried out at r.t., unless otherwise specified.
[b]At 0 °C.

using nitromethane [95], in this case the isolated FC product did not react with the aldehyde under the reaction conditions, providing further support for this mechanism. It is worth noting that other conventional Lewis acids such as BEt₃, Ti(OiPr)₄ resulted in the formation of only the FC product and the bisindole.

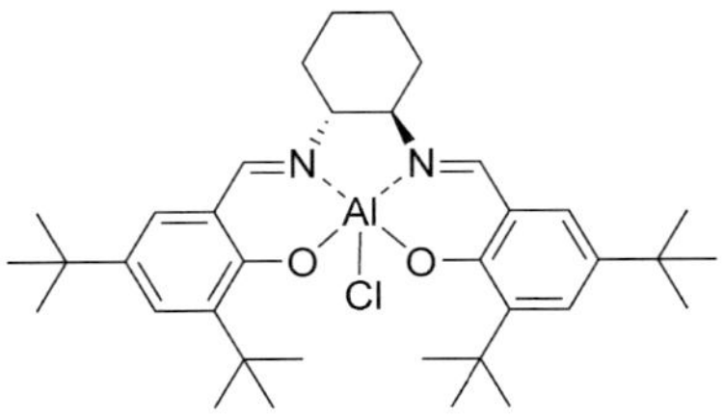

Figure 2.24 Commercially available SalenAlCl complex **14**.

By using the Salen ligand **12** a catalytic protocol was devised by Bandini and Umani-Ronchi based on catalytic amounts of the commercially available chiral [SalenAlCl] complex **14** (Figure 2.24) and pyridine as an additive, for the FC alkylation of indoles with aromatic nitroalkenes, in CH_2Cl_2 as the reaction medium [97]. The β-indolyl nitro compounds could be isolated with satisfactory yields but with moderate enantioselectivities (up to 63% *ee*) due to the occurrence of an uncatalyzed background reaction. Although the enantioselectivities obtained with this protocol are not as high as in the cases of zinc and copper BOX complexes, it must be recognized that this report represents the first example of an asymmetric FC alkylation with nitroalkenes catalyzed by chiral Lewis acids.

2.3.7.2 Friedel–Crafts Reactions of Furans and Pyrroles

Among the stereoselective alkylations of aromatic and heteroaromatic compounds, reports of the use of pyrroles and furans remain rare. Because of the relative instability of these heterocyclic compounds towards an acidic environment, classical FC reactions are unsuitable. Very recently, however, conditions have been envisaged to implement a useful FC protocol for these substrates.

2.3.7.2.1 Zinc Catalysts
The first and only catalytic asymmetric FC alkylation of a furan with nitroalkenes, specifically of the electron-rich 2-methoxyfuran, was developed by Du in 2007 [98]. Optimization of the model reaction between 2-methoxyfuran and *trans*-β-nitrostyrene led to the choice of $Zn(OTf)_2$ as the suitable metal salt, rare earth triflates and Cu(I/II) triflates leading to both low yields and low enantioselectivities, while no formation of the desired product was observed when other metal triflates were used. This salt in combination with diphenylamine-tethered BOX ligand **9b** led, on the other hand, to the formation of the most efficient catalytic complex in terms of enantioselectivity and chemical yields (Scheme 2.35). A wide range of nitroalkenes, mostly aromatic and heteroaromatic, can be engaged in this reaction and the excellent enantioselectivities observed (up to 96% *ee*) are unaffected by the electronic properties of the substituents at *para*- and *meta*-positions. A sizeable erosion of enantioselectivity can however be noticed in the case of *ortho*-substitution of the phenyl ring. The lack of reactivity of a variety of other electron-rich furans such as 2-methylfuran and phenylfuran with *trans*-β-nitrostyrene outlines, on the other hand, a serious limit to the scope of this reaction. The transition state for this reaction was assumed to be very similar to that previously outlined in Figure 2.21 in the case of the

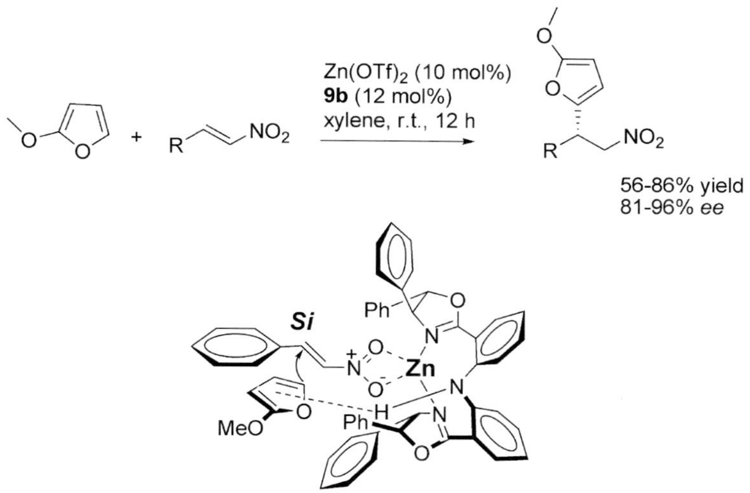

Scheme 2.35 Asymmetric FC alkylation of 2-methoxyfuran with nitroalkenes and transition state.

alkylation of the indole, with the catalyst working in a bifunctional fashion through the Zn(II) metal center and the NH group (Scheme 2.35).

The reaction of unprotected pyrroles with a variety of differently substituted nitroalkenes to give mono- and disubstituted pyrroles has been very recently disclosed by Trost [99], using as catalyst the dinuclear zinc complex **15** prepared by treating the bis-ProPhenol ligand **13** with 2 equiv. of Et$_2$Zn in THF at room temperature [100]. As shown in Table 2.26 the reaction has a wide scope, combined with excellent stereoinduction achieved with aromatic, heteroaromatic and aliphatic substituents with both branched and straight chain alkyl groups.

The access to disubstituted pyrroles with high stereoselectivities was performed through a two-step sequence by combining a MW or InCl$_3$ promoted addition of methyl- or phenylvinyl ketone to pyrrole with the **15** catalyzed FC pathway (Scheme 2.36).

To rationalize these results the authors have devised an attracting mechanism in which the dinuclear Zn-complex **15** displays double activating ability through the two

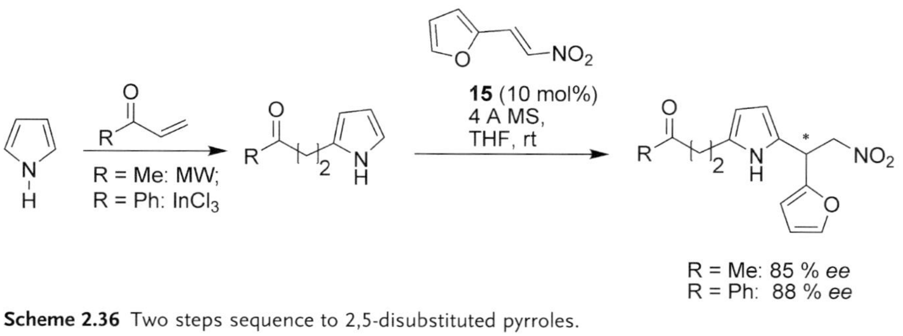

Scheme 2.36 Two steps sequence to 2,5-disubstituted pyrroles.

Table 2.26 Asymmetric FC alkylation of pyrroles with nitroalkenes.

R¹, R²	Yield (%)	ee (%)	R¹, R²	Yield (%)	ee (%)
H,4-MeC$_6$H$_4$	58	97	H,EtOOCCH$_2$	38	56
H,Ph	51	94	H,i-Pr	52	90
H,4-MeOC$_6$H$_4$	56	93	H,n-Pr	56	96
H,2-MeOC$_6$H$_4$	52	87	CH$_3$CO(CH$_2$)$_2$,2-furanyl	61	85
H,2-furanyl	90	94	CH$_3$CO(CH$_2$)$_2$,i-Pr	60	15
H,cyclohexyl	92	92	PhCO(CH$_2$)$_2$,2-furanyl	52	88
H,BnOCH$_2$	34	76	PhCO(CH$_2$)$_2$,i-Pr	67	63

metal atoms (Scheme 2.37). The proposed reaction pathway involves the deprotonation of pyrrole by the pre-catalyst accompanied by the formation of 1 equiv of ethane, followed by coordination of the nitroalkene to the second Zn atom and subsequent alkylation by the pyrrole. A proton exchange with an incoming pyrrole molecule to release the product and reform the active catalyst closes the catalytic cycle (Scheme 2.37).

Good yields and enantioselectivities for a FC alkylation of pyrroles with nitroalkenes were also obtained by Du using a BOX catalytic system, based on the combination of Zn(OTf)$_2$ with the tridentate ligand **9c** [91b], which was superior to its analogs **9a,b** previously used for the FC alkylation of indoles and of 2-methoxyfuran.

2.3.7.3 Conclusion and Outlook

From this brief overview of the chiral Lewis acid catalyzed FC alkylation of electron-rich heteroaromatic systems with nitroalkenes, the careful choice of the metal and of the reaction medium emerges as a major aspect that dictates the chemical and the stereochemical outcome. The need for the formation of stereospecific highly reactive complexes between the nitroalkene and the Lewis acid is a mandatory feature of these reactions. Although, in principle, a wide range of potential metal salts are

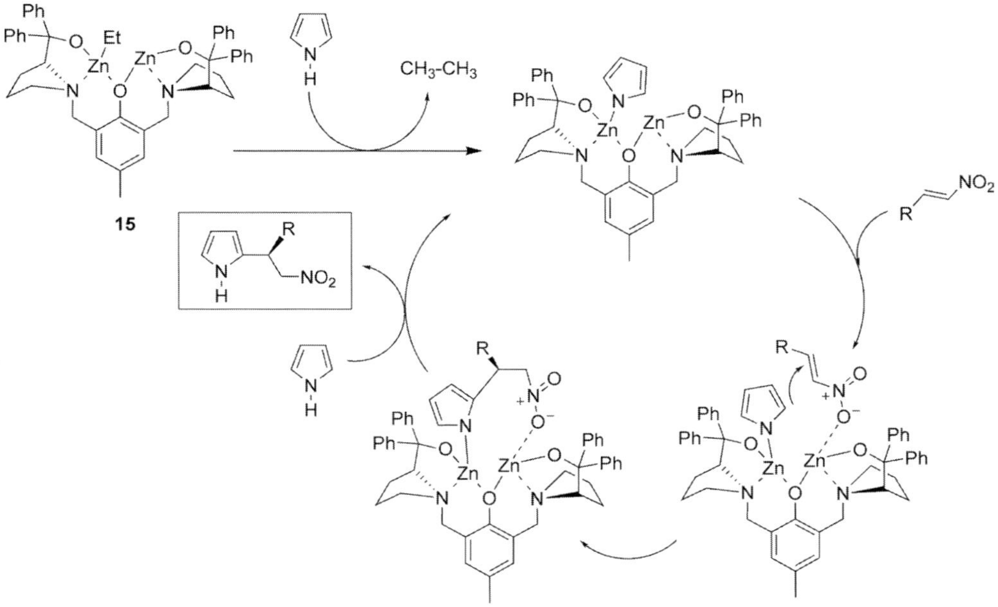

Scheme 2.37 Proposed catalytic cycle for the FC alkylation of pyrroles by catalyst **15**.

available for these reactions, at present, Zn(II) and Cu(II) appear to be the most promising candidates.

On the basis of the absolute configuration, the transition states of these reactions have been proposed, showing that the ligand–metal catalytic complex can display its activity according to a mono- or bifunctional mode. Moreover the replacement of the NH with a methyl group in the indole and pyrrole gives rise to different outcomes, depending upon the chiral complex used.

Important advances in this field can be forecast considering the possibility of applying this catalytic approach to multicomponent tandem reactions with construction of products with several contiguous stereocenters, as pioneered by Arai [96]. The use of new bifunctional catalysts generated from the metal–organic assembly of moieties, such as for example *Cinchona* alkaloids and substituted BINOLs [101], aimed at the combination of metal- and organo-catalysis and at the combinatorial construction of new chiral catalytic species can also be envisaged.

2.3.7.4 Experimental: Selected Procedures

Representative procedure for the catalytic asymmetric FC-H alkylation of indoles with nitroalkenes and aldehydes catalyzed by 11-Cu(I) (Table 2.22) [100]

A solution of **11** (14.9 mg, 19 μmol) in toluene (0.43 mL) was added to (CuOTf)$_2$·C$_6$H$_6$ (4.4 mg, 8.7 μmol) under Ar and the mixture was stirred for 2 h at room temperature. HFIP (35 μl, 0.34 mmol), benzaldehyde (35 μL, 0.34 mmol), *trans*-β-nitrostyrene (25 mg, 0.17 mmol), and indole (40 mg, 0.34 mmol) were added sequentially to the resulting clear green solution. The reaction mixture was stirred at room temperature for 16 h, then purified by column chromatography on silica gel to afford the adduct in 79% yield. The *ee* of the product was determined by HPLC using a Chiralcel OD-H column (90:10 *n*-hexane/*i*PrOH = 90:10, flow rate 0.8 mL min^{-1}, t_{maj} = 43.8 min; t_{min} = 38.6 min, 99% *ee*). $[\alpha]_D^{20}$ = +27.3 (*c* = 0.2, CHCl$_3$); ^{1}H NMR (400 MHz, CDCl$_3$) δ 8.16 (br s, 1H), 7.82 (d, *J* = 7.7 Hz, 1H), 7.12–7.45 (m, 14H), 5.67 (dd, *J* = 11.2, 3.4 Hz, 1H), 5.25 (d, *J* = 11.2 Hz, 1H), 5.03 (br s, 1H), 3.25 (br s, 1H); ^{13}C NMR (100 MHz, CDCl$_3$) δ 139.3, 138.9, 136.2, 128.8, 128.7, 128.4, 127.4, 126.3, 125.2, 122.7, 122.4, 120.3, 119.3, 114.0, 111.5, 96.0, 72.1, 43.8; IR (neat) 3419, 3029, 1548, 1369 cm^{-1}; HRMS: calc. for C$_{23}$H$_{20}$N$_2$O$_3$ (M) 372.1474, found: 372.1477.

Representative procedure for the catalytic asymmetric FC alkylation of 2-methoxyfuran with nitroalkenes catalyzed by 9b-Zn(II) (Scheme 2.35) [102]

To a flame-dried Schlenk tube were added Zn(OTf)$_2$ (9.3 mg, 0.025 mmol) and ligand **9b** (16.8 mg, 0.0275 mmol) under nitrogen, followed by addition of the xylene (3.0 mL). The solution was stirred at room temperature for 2 h and the *trans*-β-nitrostyrene (37.0 mg, 0.25 mmol) was added. The mixture was stirred for 15 min before the addition of 2-methoxyfuran (24.5 mg, 24.0 μl, 0.25 mmol). After stirring for 24 h, the mixture was separated directly by an aluminum oxide (basic, deactivated by addition of 10% w/w water) column chromatography with petroleum ether:ethyl acetate 15:1 as eluent. The (S)-2-methoxy-5-(1-phenyl-2-nitroethyl)furan was obtained as a colorless oil (52.6 mg, 85% yield). The *ee* of the product was determined by HPLC using a Daicel Chiracel OD-H column (*n*-hexane/*i*PrOH = 90:10, flow rate 1.0 mL min^{-1}, $t_{(S)}$ = 32.7 min, $t_{(R)}$ = 20.1 min). $[\alpha]_D$ = −63.2 (*c* = 1.4, CHCl$_3$); ^{1}H NMR (300 MHz, CDCl$_3$) δ 7.34–7.28 (m, 5H), 5.97 (d, *J* = 3.0 Hz, 1H), 5.04 (d, *J* = 3.0 Hz, 1H), 5.00–4.94 (m, 1H), 4.81–4.72 (m, 2H), 3.79 (s, 3H); ^{13}C NMR (75 MHz, CDCl$_3$) δ 161.4, 141.4, 136.7, 128.9, 128.0, 127.8, 108.7, 79.8, 77.9, 57.6, 43.4; HRMS: calc. for C$_{13}$H$_{13}$NO$_4$ 247.08446, found: 247.084387.

Representative procedure for the catalytic asymmetric FC alkylation of pyrroleroles with nitroalkenes using the dinuclear zinc catalyst 15-Zn(II) (Table 2.26) [103]

Preparation of the catalyst: under a nitrogen atmosphere, THF (0.5 mL) was syringed into a test tube capped with a rubber septum containing the Bis-ProPhenol ligand **13** (32 mg, 0.05 mmol). The test tube was evacuated using high vacuum (1 torr) and flushed with nitrogen. This procedure was repeated three times. A solution of diethylzinc (100 μl of 1.0 M solution in hexane, 0.01 mmol) was added dropwise at r.t. to give a solution of catalysts **15-Zn(II)** (0.1 M in THF).

Asymmetric FC reaction: under an argon atmosphere, the solution of catalyst **15-Zn(II)** was added via syringe to a mixture of pyrrole (100 mg, 1.50 mmol), *trans*-β-nitrostyrene (75 mg, 0.50 mmol) and 4 Å molecular sieve (100 mg) in THF (0.5 mL) at r.t.. The reaction was stirred for 24 h at r.t. After the reaction, the mixture was quenched with water (1 mL) and diethyl ether (1 mL), and the aqueous phase was extracted with diethyl ether (2 mL × 4). The combined organic layers were dried over magnesium sulfate. The solvent was removed under reduced pressure using a rotary evaporator with ice-bath. The residue was purified by silica gel column chromatography to give the expected compound as brown crystals (65.9 mg, 0.31 mmol, 61%). The ee of the product was determined by HPLC using a Daicel Chirapak AD column (heptane/iPrOH = 90:10, flow rate 1.0 mL min^{-1}, t = 10.2, 11.2 min, 94% ee). [α]$_D$ = +73.6 (c = 1.02, CHCl$_3$); m.p. = 65–66 °C. ^{1}H NMR (400 MHz, CDCl$_3$) δ 7.82 (bs, 1H), 7.36–7.27 (m, 3 H), 7.24–7.20 (m, 2H), 6.67–6.65 (m, 1H), 6.17–6.14 (m, 1H), 6.08–6.06 (m, 1H), 4.96 (dd, J = 11.9, 7.3 Hz, 1H), 4.89–4.86 (m, 1 H), 4.78 (dd, J = 11.9, 7.6 Hz, 1H); ^{13}C NMR (100 MHz, CDCl$_3$) δ 137.9, 129.2 (2C), 128.9, 128.1, 127.9 (2C), 118.2, 108.6, 105.8, 79.2, 42.9. Elemental analysis calc. for $C_{12}H_{12}N_2O_2$: C, 66.65; H, 5.59; N, 12.96. Found: C, 66.81, H, 5.40; N, 13.07.

Acknowledgments

We are indebted to our coworkers Valentina Sgarzani, Gabriella Dessole, and Dr Raquel P. Herrera who contributed to the development of some of the work presented in this chapter. We acknowledge financial support from "Stereoselezione in Sintesi Organica Metodologie e Applicazioni" 2005. Financial support by the Merck-ADP grant 2007 is also gratefully recognized.

Abbreviations

Ac	acetyl
Alk	alkyl
Ar	aryl
BINAM	1,1′-binaphthyl-2,2′-diamine
BINOL	1,1′-binaphthyl-2,2′-diol
BOX	bisoxazoline
Bn	benzyl
Bs	benzensulfonyl
CD	circular dichroism
DCE	1,2-dichloroethane
ee	enantiomeric excess
FC	Friedel–Crafts
FC-H	Friedel–Crafts–Henry
HFIP	1,1,1,3,3,3,-hexafluoro-2-propanol
HTS	high throughput screening
L	ligand
MS	molecular sieves
MW	microwave
TFA	trifluoroacetic acid
THF	tetrahydrofuran
Tf	trifluoromethanesulfonyl
Ts	p-toluensulfonyl

References

1 (a) Jørgensen, K.A. (2003) *Synthesis*, 1117–1125; (b) Bandini, M., Melloni, A. and Umani-Ronchi, A. (2004) *Angewandte Chemie – International Edition*, **43**, 550–556; (c) Bandini, M., Melloni, A., Tommasi, S. and Umani-Ronchi, A. (2005) *Synlett*, 1199–1222; (d) Poulsen, T.B. and Jørgensen, K.A. (2008) *Chemical Reviews*, **108**, 2903–2915.

2 (a) Tsogoeva, S.B. (2007) *European Journal of Organic Chemistry*, 1701; (b) Bandini, M., Eichholzer, A. and Umani-Ronchi, A. (2007) *Mini-Reviews in Organic Chemistry*, **4**, 115; For examples of potential medicinal agents, see: (c) Rawson, D.J., Dack, K.N., Dickinson, R.P. and James, K. (2002) *Bioorganic & Medicinal Chemistry Letters*, **12**, 125–128; (d) Dillard, R.D. *et al.* (1996) *Journal of Medicinal Chemistry*, **39**, 5119–5136; (e) Chang-Fong, J., Rangisetty, J.B., Dukat, M., Setola, V., Raffay, T., Roth, B. and Glennon, R. (2004) *Bioorganic & Medicinal Chemistry Letters*, **14**, 1961–1964; For examples of natural products, see: (f) Saxton, J.E. (1997) *Natural Product Reports*, **14**, 559–590; (g) Toyota, M. and Ihara, M. (1998) *Natural Product Reports*, **15**, 327–340; cycloaplysinopsin: (h) Mancini, I., Guella, G., Zibrowius, H. and Pietra, F. (2003) *Tetrahedron*, **59**, 8757–8762; 10,11-Dimethoxynareline: (i) Kam, T. and Choo, Y. (2004) *Journal of Natural Products*, **67**, 547–552; Hapalindoles: (j) Kinsman, A.C. and Kerr, M.A. (2003) *Journal of the American Chemical Society*, **125**, 14120–14125; (k) Huber, U., Moore, R.E. and Patterson, G.M.L. (1998) *Journal of Natural Products*, **61**, 1304–1306; Indolmycin: (l) Shinohara, T. and Suzuki, K. (2002) *Tetrahedron Letters*, **43**, 6937–6940; Ambiguine H, hapalindole U, fischerindole I, welwitindolinone A: (m) Baran, P.S., Maimone, T.J. and Richter, J.M. (2007) *Nature*, **446**, 404–408.

3 (a) Paras, N.A. and MacMillan, D.W.C. (2001) *Journal of the American Chemical Society*, **123**, 4370–4371; (b) Austin, J.F. and MacMillan, D.W.C. (2002) *Journal of the American Chemical Society*, **124**, 1172–1173; (c) Paras, N.A. and MacMillan, D.W.C. (2002) *Journal of the American Chemical Society*, **124**, 7894–7895; For a review, see: (d) Lelais, G. and MacMillan, D.W.C. (2006) *Aldrichimica Acta*, **39**, 79–87.

4 Reviews: (a) Desimoni, G., Faita, G. and Jørgensen, K.A. (2006) *Chemical Reviews*, **106**, 3561–3651; (b) Rasappan, R., Laventine, D. and Reiser, O. (2008) *Coordination Chemistry Reviews*, **252**, 702–714.

5 Review: Desimoni, G., Faita, G. and Quadrelli, P. (2003) *Chemical Reviews*, **103**, 3119–3154.

6 Review: Gade, L.H. and Bellemin-Laponnaz, S. (2008) *Chemistry – A European Journal*, **14**, 4142–4152.

7 Reviews: (a) Berthod, M., Mignani, G., Woodward, G. and Lemaire, M. (2005) *Chemical Reviews*, **105**, 1801–1836; (b) Noyori, R. and Takaya, H. (1990) *Accounts of Chemical Research*, **23**, 345–350.

8 Zhuang, W., Hansen, T. and Jørgensen, K.A. (2001) *Chemical Communications*, 347–348.

9 Krapcho, A.P. (1982) *Synthesis*, 805–822.

10 Yamazaki, S. and Iwata, Y. (2006) *The Journal of Organic Chemistry*, **71**, 739–743.

11 Zhou, J. and Tang, Y. (2004) *Chemical Communications*, 432–433.

12 (a) Rasappan, R., Hager, M., Gissibl, A. and Reiser, O. (2006) *Organic Letters*, **8**, 6099–6102; (b) Schätz, A., Rasappan, R., Hager, M., Gissibl, A. and Reiser, O. (2008) *Chemistry – A European Journal*, **14**, 7259–7265.

13 Zhou, J. and Tang, Y. (2002) *Journal of the American Chemical Society*, **124**, 9030–9031.

14 (a) Zhou, J., Ye, M.-C., Huang, Z.-Z. and Tang, Y. (2004) *The Journal of Organic Chemistry*, **69**, 1309–1320; (b) Ye, M.-C.,

Li, B., Zhou, J., Sun, X.-L. and Tang, Y. (2005) *The Journal of Organic Chemistry*, **70**, 6108–6110.

15 Gibson, S.E., Guillo, N. and Tozer, M.J. (1999) *Tetrahedron*, **55**, 585–615.

16 Jensen, K.B., Thorhauge, J., Hazell, R.G. and Jørgensen, K.A. (2001) *Angewandte Chemie – International Edition*, **40**, 160–163.

17 For more information about the structure of Cu-bis(oxazoline) complexes, see: (a) Johnson, J.S. and Evans, D.A. (2000) *Accounts of Chemical Research*, **33**, 325–335; (b) Thorhauge, J., Roberson, M., Hazell, R.G. and Jørgensen, K.A. (2002) *Chemistry – A European Journal*, **8**, 1888–1898.

18 Rueping, M., Nachtsheim, B.J., Moreth, S.A. and Bolte, M. (2008) *Angewandte Chemie – International Edition*, **47**, 593–596.

19 (a) Palomo, C., Oiarbide, M., García, J.M., González, A. and Arceo, E. (2003) *Journal of the American Chemical Society*, **125**, 13942–13943; (b) Palomo, C., Oiarbide, M., Halder, R., Kelso, M., Gómez-Bengoa, E. and García, J.M. (2004) *Journal of the American Chemical Society*, **126**, 9188–9189; (c) Palomo, C., Oiarbide, M., Arceo, E., García, J.M., López, R., González, A. and Linden, A. (2005) *Angewandte Chemie – International Edition*, **44**, 6187–6190; (d) Palomo, C., Oiarbide, M., Kardak, B.G., García, J.M. and Linden, A. (2005) *Journal of the American Chemical Society*, **127**, 4154–4155; (e) Palomo, C., Pazos, R., Oiarbide, M. and García, J.M. (2006) *Advanced Synthesis and Catalysis*, **348**, 1161–1164.

20 Palomo, C., Oiarbide, M., Kardak, B.G., García, J.M. and Linden, A. (2005) *Journal of the American Chemical Society*, **127**, 4154–4155.

21 Yang, H., Hong, Y.-T. and Kim, S. (2007) *Organic Letters*, **9**, 2281–2284.

22 Bandini, M., Fagioli, M., Garavelli, M., Melloni, A., Trigari, V. and Umani-Ronchi, A. (2004) *The Journal of Organic Chemistry*, **69**, 7511–7518.

23 (a) Evans, D.A., Scheidt, K.A., Fandrick, K.R., Lam, H.W. and Wu, J. (2003) *Journal of the American Chemical Society*, **125**, 10780–10781; (b) Evans, D.A., Fandrick, K.R., Song, H.-J., Scheidt, K.A. and Xu, R. (2007) *Journal of the American Chemical Society*, **129**, 10029–10041.

24 (a) Evans, D.A., Fandrick, K.R. and Song, H.-J. (2005) *Journal of the American Chemical Society*, **127**, 8942–8943; (b) Evans, D.A. and Fandrick, K.R. (2006) *Organic Letters*, **8**, 2249–2252.

25 Cavdar, H. and Saracoğlu, N. (2005) *Tetrahedron*, **61**, 2401–2405.

26 Davies, H.W., Matasi, J.J. and Ahmed, G. (1996) *The Journal of Organic Chemistry*, **61**, 2305–2313.

27 For a partial list of syntheses of heliotridane, see: (a) Kim, S.-H., Kim, S.-I. and Cha, J.K. (1999) *The Journal of Organic Chemistry*, **64**, 6771–6775; (b) Pandey, G., Reddy, G.D. and Chakrabarti, D. (1994) *Journal of the Chemical Society – Perkin Transactions 1*, 219–224; (c) Keusenkothen, P.F. and Smith, M.B. (1994) *Journal of the Chemical Society – Perkin Transactions 1*, 2485–2492; (d) Doyle, M.P. and Kalinin, A.V. (1996) *Tetrahedron Letters*, **37**, 1371–1374.

28 Bandini, M., Melloni, A., Tommasi, S. and Umani-Ronchi, A. (2003) *Helvetica Chimica Acta*, **86**, 3753–3763.

29 Sui, Y., Liu, L., Zhao, J.-L., Wang, D. and Chen, Y.-J. (2007) *Tetrahedron*, **63**, 5173–5183.

30 Barroca, N. and Schmidt, R.R. (2004) *Organic Letters*, **6**, 1551–1554.

31 For recent reviews on catalytic, asymmetric Friedel–Crafts reactions, see: Poulsen, T.B. and Jørgensen, K.A. (2008) *Chemical Reviews*, **108**, 2903–2915.

32 (a) Gupta, R.R., Kumar, M. and Gupta, V. (1999) *Heterocyclic Chemistry*, Vol. 3, Springer-Verlag, Heidelberg, (b) Strell, M. and Kalojanoff, A. (1954) *Chemische Berichte*, **87**, 1019–1024.

33 For selected examples, see: (a) Jensen, K.B., Thorhauge, J., Hazell, R.G. and Jørgensen, K.A. (2001) *Angewandte*

Chemie – International Edition, **40**, 160–163; (b) Zhuang, W., Hansen, T. and Jørgensen, K.A. (2001) Chemical Communications, 347–348; (c) Zhou, J. and Tang, Y. (2002) Journal of the American Chemical Society, **124**, 9030–9031; (d) Evans, D.A., Scheidt, K.A., Fandrick, K.R., Lam, H.W. and Wu, J. (2003) Journal of the American Chemical Society, **125**, 10780–10781; (e) Palomo, C., Oiarbide, M., Kardak, B.G., Garcia, J.M. and Linden, A. (2005) Journal of the American Chemical Society, **127**, 4154–4155; (f) Evans, D.A., Fandrick, K.R. and Song, H.-J. (2005) Journal of the American Chemical Society, **127**, 8942–8943; (g) Evans, D.A., Fandrick, K.R., Song, H.-J., Scheidt, K.A. and Xu, R. (2007) Journal of the American Chemical Society, **129**, 10029–10041.

34 For reviews on asymmetric iminium ion catalysis, see: Erkkilä, A., Majander, I. and Pihko, P.M. (2007) Chemical Reviews, **107**, 5416–5470.

35 Bandini, M., Fagioli, M., Melchiorre, P. and Umani-Ronchi, A. (2003) Tetrahedron Letters, **44**, 5846–5849.

36 General reviews on asymmetric organocatalysis, see: (a) Dalko, P.I.(ed.) (2007) Enantioselective Organocatalysis, Wiley-VCH, Weinheim; (b) Dondoni, A. and Massi, A. (2008) Angewandte Chemie – International Edition, **47**, 4638–4660; (c) Gaunt, M.J., Johansson, C.C.C., McNally, A. and Vo, N.C. (2007) Drug Discovery Today, **12**, 8–27; (d) List, B. and Yang, J.W. (2006) Science, **313**, 1584–1586.

37 (a) Barbas, C.F. III, (2008) Angewandte Chemie – International Edition, **47**, 42–47; (b) Melchiorre, P., Marigo, M., Carlone, A. and Bartoli, G. (2008) Angewandte Chemie – International Edition, **47**, 6138–6171.

38 (a) Ahrendt, K.A., Borths, C.J. and MacMillan, D.W.C. (2000) Journal of the American Chemical Society, **122**, 4243–4244; (b) Jen, W.S., Wiener, J.J.M. and MacMillan, D.W.C. (2000) Journal of the American Chemical Society, **122**, 9874–9875.

39 (a) Sundberg, R.J. (1996) Indoles, Academic Press, San Diego, p. 175; (b) Nicolaou, K.C. and Snyder, S.A. (2003) Classics in Total Synthesis II, Wiley-VCH, Weinheim.

40 For theoretical explorations of pyrrole and indole asymmetric alkylations organocatalyzed by imidazolidinone catalysts, see: (a) Gordillo, R., Carter, J. and Houk, N.K. (2004) Advanced Synthesis and Catalysis, **346**, 1175–1185; For the use of chiral aziridines to catalyze the asymmetric pyrrole and indole alkylations with enals, see: (b) Bonini, B.F., Capitò, E., Comes-Franchini, M., Fochi, M., Ricci, A. and Zwanenburg, B. (2006) Tetrahedron: Asymmetry, **17**, 3135–3143.

41 Brown, S.P. (2005) Doctoral Dissertation, California Institute of Technology, Pasadena, CA.

42 (a) King, H.D., Meng, Z., Denhart, D., Mattson, R., Kimura, R., Wu, D., Gao, Q. and Macor, J.E. (2005) Organic Letters, **7**, 3437–3440; (b) Denhart, D., Ditta, J., King, H.D., Macor, J.E., Marcin, L., Mattson, R. and Meng, Z. (2004) WO 026237. (2004) Chemical Abstracts, **140**, 303528; For the application of aminocatalytic FC alkylation of anilines with enals in the total synthesis of (+)-curcuphenol, see: (c) Kim, S.-G., Kim, J. and Jung, H. (2005) Tetrahedron Letters, **46**, 2437–2439.

43 Fleming, I. (1976) Frontier Orbitals and Organic Chemical Reactions, John Wiley & Sons, Chichester, UK, pp. 33–85.

44 (a) Lee, S. and MacMillan, D.W.C. (2007) Journal of the American Chemical Society, **129**, 15438–15439; for a different approach, leading to the enantioselective Friedel–Crafts alkylation at the 2-position of indole with simple enones, see: (b) Blay, G., Fernández, I., Pedro, J.R. and Vila, C. (2007) Tetrahedron Letters, **48**, 6731–6734.

45 (a) Walji, A.M. and MacMillan, D.W.C. (2007) Synlett, 1477–1489; (b) Nicolaou, K.C., Edmonds, D.J. and Bulger, P.G. (2006) Angewandte Chemie – International Edition, **45**, 7134–7186;(c) Tietze, L.F.,

Brasche, G. and Gerike, K. (2006) *Domino Reactions in Organic Chemistry*, Wiley-VCH, Weinheim; (d) Ramón, D.J. and Yus, M. (2005) *Angewandte Chemie – International Edition*, **44**, 1602–1634.

46 Austin, J.F., Kim, S.G., Sinz, C.J., Xiao, W.J. and MacMillan, D.W.C. (2004) *Proceedings of the National Academy of Sciences of the United States of America*, **101**, 5482–5487.

47 Huang, Y., Walji, A.M., Larsen, C.H. and MacMillan, D.W.C. (2005) *Journal of the American Chemical Society*, **127**, 15051–15053.

48 Reviews on organocatalytic domino reactions: (a) Enders, D., Grondal, C. and Hüttl, M.R.M. (2007) *Angewandte Chemie – International Edition*, **46**, 1570–1581; (b) Guillena, G., Ramón, D.J. and Yus, M. (2007) *Tetrahedron: Asymmetry*, **18**, 693–700; Relevant examples, see: (c) Enders, D., Hüttl, M.R.M., Grondal, C. and Raabe, G. (2006) *Nature*, **441**, 861–863; (d) Marigo, M., Schulte, T., Franzen, J. and Jørgensen, K.A. (2005) *Journal of the American Chemical Society*, **127**, 15710–15711; (e) Carlone, A., Cabrera, S., Marigo, M. and Jørgensen, K.A. (2007) *Angewandte Chemie – International Edition*, **46**, 1101–1104; (f) Penon, O., Carlone, A., Mazzanti, A., Locatelli, M., Sambri, L., Bartoli, G. and Melchiorre, P. (2008) *Chemistry – A European Journal*, **14**, 4788–4791.

49 Chi, Y., Scroggins, S.T. and Fréchet, J.M.J. (2008) *Journal of the American Chemical Society*, **130**, 6322–6323.

50 (a) Chen, W., Du, W., Yue, L., Li, R., Wu, Y., Ding, L.-S. and Chen, Y.-C. (2007) *Organic and Biomolecular Chemistry*, **5**, 816–821; (b) Bartoli, G., Bosco, M., Carlone, A., Pesciaioli, F., Sambri, L. and Melchiorre, P. (2007) *Organic Letters*, **9**, 1403–1405.

51 Taylor, M.S., Zalatan, D.N., Lerchner, A.M. and Jacobsen, E.N. (2005) *Journal of the American Chemical Society*, **127**, 1313–1317.

52 Blay, G., Fernández, I., Pedro, J.R. and Vila, C. (2007) *Organic Letters*, **9**, 2601–2601.

53 For meaningful examples of the potential of chiral primary amines in iminium activation of enones, see: (a) Kim, H., Yen, C., Preston, P. and Chin, J. (2006) *Organic Letters*, **8**, 5239–5242; (b) Xie, J.-W., Chen, W., Li, R., Zeng, M., Du, W., Yue, L., Chen, Y.-C., Wu, Y., Zhu, J. and Deng, J.-G. (2007) *Angewandte Chemie – International Edition*, **46**, 389–392; (d) Singh, R.P., Bartelson, K., Wang, Y., Su, H., Lu, X. and Deng, L. (2008) *Journal of the American Chemical Society*, **130**, 2422–2423; (e) Wang, X., Reisinger, C.M. and List, B. (2008) *Journal of the American Chemical Society*, **130**, 6070–6071. Application of chiral primary amines in enamine activation of ketones, see: (f) Tsogoeva, S.B. and Wei, S. (2006) *Chemical Communications*, 1451–1453; (g) Huang, H. and Jacobsen, E.N. (2006) *Journal of the American Chemical Society*, **128**, 7170–7171; (h) McCooey, S.H. and Connon, S.J. (2007) *Organic Letters*, **9**, 599–602; (i) Liu, T.-Y., Cui, H.-L., Zhang, Y., Jiang, K., Du, W., He, Z.-Q. and Chen, Y.-C. (2007) *Organic Letters*, **9**, 3671–3674.

54 For an organocatalytic indole alkylation with enones promoted by an achiral amine, see: Li, D.-P., Guo, Y.-C., Ding, Y. and Xiao, W.-J. (2006) *Chemical Communications*, pp. 799–801. In this report, an initial attempt to perform an asymmetric version using the MacMillan imidazolidinone **6** afforded poor selectivity (28% ee).

55 The chiral primary amines, directly derived from natural Cinchona alkaloids, were also successfully employed as highly efficient enamine catalysts for the α-functionalization of ketones, see Ref [56 h–i].

56 Zhou, W., Xu, L.-W., Li, L., Yang, L. and Xia, C.-G. (2006) *European Journal of Organic Chemistry*, 5225–5227.

57 Tang, H.-Y., Lu, A.-D., Zhou, Z.-H., Zhao, G.-F., He, L.-N. and Tang, C.-C. (2008) *European Journal of Organic Chemistry*, 1406–1410.

58 For reviews: (a) Brown, R.T. (1983) *Indoles* (ed. J.E. Saxton), Wiley-Interscience, New York, Part 4 (The Monoterpenoid Indole Alkaloids); (b) Bentley, K.W. (2004) *Natural Product Reports*, **21**, 395–424, and references cited therein.

59 Angeli, M., Bandini, M., Garelli, A., Piccinelli, F., Tommasi, S. and Umani-Ronchi, A. (2006) *Organic and Biomolecular Chemistry*, **4**, 3291–3296.

60 (a) Li, C.-F., Liu, H., Liao, J., Cao, Y.-J., Liu, X.-P. and Xiao, W.-J. (2007) *Organic Letters*, **9**, 1847–1850; For related studies, see: (b) Banwell, M.G., Beck, D.A.S. and Smith, J.A. (2004) *Organic and Biomolecular Chemistry*, **2**, 157–159.

61 (a) Berner, O.M., Tedeschi, L. and Enders, D. (2002) *European Journal of Organic Chemistry*, 1877–1894.(b) Ballini, R., Marcantoni, E. and Petrini, M. (2008) *Amino Group Chemistry* (ed. A. Ricci), Wiley-VCH, Weinheim, pp. 93–148, Ch. 3.

62 Ono, N. (2001) *The Nitro Group in Organic Synthesis*, Wiley-VCH, New York.

63 Noland, W.E. and Hartman, P.J. (1954) *Journal of the American Chemical Society*, **76**, 3227–3228.

64 Early examples: (a) Bigi, F., Casiraghi, G., Casnati, G., Sartori, G., Fava, G.G. and Ferrari Belicchi, M. (1985) *The Journal of Organic Chemistry*, **50**, 5018–5022; (b) Erker, G. and van der Zeijden, A.A.H. (1990) *Angewandte Chemie*, **102**, 543–545; (1990) *Angewandte Chemie – International Edition*, **29**, 512–514; (c) Johannsen, M. (1999) *Chemical Communications*, 2233–2234; (d) Kawate, T., Yamada, H., Soe, T., and Nakagawa, M. (1996) *Tetrahedron: Asymmetry*, **7**, 1249–1252.

65 Reviews: (a) Taylor, M.S. and Jacobsen, E.N. (2006) *Angewandte Chemie*, **118**, 1550–1572; (2006) *Angewandte Chemie – International Edition*, **45**, 1520–1543; (b) Doyle, A.G. and Jacobsen, E.N. (2007) *Chemical Reviews*, **107**, 5713–5743; (c) Takemoto, Y. (2005) *Organic and Biomolecular Chemistry*, **3**, 4299–4306.

66 Etter, M.C., Urbañczyk-Lipkowska, Z., Zia-Ebrahimi, M. and Panunto, T.W. (1990) *Journal of the American Chemical Society*, **112**, 8415–8426.

67 Dessole, G., Herrera, R.P. and Ricci, A. (2004) *Synlett*, 2374–2378.

68 Schreiner, P.R. and Wittkopp, A. (2002) *Organic Letters*, **4**, 217–220.

69 Herrera, R.P., Sgarzani, V., Bernardi, L. and Ricci, A. (2005) *Angewandte Chemie*, **117**, 6734–6737; (2005) *Angewandte Chemie – International Edition*, **44**, 6576–6579.

70 Okino, T., Hoashi, Y. and Takemoto, Y. (2003) *Journal of the American Chemical Society*, **125**, 12672–12673.

71 Simón, L., Muñiz, F.M., Sáez, S., Raposo, C. and Morán, J.R. (2008) *European Journal of Organic Chemistry*, 2397–2403.

72 Fleming, E.M., McCabe, T. and Connon, S.J. (2006) *Tetrahedron Letters*, **47**, 7037–7042.

73 Corey, E.J., Imwinkelried, R., Pikul, S. and Xiang, Y.B. (1989) *Journal of the American Chemical Society*, **111**, 5493–5495.

74 Zhuang, W., Hazell, R.G. and Jørgensen, K.A. (2005) *Organic and Biomolecular Chemistry*, **3**, 2566–2571.

75 Reviews: (a) Connon, S.J. (2006) *Angewandte Chemie*, **118**, 4013–4016; (2006) *Angewandte Chemie – International Edition*, **45**, 3909–3912; (b) Akiyama, T. (2007) *Chemical Reviews*, **107**, 5744–5758.

76 Itoh, J., Fuchibe, K. and Akiyama, T. (2008) *Angewandte Chemie*, **120**, 4080–4082; (2008) *Angewandte Chemie – International Edition*, **47**, 4016–4018.

77 Storer, R.I., Carrera, D.E. and MacMillan, D.W.C. (2006) *Journal of the American Chemical Society*, **128**, 84–86.

78 Liu, T.-Y., Cui, H.-L., Chai, Q., Long, J., Li, B.-J., Wu, Y., Ding, L.-S. and Chen, Y.-C. (2007) *Chemical Communications*, 2228–2230.

79 Li, B.-J., Jiang, L., Liu, M., Chen, Y.-C., Ding, L.-S. and Wu, Y. (2005) *Synlett*, 603–606.

References

80 Li, H., Wang, Y., Tang, L. and Deng, L. (2004) *Journal of the American Chemical Society*, **126**, 9906–9907.

81 Wang, J., Li, H., Zu, L. and Wang, W. (2006) *Organic Letters*, **8**, 1391–1394.

82 Gandelman, M. and Jacobsen, E.N. (2005) *Angewandte Chemie*, **117**, 2445–2449; (2005) *Angewandte Chemie – International Edition*, **44**, 2393–2397.

83 (a) Kidwai, M., Saxena, S., Khan, M.K.R. and Thukral, S.S. (2005) *Bioorganic & Medicinal Chemistry Letters*, **15**, 4295–4298; (b) Brandes, S., Bella, M., Kjærsgaard, A. and Jørgensen, K.A. (2006) *Angewandte Chemie*, **118**, 1165–1169; (2006) *Angewandte Chemie – International Edition*, **45**, 1147–1151.

84 Harrington, P.F. and Kerr, M.A. (1996) *Synlett*, 1047–1048.

85 Bandini, M., Melchiorre, P., Melloni, A. and Umani-Ronchi, A. (2002) *Synthesis*, 1110–1114.

86 Alam, M.M., Varala, R., Srinivas, S. and Adapa, R. (2003) *Tetrahedron Letters*, **44**, 5115–5119.

87 Lin, C., Hsu, J., Sastry, M.N.V., Fang, H., Tu, Z., Liu, J.-T. and Ching-Fa, Y. (2005) *Tetrahedron*, **61**, 11751–11757.

88 (a) Bandini, M., Fagioli, M. and Umani-Ronchi, A. (2004) *Advanced Synthesis and Catalysis*, **346**, 545–548; (b) Azizi, N., Arynasab, F. and Saidi, M.R. (2006) *Organic and Biomolecular Chemistry*, **4**, 4275–4277.

89 Guiry, P.J. and Saunders, C.P. (2004) *Advanced Synthesis and Catalysis*, **346**, 497–537.

90 Jia, Y.-X., Zhu, S.-F., Yang, Y. and Zhou, Q.-L. (2006) *The Journal of Organic Chemistry*, **71**, 75–80.

91 (a) Lu, S.-F., Du, D.-M. and Xu, J. (2006) *Organic Letters*, **8**, 2115–2118; (b) Liu, H., Lu, S.-F., Xu, J. and Du, D.-M. (2008) *Chemistry, an Asian Journal*, **3**, 1111–1121.

92 Yuan, Z.-L., Lei, Z.-Y. and Shi, M. (2008) *Tetrahedron: Asymmetry*, **19**, 1339–1346.

93 Tsuzuki, S., Honda, K., Uchimaru, T., Mikami, M. and Tanabe, K. (2000) *Journal of the American Chemical Society*, **122**, 11450–11458.

94 Singh, P.K., Bisai, A. and Singh, V.K. (2007) *Tetrahedron Letters*, **48**, 1127–1129.

95 Arai, T., Yokoyama, N. and Yanagisawa, A. (2008) *Chemistry – A European Journal*, **14**, 2052–2059.

96 Arai, T. and Yokoyama, N. (2008) *Angewandte Chemie*, **120**, 5067–5070; (2008) *Angewandte Chemie – International Edition*, **47**, 4989–4992.

97 Bandini, M., Garelli, A., Rovinetti, M., Tommasi, S. and Umani-Ronchi, A. (2005) *Chirality*, **17**, 522–529.

98 Liu, H., Xu, J. and Du, D.-M. (2007) *Organic Letters*, **9**, 4725–4728.

99 Trost, B.M. and Müller, C. (2008) *Journal of the American Chemical Society*, **130**, 2438–2439.

100 Trost, B.M. and Ito, H. (2000) *Journal of the American Chemical Society*, **122**, 12003–12004.

101 Yang, F., Zhao, D., Lan, J., Xi, P., Yang, L., Xiang, S. and You, J. (2008) *Angewandte Chemie*, **120**, 5728–5731; (2008) *Angewandte Chemie – International Edition*, **47**, 5646–5649.

3
Addition to Carbonyl Compounds

3.1
Aldehydes/Ketones

Jia-Rong Chen and Wen-Jing Xiao

Summary

The Friedel–Crafts reaction of aromatic compounds with carbonyl compounds is one of the most fundamental C—C bond-forming reactions in organic chemistry. In particular, the catalytic enantioselective variant of this reaction provides a simple and straightforward access to enantiomerically enriched aromatic compounds bearing benzylic stereocenters, most of which possess biologically important properties. This chapter puts particular emphasis on the recent advances in catalytic and enantioselective alkylation of aromatics via direct 1,2-addition to carbonyl compounds.

3.1.1
Introduction

The Friedel–Crafts (FC) reaction of aromatic compounds with carbonyl compounds is one of the most fundamental C—C bond-forming reactions in organic chemistry [1]. In particular, the catalytic enantioselective variant of this reaction provides a simple and straightforward approach to enantiomerically enriched aromatic compounds bearing benzylic stereocenters, most of which possess biologically important properties (Scheme 3.1). In the mid-1980s, Casiraghi and coworkers reported the first example of the asymmetric addition of aromatic C—H bonds to carbonyl compounds with stoichiometric amounts of chiral alkoxyaluminum chlorides, though stoichiometric amounts of catalyst were needed (Scheme 3.2) [2]. Shortly after, Erker developed the first catalytic enantioselective addition of pyruvic ester with 1-naphthol using a chiral zirconium complex [3]. Since then, catalytic enantioselective FC reactions of reactive aromatic compounds with a variety of C=O bonds have attracted extensive attention in the field of asymmetric catalysis [4]. This chapter puts particular emphasis on the recent advances in catalytic and

---

Catalytic Asymmetric Friedel–Crafts Alkylations. Edited by M. Bandini and A. Umani-Ronchi
Copyright © 2009 WILEY-VCH Verlag GmbH & Co. KGaA, Weinheim
ISBN: 978-3-527-32380-7

Scheme 3.1 Asymmetric and enantioselective alkylation of aromatic compounds via 1,2-addition to carbonyl compounds.

enantioselective alkylation of aromatics via direct 1,2-addition to carbonyl compounds. The selected examples are roughly organized according to the catalysts employed in the reaction, namely organometallic catalysis, organocatalysis, and heterogeneous catalysis.

3.1.2
Organometallic Catalysis

3.1.2.1 Fundamental Examples and Mechanism

The addition of electron-rich aromatic and heteroaromatic compounds to carbonyl compounds, such as aldehydes and ketones, results in the formation of versatile functionalized compounds [1f]. Numerous examples of this transformation involving activated carbonyl [5, 6] and α-dicarbonyl compounds [7, 8] with the use of Lewis acids such as $AlCl_3$ have been described. Diastereoselective versions have also been reported with the use of glyoxylate derivatives bearing a chiral auxiliary [9], chiral ketoesters [10], and aldehydes [11] to react with electron-rich substrates. There are also a few examples of FC-type reactions of reactive arenes with α-dicarbonyl compounds attached to a chiral auxiliary, affording optically active products [12]. Recently, there has been considerable interest in developing asymmetric FC reactions by using various chiral metal-based catalysts. The following section provides a brief summary of the development in organometallic catalyzed enantioselective reactions, along with a discussion of mechanisms and selectivity models.

In 1990, Erker reported the first catalytic asymmetric addition of 1-naphthol **5** to pyruvic esters **6** promoted by the chiral zirconocene complex **4** (Scheme 3.3) [3].

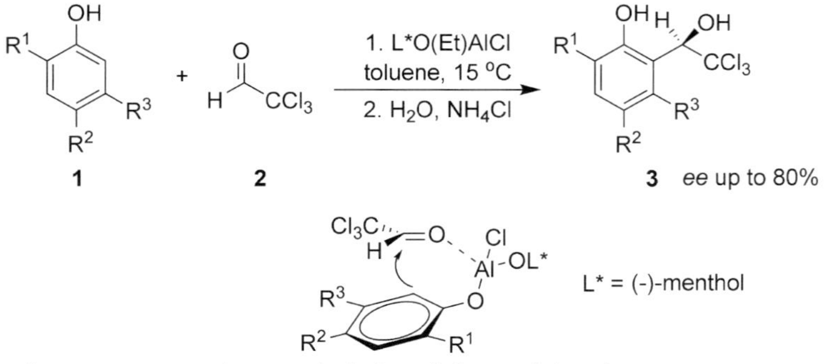

Scheme 3.2 Enantioselective ortho-hydroxyalkylation of phenols with chloral mediated by chiral alkoxyaluminum chloride.

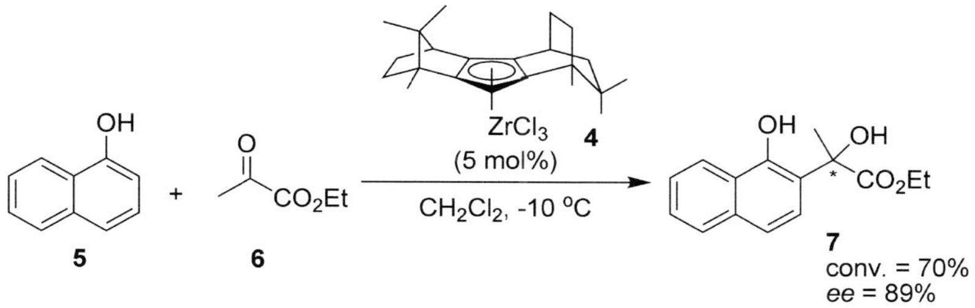

Scheme 3.3 Enantioselective addition of 1-hydroxynaphthalene **5** to the pyruvic ester **6** catalyzed by a Zr-dibornacyclopentadienyl complex.

Under optimized conditions, up to 89% *ee* was obtained with the addition of small amounts of water to the reaction mixture. Though they did not obtain a defined adduct of **8** with the individual component 1-naphthol **5** or pyruvic esters **6**, the zirconium complex **9** was isolated by adding a mixture of **5** and **6** to CpZrCl$_3$ (Scheme 3.4). Complex **9** could catalyze the reaction of **5** and **6** to give **7**. Accordingly, the authors proposed that the enantioselective C–C coupling reaction involving the chiral organometallic Lewis acid **4** proceeded through a similar reaction route.

3.1.2.1.1 Chiral Titanium (IV) Catalysis
Asymmetric synthesis of organofluorine compounds has received much attention in the pharmaceutical industry and in materials science owing to the unique properties of fluorinated compounds [13].

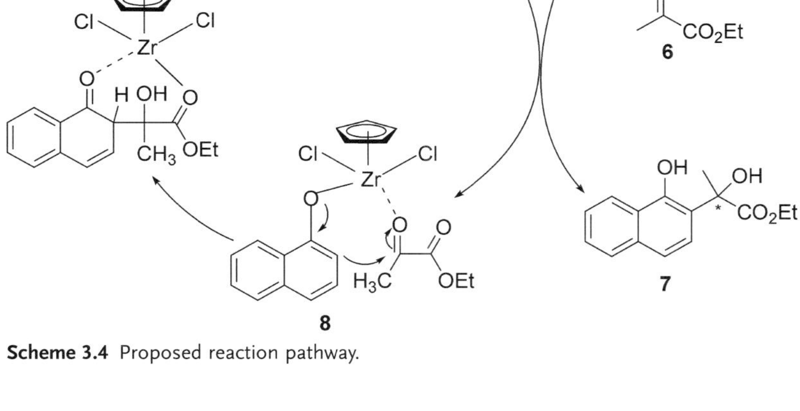

Scheme 3.4 Proposed reaction pathway.

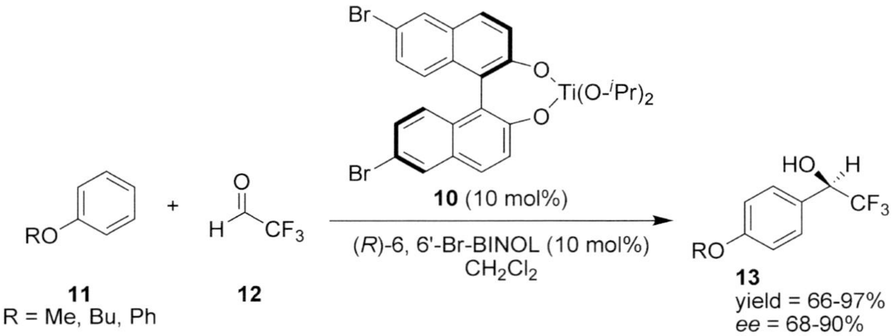

Scheme 3.5 Catalytic enantioselective FC reaction of electron-rich arenes with fluoral catalyzed by a BINOL-titanium complex **10**.

Mikami and coworkers first reported a practical synthetic route to chiral 1-aryl-2,2,2-trifluoroethanol derivatives through the FC reaction of alkoxyarenes **11** with 2,2,2-trifluoroacetaldehyde using chiral BINOL-Ti complex as the catalyst (Scheme 3.5) [14, 15]. In this type of FC alkylation, the substitution pattern of the BINOL is critical to the catalytic activity and the asymmetric induction of BINOL-Ti catalysts. The (R)-6,6'-Br$_2$-BINOL-Ti complex, which was prepared from [Ti(O-Pr^i)$_4$] and (R)-6,6'-Br$_2$-BINOL, was found to be the most effective catalyst. It is worth noting that the bis-adduct with 2,2,2-trifluoroacetaldehyde was not observed, even when a large excess of fluoral was used in the reaction of diphenyl ether [16]. They also found that the catalytic efficiency of BINOL-Ti complexes could be further improved through asymmetric activation by the addition of acidic activators, such as (R)-6,6'-Br$_2$-BINOL [17]. This study represented the first example of asymmetric catalytic FC reactions with fluoral, providing direct access to chiral synthetically useful 1-aryl-2,2,2-trifluoroethanol derivatives **13**.

As a part of their continuing work on titanium chemistry [18], Ding and coworkers extended the efficiency of chiral titanium complexes to the FC reaction of N,N-dialkylamino aromatics with ethyl glyoxylate **15**. Here, the practical synthesis of amino-mandelic acid derivatives was accomplished in a highly enantioselective manner (Table 3.1) [19]. The authors found that the substitution pattern on the binaphthyl backbone in the ligands is critical to get high reactivity and enantioselectivity in this asymmetric FC reaction. After carefully tuning the steric and electronic modifications on the diol ligands, they identified (R)-6,6'-Br$_2$-BINOL as the best ligand. Under the optimized conditions, the reaction of a variety of meta-substituted N,N-dimethylanilines with ethyl glyoxylate proceeded smoothly to give the corresponding p-amino-mandelic acid derivatives in high yields (85–99%) and ees (80–96%). Moreover, the substrate scope was extended to heterocyclic aromatic amines, with ees of 90% and 97%, respectively (Scheme 3.6). The preparative utility of this methodology was demonstrated by conducting the reaction of ethyl glyoxylate with N,N-dimethylaniline on a gram scale using only 1 mol% of (R)-6,6'-Br$_2$-BINOL/Ti catalyst, affording the desired adduct **16a** in 91% yield with 85% ee.

Table 3.1 Enantioselective FC reaction of aromatic amines with ethyl glyoxylate catalyzed by a BINOL-titanium complex **10**.

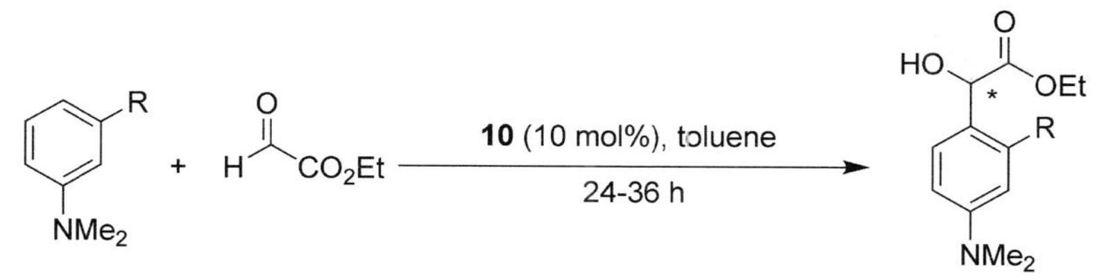

14 **15** **16**

a: R = H; **b:** R = F
c: R = Me; **d:** R = OMe

Entry	*N,N*-Dimethylaniline	Temp. (°C)	Product	Yield (%)	*ee* (%)
1	**14a**	−10	(*R*)-**16a**	99	96
2	**14b**	−30	(−)-**16b**	85	88
3	**14c**	−30	(−)-**16c**	96	90
4	**14d**	−10	(−)-**16d**	99	80

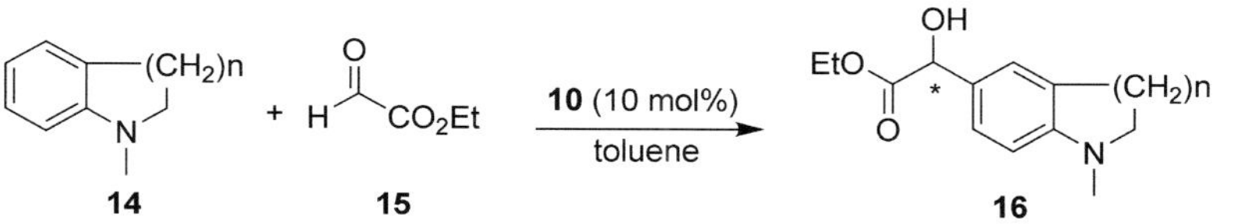

14 **15** **16**

e: n = 1
f: n = 2

e: n = 1; yield = 99%, *ee* = 91% (-30 °C)
f: n = 2; yield = 99%, *ee* = 97% (-10 °C)

Scheme 3.6 Asymmetric FC reaction of *N*-methylindoline and *N*-methyltetrahydroquiniline with ethyl glyoxylate catalyzed by **10**.

Based on the observed absolute configuration of the FC product and Corey's transition-state model for carbonyl-ene reaction (a) [20], they proposed a possible asymmetric induction model (b), in which the bottom (*Re*) face of the formyl group is much more favorable for attack by the nucleophile than the top (*Si*) face because of the steric interaction (Scheme 3.7).

As a part of the continuous research interest in functionalization of indole [21], and to broaden the spectrum of BINOL-Ti complexes catalyzed asymmetric reactions, Xiao and coworkers demonstrated that BINOL-Ti(IV) can efficiently catalyze FC reactions of indoles **18** with ethyl glyoxylate **15** [22]. Under optimized conditions, various structural alterations of indoles were tolerated and good yields (64–88%) and high enantiomeric excess (62–96%) were obtained for the desired chiral indole derivatives **19** (Scheme 3.8).

It was found that the carbonyl electrophile played an important role in the success of this strategy. For example, methyl 3,3,3-trifluoropyruvate **20a** only afforded the FC

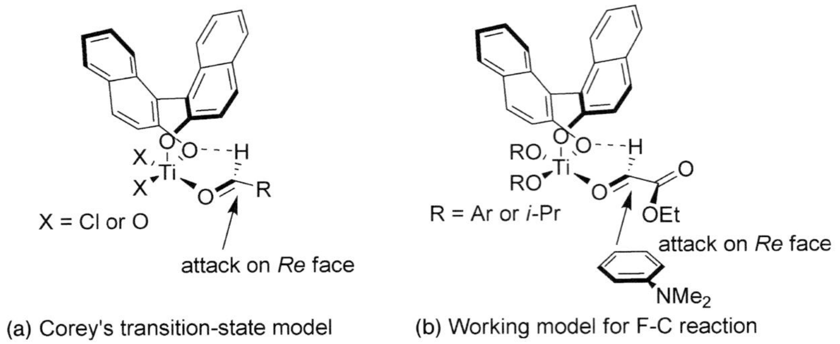

Scheme 3.7 Proposed working model for the stereo-induction.

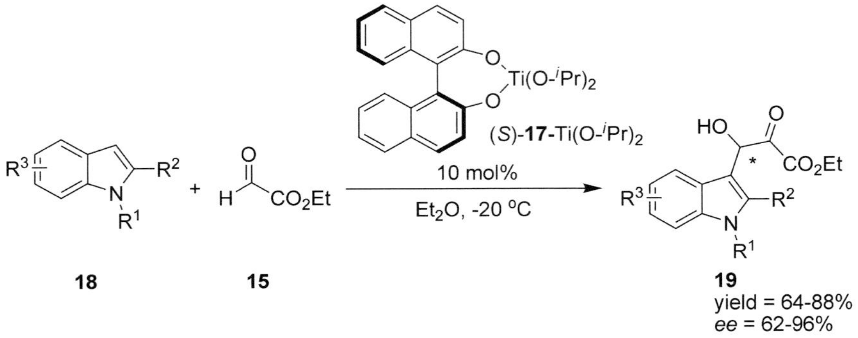

(a) Corey's transition-state model

(b) Working model for F-C reaction

Scheme 3.8 Asymmetric FC alkylation of indoles with ethyl glyoxylate catalyzed by (s)-BINOL-titanium complex.

product in 10% ee, while the bisindole derivatives were formed exclusively for the methyl pyruvate **22** and p-chlorophenylglyoxal **24** (Scheme 3.9).

The authors postulated a possible mechanism as outlined in Scheme 3.10. First, the chiral titanium catalyst preferentially coordinates to ethyl glyoxylate **15** due to its high oxophilicity. Then the activated carbonyl is attacked by N-methylindole forming the intermediate **26**. In the following steps, intermediate **26** may follow pathway **I** to give the bisindole **19a′** or pathway **II**, providing the expected product **19a** and releasing the chiral catalyst for the next catalytic cycle. Note that the formation of bisaryl compounds was almost completely suppressed when the reactions were carried out in Et_2O at sub-zero temperatures.

While a precise reaction mechanism and the exact structure of the active catalyst species await further studies, a stereo-controlling model for the transition state was proposed based on Corey [20] and Ding's [19] previous models. As shown in Scheme 3.11, the H atom of the formyl group forms a hydrogen bond with the oxygen atom of the BINOL ligand, which plays an important role in stabilizing the transition state of the process. As result, the *Si* face of the formyl group is more favorable to attack by indole derivatives than the *Re* face, since the latter is blocked by the nearby naphthyl subunit. Accordingly, the absence of the hydrogen bond

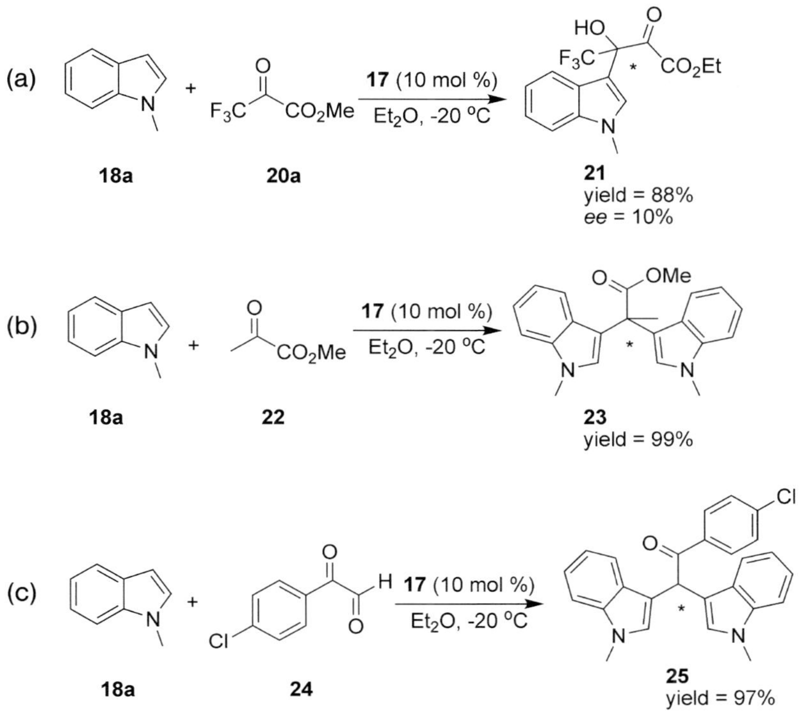

Scheme 3.9 F-C alkylation of 1-methylindole with methyl 3,3,3-trifluoropyruvate (**20a**), methyl pyruvate (**22**) and p-chlorophenylglyoxal (**24**).

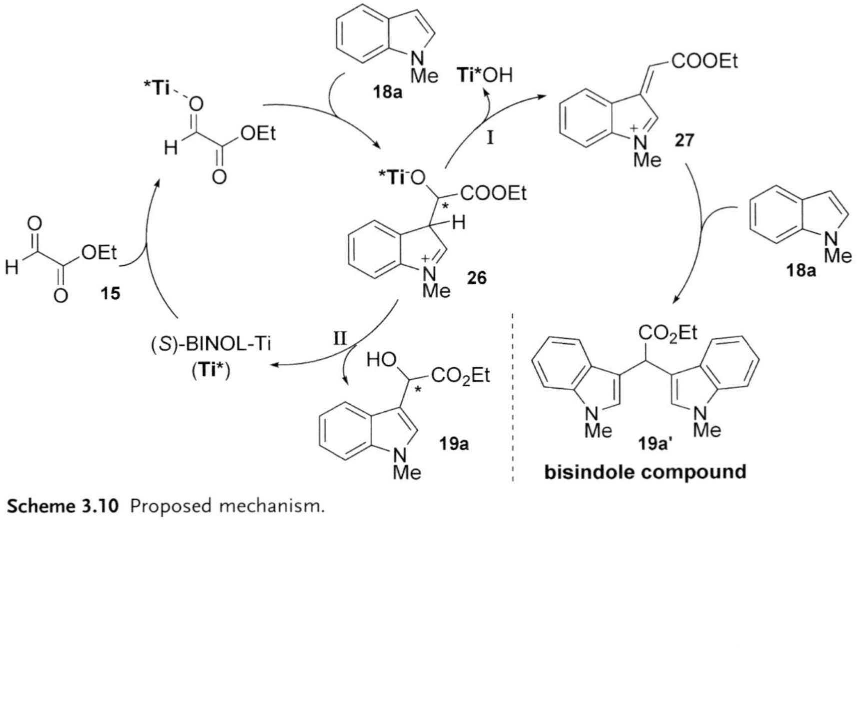

Scheme 3.10 Proposed mechanism.

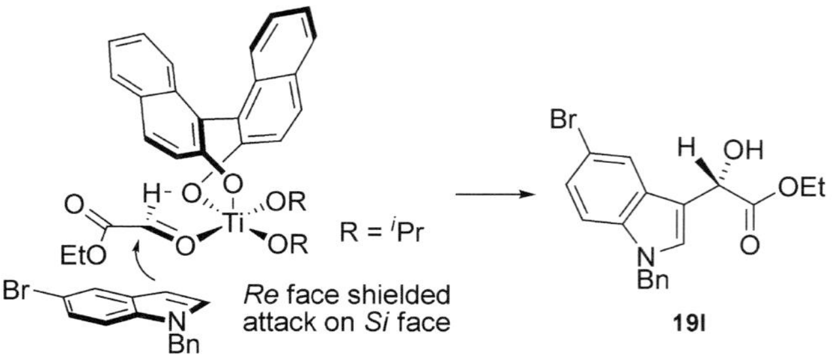

Scheme 3.11 Proposed working model.

interaction in the case of methyl trifluoropyruvate **20b** might explain why a lower *ee* was obtained. Finally, other 1,2-dicarbonyl derivatives such as methyl pyruvate **22** and *p*-chlorophenylglyoxal **24** afforded only the undesired bisindole products (Scheme 3.9).

3.1.2.1.2 Chiral Bisoxazoline-Cu(II) Catalysis
The Jørgensen group reported the first bisoxazoline-copper(II) catalyzed FC reaction of aromatic amines with glyoxylate, affording various optically active mandelic acid derivatives (Table 3.2) [23]. The combination of Cu(OTf)$_2$ with *tert*-butyl bisoxazoline (*S*)-**28a** was found to be the best

Table 3.2 Enantioselective FC reaction of aromatic amines with ethyl glyoxylate catalyzed by (*S*)-**28a**-Cu(OTf)$_2$.

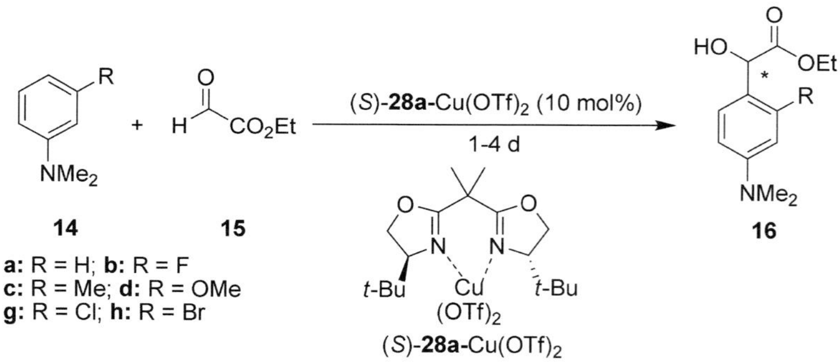

Entry	Substrate	Temp. (°C)	Solvent	Product	Yield (%)	ee (%)
1	14a	0	THF	(+)-(*S*)-**16a**	82	94
2	14a	r.t.	CH$_2$Cl$_2$	(+)-(*S*)-**16a**	72	90
3	14b	r.t.	CH$_2$Cl$_2$	(+)-**16b**	80	85
4	14c	r.t.	THF	(+)-**16c**	76	92
5	14d	r.t.	THF	(+)-**16d**	19	86
6	14g	r.t.	CH$_2$Cl$_2$	(+)-**16g**	84	93
7	14h	r.t.	CH$_2$Cl$_2$	(+)-**16h**	68	88

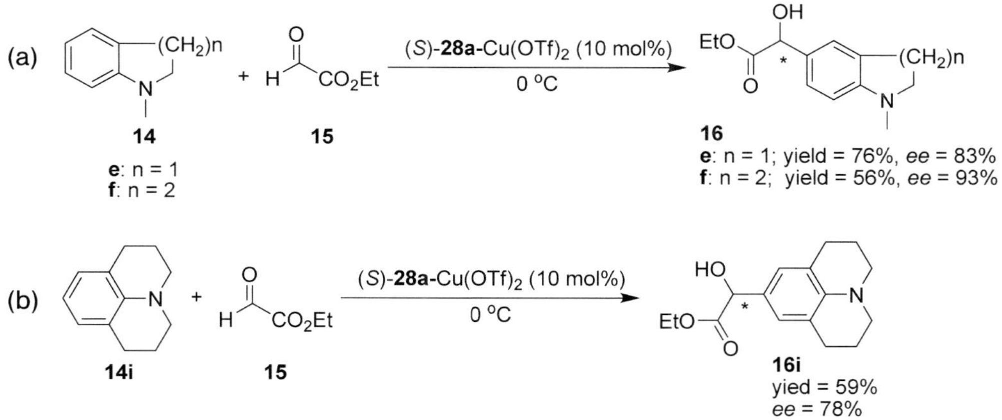

Scheme 3.12 Enantioselective FC reaction of ethyl glyoxylate (**15**) with N-methylindoline (**14e**), N-methyltetrahydroquinoline (**14f**) and Juloindine (**14i**).

choice. Both electron-donating and electron-withdrawing substituents in the meta-position of the benzene ring are well tolerated in this catalytic system (21–84% yield, 86–94% ee). This strategy was also found to be applicable to other aromatic systems, such as N-methylindoline (**14e**), N-methyltetrahydroquinoline (**14f**), and julolidine (**14i**). In these cases, moderate to good enantioselectivities were obtained (Scheme 3.12), while reduced yields and ee values were observed if furans were employed as substrates. On the basis of the experimental results and the absolute configuration of the products, the authors proposed a square-planar adduct formed by **15** coordinated to (S)-**28a**-Cu(OTf)$_2$ in a two-site binding mode (Scheme 3.13). This is the first catalytic enantioselective FC reaction of aromatic amines with glyoxylate.

The synthetic versatility of the bisoxazoline-copper(II)-based catalyst system was further explored by the Jørgensen group. They found that (S)-**28a**-Cu(OTf)$_2$ could also efficiently catalyze the FC reaction of heteroaromatic and aromatic compounds with ethyl trifluoropyruvate **20b**, leading to a simple synthetic procedure for the introduction of an optically active hydroxy-trifluoro-methyl ethyl ester group in aromatic and heteroaromatic compounds (Table 3.3) [24]. For example, the reaction proceeds with high yield (88–94%) and high enantioselectivity (83–94%) for sterically

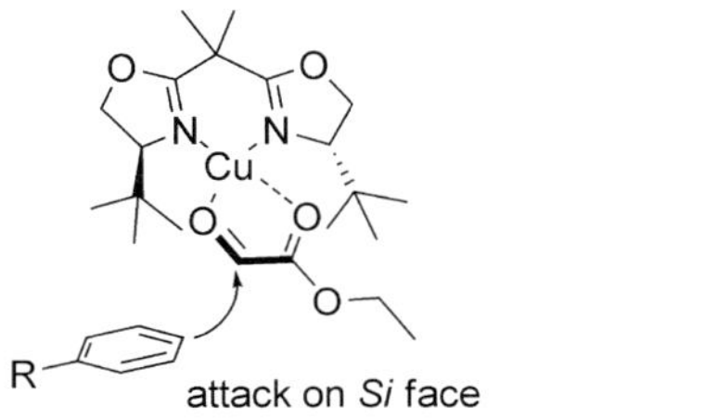

Scheme 3.13 Proposed transition state.

Table 3.3 Enantioselective FC reaction of aromatic compounds with ethyl trifluoropyruvate **20b** catalyzed by (S)-**28a**-Cu(OTf)$_2$.

Entry	Aromatic compound	Electrophile	Product	Yield (%)	ee (%)
1	**18**, R^1 = Me, R^2 = R^3 = H	**20b**	**21**, R^1 = Me, R^2 = R^3 = H	94	89
2	R^1 = R^2 = H, R^3 = Cl		R^1 = R^2 = H, R^3 = Cl	70	89
3	R^1 = R^3 = H, R^2 = Ph		R^1 = R^3 = H, R^2 = Ph	61	87
4	R^1 = R^2 = R^3 = H		R^1 = R^2 = R^3 = H	93	83
5	R^1 = R^2 = Me, R^3 = H		R^1 = R^2 = Me, R^3 = H	88	94

Entry	Aromatic compound	Electrophile	Product	Yield (%)	ee (%)
	29 (pyrrole, R^1, R^2)	F_3C-CO-CO_2Et **20b**	**30**		
6	$R^1 = R^2 = H$		$R^1 = R^2 = H$	80	83
7	$R^1 = Me, R^2 = H$		$R^1 = Me, R^2 = H$	42	93
8	$R^1 = Me, R^2 = (CO)Me$		$R^1 = Me, R^2 = (CO)Me$	69	89
	31 (R^1, X)	F_3C-CO-CO_2Et **20b**	**32**		
9	X = O, R^1 = Me		X = O, R^1 = Me	65	93
10	X = O, R^1 = H		X = O, R^1 = H	15	81
11	X = O, R^1 = TMS		X = O, R^1 = TMS	32	76
12	X = S, R^1 = Me		X = S, R^1 = Me	16	79

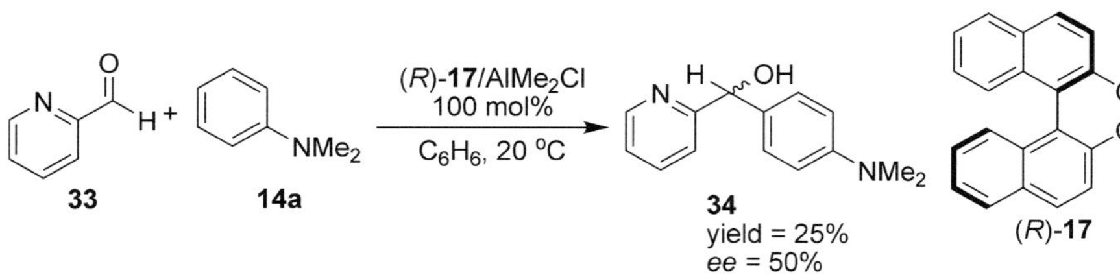

Scheme 3.14 Enantioselective FC reaction of pyridine-2-carbaldehyde with N,N-dimethylaniline mediated by chiral aluminum complex.

unhindered indoles and N-methyl-protected indoles. Other heteroaromatic compounds, such as pyrroles, furans and thiophene, react with ethyl trifluoropyruvate in moderate to good yield (16–80%) and high enantioselectivity (76–93%).

Meanwhile, the Jørgensen group also made a preliminary investigation of the enantioselective hydroxalkylation reaction of N,N-dimethylaniline **14a** with pyridine-2-carbaldehyde **33** in the presence of chiral aluminium catalyst (Scheme 3.14) [25]. It was found that pyridine-2-carbaldehyde activated by (R)-BINOL-AlCl reacts with for example, N,N-dimethylaniline, to give the corresponding alcohol **34** in moderate yields and enantiomeric excess, whereas the pyridine-3- and -4-carbaldehydes give a diarylated product. Significantly, this is the first example of a simple aromatic aldehyde reacting with an aromatic C—H bond to give an optically active FC product in the presence of a chiral Lewis acid.

In order to explore the versatility of a new chiral nonracemic 2,2′-bipyridyl ligand (+)-**35** [26], Wilson et al. recently described that C_2-symmetric 2,2′-bipyridyl copper (II) triflate complex proved to be a suitable catalyst for the asymmetric FC reaction of a series of substituted indoles with methyl trifluoropyruvate (Table 3.4) [27]. Under the optimized conditions, the Cu(OTf)$_2$-(+)-**35** complex displayed excellent catalytic activity with good to high yields (62–79%) and ee values (60–90%). On the contrary, 1-methylindole and 1-methyl-2-phenylindole afforded the corresponding products in low enantiomeric excess (18%) (Table 3.4, entries 7 and 8). It should be noted that this is the first report of the application of a 2,2′-bipyridyl ligand in catalytic enantioselective FC reactions.

To rationalize the stereochemical induction of this copper(II)-catalyzed asymmetric FC reaction of indoles with the methyl ester of 3,3,3-trifluoropyruvate **20b**, they postulated a transition state with an approximately square-planar geometry, in which the relatively small α-ketoester **20b** coordinates the copper center in a bidentate fashion (Scheme 3.15). Accordingly, the indole **18** would then be expected to approach the carbonyl moiety from the less sterically accessible Re-face.

The enantioselective FC alkylations of electron-rich aromatic and heteroaromatic compounds such as aniline and indole derivatives with glyoxylate or trifluoropyruvate have been extensively investigated. However, there are very few studies on the catalytic enantioselective FC alkylation of the less reactive aromatic ethers with carbonyl compounds. Recently, Chen and coworkers reported a highly enantioselective FC alkylation of simple aromatic ethers with trifluoropyruvate catalyzed by (4R,5S)-Diph-BOX (**28b**)-Cu(OTf)$_2$ under solvent-free conditions (Scheme 3.16) [28].

Table 3.4 Enantioselective FC alkylation reactions catalyzed by a chiral nonracemic C2-symmetric 2,2′-bipyridyl copper(II) complex.

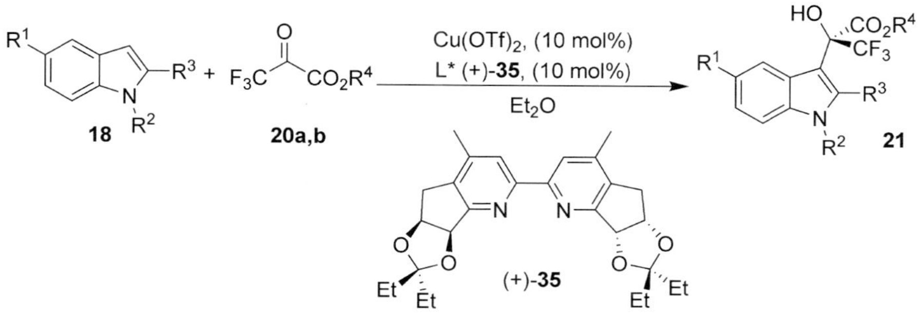

Entry	R^1	R^2	R^3	R^4	Temp. (°C)	Yield (%)	ee (%)
1	H	H	H	Et	0	68	74
2	H	H	H	Me	0	77	90
3	H	H	H	Me	−10	62	90
4	H	H	Me	Me	0	79	86
5	OMe	H	H	Me	0	69	72
6	NO_2	H	H	Me	0	75	60
7	H	Me	H	Me	0	74	18
8	H	Me	Ph	Me	0	65	18

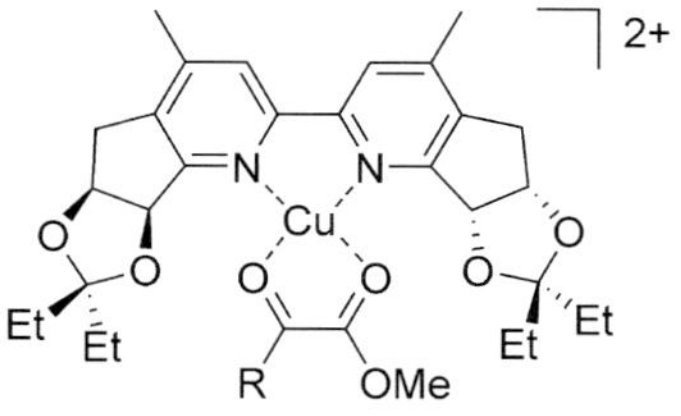

Square-planar coordination
Scheme 3.15 Proposed coordination geometry of the ketoester with the chiral catalyst.

After screening several representative BOX-ligands and optimization of other reaction parameters, they found that the ligand (4R,5S)-**28b**, in combination with Cu(OTf)$_2$, showed the best catalytic performance for the reaction of a variety of aromatic ethers **36** with **20a** under solvent-free conditions. Moderate to high yields (55–98%) and high enantioselectivities (90–93%) were achieved in all cases. It is noteworthy that only 1 mol% of catalyst loading was used in this process.

As for the stereochemical course of the reaction, a tetrahedral transition structure involving the coordination of trifluoropyruvate to (4R,5S)-Diph-BOX-Cu(II) was proposed, wherein, the π–π stacking interaction of a phenyl group with the carbonyl moiety of trifluoropyruvate played an important role, leading predominantly to Re face attack (Scheme 3.17).

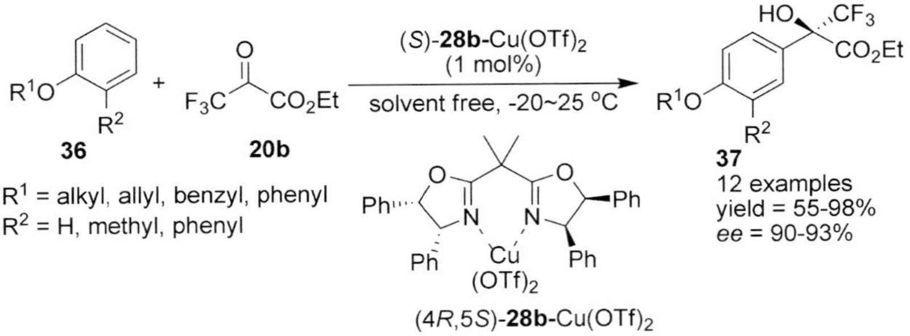
(S)-28b-Cu(OTf)2
(1 mol%)
solvent free, -20~25 °C
36
20b
37
R1 = alkyl, allyl, benzyl, phenyl
R2 = H, methyl, phenyl
12 examples
yield = 55-98%
ee = 90-93%
(4R,5S)-28b-Cu(OTf)2

Scheme 3.16 Catalytic enantioselective FC reaction of aromatic ethers **36** with ethyl trifluoropyruvate **20b** under solvent-free conditions.

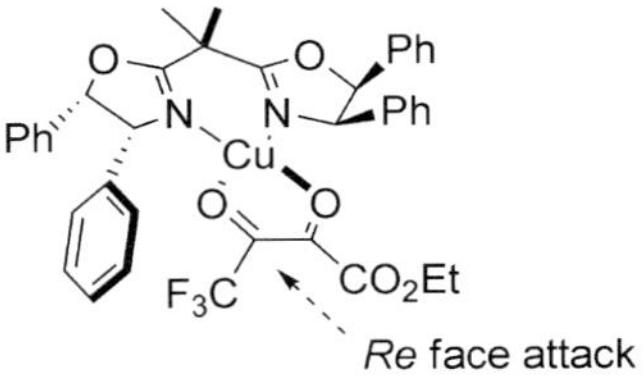
Re face attack

Scheme 3.17 Proposed model for the asymmetric FC alkylation.

3.1.2.2 Experiments: Selected Representative Procedures

General procedure for the Friedel–Crafts reactions with fluoral catalyzed by BINOL-Ti complexes 10 (Scheme 3.5) [15]

To a solution of Ti(O-^{i}Pr)$_4$ (28.4 mg, 0.1 mmol) in dehydrated DCM (1 mL) was added (*R*)-6,6′-Br$_2$-BINOL (44.4 mg, 0.1 mmol) at room temperature under an argon atmosphere. After stirring for 1 h, the additive (0.1 mmol) in dehydrated DCM (1 mL) was added to the mixture. After stirring for an additional 1 h, aromatic substrate (1 mmol) in dehydrated DCM (1 mL) was added, and then an excess amount of freshly dehydrated and distilled fluoral was added to the catalyst solution at 0 °C. The reaction mixture was stirred for 12 h at the same temperature. DCM (5 mL) and H_2O (3 mL) were added to the mixture. Insoluble material was filtered off through a pad of Celite, and the filtrate was extracted three times with DCM. The combined organic layer was washed with brine, dried over $MgSO_4$, and evaporated under reduced pressure. Chromatographic separation by silica gel (DCM : *n*-hexane 3 : 2) gave the product.

General procedure for titanium-catalyzed enantioselective Friedel–Crafts reaction of aromatic amines 14 with ethyl glyoxylate 15 (Table 3.1) [19]

A 10 mL Schlenck tube was equipped with a magnetic stirrer. The air in the tube was replaced by argon. Ligand (*R*)-6,6′-Br$_2$-BINOL (22 mg, 0.05 mmol) was dissolved in 1 mL of toluene, and then Ti(O-^{i}Pr)$_4$ (0.5 M in DCM, 50 μL, 0.025 mmol) was added to the solution. The mixture was stirred at room temperature for 2 h. The resulting orange solution was cooled to the specific temperature, and then aromatic amine **14** (0.25 mmol) and freshly distilled ethyl glyoxylate **15** (40 µL, 0.5 mmol) were introduced into the reaction system. The reaction process was monitored by TLC. After the completion of the reaction, the solvent was removed under reduced pressure and the residue was submitted to flash chromatography separation on silica gel using hexanes–ethyl acetate (3 : 1) as an eluent to give the corresponding FC reaction product **16**.

General procedure for enantioselective FC reaction of indoles 18 with ethyl glyoxylate 15 catalyzed by (S)-BINOL-Ti complex (Scheme 3.8) [22]

To a 5-mL flask equipped with a magnetic stirrer, in which the air was replaced by nitrogen, was added (S)-BINOL (28.6 mg, 0.1 mmol), diethyl ether (1 mL), and Ti(O-^{i}Pr)$_4$ (14.9 µL, 0.05 mmol). The mixture was stirred at room temperature for 1 h. Then the resulting orange solution was cooled to the specified temperature, the indole **18** (0.5 mmol) and ethyl glyoxylate **15** (0.15 mL, 50% in toluene, 0.75 mmol) were introduced into the reaction system. After completion of the reaction (monitored by TLC), H$_2$O (3 mL) and DCM (5 mL) were added to the mixture. Insoluble material was filtered off through a pad of Celite, and the filtrate was extracted three times with DCM. The combined organic layer was washed with brine, dried over MgSO$_4$, and the solvent was removed under reduced pressure. The residue was submitted to flash chromatographic separation on silica gel using petroleum ether–ethyl acetate (3 : 1) as an eluent to give the corresponding FC reaction product **19**.

General procedure for the catalytic FC reaction between aromatic amines and ethyl glyoxylate catalyzed by chiral bisoxazoline-copper (II) complex (Table 3.2) [23]

To a flame-dried Schlenk tube was added Cu(OTf)$_2$ (18.1 mg, 0.05 mmol, 10 mol%) and (S)-*tert*-butyl-bisoxazoline **28a** (15.5 mg, 0.055 mmol, 11 mol%). The mixture was dried under vacuum for 1–2 h and freshly distilled anhydrous solvent (2.0 mL) was added and the solution was stirred for 0.5–1 h. Freshly distilled ethyl glyoxylate **15** (255 mg, 2.5 mmol) and amine **14** (0.5 mmol, 1 equiv.) were added. After stirring for a specific period (4 h to 4 d) at the indicated temperature, the reaction mixture was filtered through a pad of silica with Et$_2$O, concentrated *in vacuo*, and the product was purified by flash chromatography.

General procedure for the copper (II)-catalyzed asymmetric FC reactions using ligand (+)-35 (Table 3.4) [27]

To a flame-dried Schlenk tube was added Cu(OTf)$_2$ (3.1 mg, 8.6 µmol), ligand (+)-**35** (4.1 mg, 8.8 µmol) and anhydrous ether (1.5 mL) and the resultant solution was stirred at room temperature for 1 h. The catalyst solution was then cooled to 0 °C and treated with the appropriate indole **18** (94 µmol) and the ethyl or methyl 3,3,3-trifluoropyruvate **20a,b** (86 µmol). The Schlenk tube was then sealed and stirred at 0 °C for 16 h. The reaction mixture was then concentrated *in vacuo* to afford the crude product. Flash chromatography using hexanes-ether (1 : 1) as the eluent afforded the desired substituted indoles **21**.

Typical experimental procedure for enantioselective FC reaction of aromatic ethers 36 with ethyl trifluoropyruvate 20b under solvent-free conditions (Scheme 3.16) [28]

To a solid sample of pre-prepared chiral Lewis acid catalyst (4R,5S)-**28b**-Cu(OTf)$_2$ (2.0 mg, 0.002 mmol) in a dry test tube, ethyl trifluoropyruvate **20b** (32 µL, 0.24 mmol, 1.2 equiv.) was added under the protection of argon at 0 °C with an ice bath, followed by stirring for 10 min. Anisole **36** (0.2 mmol) was added by syringe. After stirring for 13 h at the same temperature, ethyl acetate (0.5 mL) was added to quench the reaction and the crude product was purified by flash chromatography on silica gel (eluent: petroleum ether–ethyl acetate = 10 : 1 v/v) to give the desired product.

3.1.3
Organocatalysis

3.1.3.1 Fundamental Examples and Mechanism

Recently, asymmetric organocatalysis [29, 30] has become an interesting important subfield of organic chemistry, due to the operational simplicity, the "green" aspect of this chemistry and the ready availability of the inexpensive bench-stable catalysts. In this rapidly expanding field, the application of chiral hydrogen bond donors as the catalyst for electrophilic activation of unsaturated polarized molecular entities, including the carbonyl group, has received remarkable attention [31].

In 2005, Jørgensen and coworkers applied the concept of hydrogen bond activation to the asymmetric FC reaction of substituted indoles with α-dicarbonyl compounds, ethyl glyoxylate **15** and ethyl 3,3,3-trifluoropyruvate **20b** [32]. They demonstrated that the catalyst, bis-sulfonamide **38**, showed superior catalytic activity for a variety of indoles and the corresponding products were obtained in good to excellent yields (73–99%) with moderate *ee*s (23–63%) (Scheme 3.18). A double hydrogen-bonding interaction model was proposed for the stereo-induction of this reaction.

Almost at the same time, the Török group described the use of simple cinchona alkaloids, such as cinchonidine **39** and cinchonine **32**, in the asymmetric addition

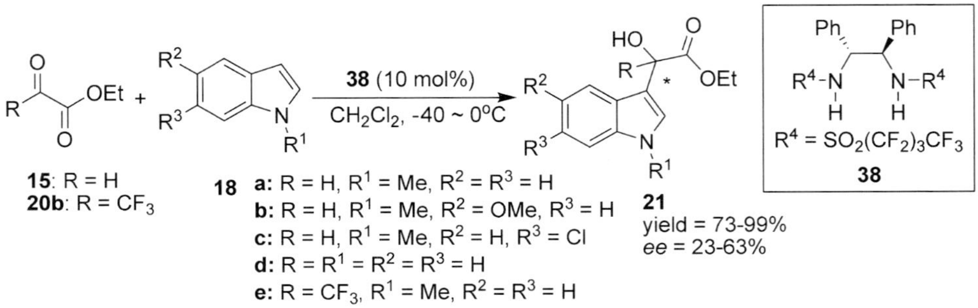

Scheme 3.18 Hydrogen bonding catalyzed enantioselective FC reaction of indole.

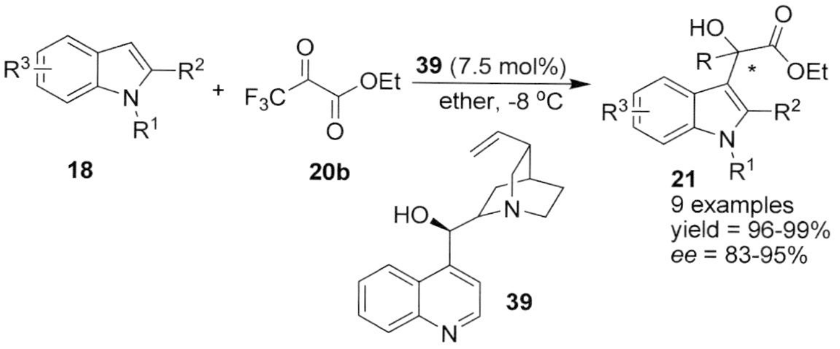

Scheme 3.19 Highly enantioselective organocatalytic FC reaction of indole **18** with ethyl trifluoropyruvate **20b**.

of indoles to ethyl 3,3,3-trifluoropyruvate **20b** (Scheme 3.19) [33]. Both catalysts provided excellent yields (96–99%) and remarkable enantioselectivity (83–95%) for a wide range of substituted indoles, with opposite stereoinduction. As for 1-methylindole derivatives or 2-subtituted indole, racemic or low *ee* of the corresponding product were obtained.

Prompted by the additional catalyst structure-enantiodifferentiation profile that the indole NH, cinchona 9-OH, and quinuclidine N groups are essential to the catalytic performance, Török and coworkers proposed that the alkaloid forms a weak hydrogen-bonded complex with indole and then anchors ethyl 3,3,3-trifluoropyruvate **20b** to form another active hydrogen-bonded intermediate. This is the first highly enantioselective organocatalytic FC hydroxyalkylation reaction.

In 2006, the Deng group demonstrated that 6′-OH cinchona alkaloid **40** is able to promote highly enantioselective FC reactions of indole with carbonyl compounds (Scheme 3.20) [34]. Impressively, this catalytic strategy is applicable not only to a wide variety of indoles but also to a substantial range of α-ketoesters **42** and **43** (52–97% yield, 81–99% *ee*) and aldehydes **15** and **41** (60–96% yield, 82–93% *ee*). It is particularly noteworthy that this study documents the first highly enantioselective catalytic FC reaction with simple aldehydes. Thus, this new asymmetric

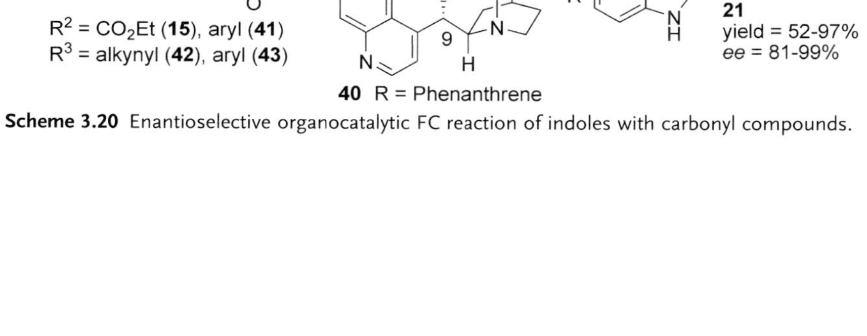

Scheme 3.20 Enantioselective organocatalytic FC reaction of indoles with carbonyl compounds.

reaction expands substantially the range of optically active indole derivatives which can be directly generated from readily available prochiral precursors with useful enantioselective excess.

3.1.3.2 Experiments: Selected Representative Procedures

A typical experimental procedure for cinchona alkaloid catalyzed hydroxyalkylation of indoles with ethyl 3,3,3-trifluropyruvate (Scheme 3.19) [33]

Indole (0.5 mmol) and cinchonidine (0.0375 mmol) were placed in a glass reaction vessel and 3 mL Et$_2$O was added. The mixture was stirred at $-8\,°C$ (salt–ice cooling bath) for 30 min. 0.75 mmol of ethyl 3,3,3-trifluoropyruvate was then added and the mixture was stirred at $-8\,°C$ (salt–ice cooling bath) for 3 h, the progress was monitored by TLC. After the reaction was completed, the solvent and excess of ethyl trifluoropyruvate were removed by evaporation. The mixture was dissolved in ether and the catalyst was removed by treatment with 500 mg of K-10 montmorillonite (a solid acid). After the treatment cinchonidine-K-10 complex was removed by filtration and the solvent was evaporated. A colorless solid was obtained in 98% yield. The enantiomeric excess of the product has been determined by chiral HPLC.

General procedure for FC addition of indoles to α-ketoester (Scheme 3.20) [34]

At room temperature, to the solution of pyruvate (0.2 mmol) in Et$_2$O (0.4 mL) was added QD-**40** (10 mol%), followed by addition of indoles (0.4 mmol). The mixture was then left to stand at room temperature for 40–72 h. The reaction mixture was subjected to flash chromatography to afford the desired product.

3.1.4
Heterogeneous Catalysis

Recently, Corma *et al.* have reported two solid catalysts, in which a chiral copper(II) bisoxazoline has been covalently anchored on silica and MCM-41, and have studied their catalytic efficiency in the asymmetric FC reaction of 1,3-dimethoxybenzene with ethyl 3,3,3-trifluoropyruvate (Scheme 3.21) [35]. With the supported catalysts, higher enantioselectivities (up to 92%) were observed than that observed under homogeneous conditions [23]. More importantly, it was found that the chiral catalyst **45**-MCM-41 can be reused and the second catalytic reaction provided the same level of enantioselectivity (84%) with only a slight decrease in conversion (73%). This is the first example of heterogeneous catalytic asymmetric FC alkylation.

Acknowledgments

Financial support was provided by the National Science Foundation of China (20672040), the National Basic Research Program of China (2004CCA00100), the Program for New Century Excellent Talents in University (NCET-05-0672), and the

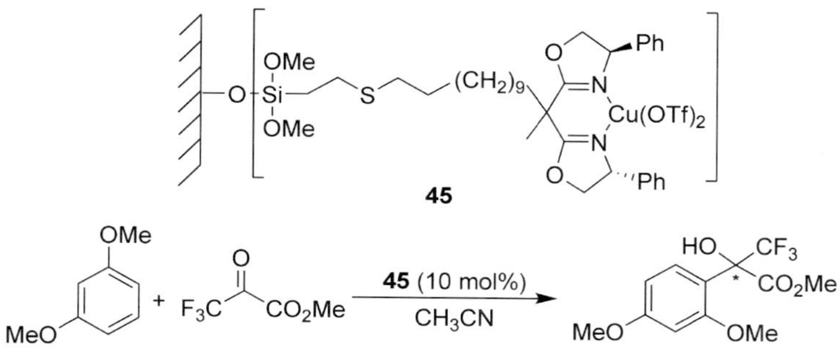

Scheme 3.21 Solid-supported catalyst for the enantioselective alkylation of 1,3-dimethoxybenzene with methyl 3,3,3-trifluoropyruvate.

Cultivation Fund of the Key Scientific and Technical Innovation Project (Ministry of Education of China, No 705039).

Abbreviations

FC   Friedel–Crafts
DCM   dichloromethane
OTf   triflate

3.2 Imines

Shu-Li You

Summary

Asymmetric Friedel–Crafts reaction of aromatic compounds with imines provides a class of highly desirable chiral amines bearing a benzylic carbon stereocenter. With a catalytic amount of $CuPF_6$/Tol-BINAP or $Cu(OTf)_2$/(S)-Bn-bisoxazoline, asymmetric Friedel–Crafts reactions of α-imino esters or N-sulfonyl aldimines with electron-rich arenes, indoles, pyrroles, furans, or thiophenes have been realized in good yields with up to excellent enantioselectivities. Cinchona alkaloid-derived thioureas and chiral phosphoric acids have been demonstrated to be efficient catalysts for the asymmetric Friedel–Crafts reaction of indoles with imines. This methodology could be extended to the synthesis of chiral trifluoromethyl-containing compounds, 2-indolyl methanamines and the 1,3-disubstituted isoindolines. As the intramolecular version of FC reactions of imines, Pictet–Spengler-type reactions were achieved in high *ee*s with chiral thioureas or chiral phosphoric acids as catalysts. The utility of the methodology has been demonstrated by the efficient total synthesis of (+)-yohimbine and (+)-harmicine.

3.2.1
Catalytic Asymmetric Friedel–Crafts Reaction of Imines

Asymmetric FC reaction is one of the most powerful methods for the synthesis of optically active aromatic compounds [36, 37]. In particular, with imines as the electrophilic partners, asymmetric FC reaction provides a class of optically active amines bearing a benzylic carbon stereocenter, which exists widely in biologically active natural products and pharmaceutically relevant compounds, Equation (3.1) [38].

$$R^1\text{-Ar} + \underset{R^2}{N^{-R^3}} \xrightarrow{\text{cat*}} R^1\text{-Ar}\overset{NHR^3}{\underset{R^2}{\overset{*}{C}}} \tag{3.1}$$

The utilization of imines offers the opportunity for them to be activated by chiral Lewis acid and chiral Brønsted acids to realize the enantioselective FC reaction. In addition, this reaction is also very attractive due to its perfect atom economy. This chapter focuses on the catalytic asymmetric FC reaction of imines with different electron-rich aromatic compounds. It should be noted that the easy access to optically pure imine substrates from chiral amines has inspired many studies on the chiral auxiliary based diastereoselective FC reaction of imines. Although, in many cases, high diastereoselectivities were realized, they will not be discussed in this chapter [39].

3.2.1.1 Intermolecular Approach

3.2.1.1.1 Organometallic Catalysis
Synthesis of optically pure α-amino acids and their derivatives constitutes an important task in organic synthesis because of their broad utility in all disciplines of biology, medicine and chemistry [40]. Aryl glycines and heteroaromatic glycines are a particularly interesting and important class of α-amino acids that have found wide applications in the synthesis of biologically active compounds. With several successful methods in this field, such as the asymmetric Strecker reaction, the Sharpless amino-hydroxylation, hydrogenation of the α-imino acids, and so on, a general catalytic asymmetric synthesis is still in great demand. The catalytic enantioselective addition of aromatic and heteroaromatic compounds to α-imino esters, commonly referred to as aza-Friedel–Crafts reaction, is a direct approach to optically pure aryl and heteroaryl glycines and has attracted great attention for a long time.

In 1999, the first study of asymmetric FC reaction of α-imino esters was reported by Johannsen, utilizing $CuPF_6$/Tol-BINAP as the catalyst [41]. With 1–5 mol% of the catalyst, the reaction of indoles with N-Ts α-imino esters afforded the addition products **3** in 67–89% yields with up to 97% ee (Scheme 3.22). When 2-acetylpyrrole was used, the 4-position alkylation product was obtained with a high enantioselectivity of 94% ee.

Besides indole and pyrrole, the highly enantioselective FC reactions of other electron-rich aromatic compounds with α-imino ester were achieved by using the

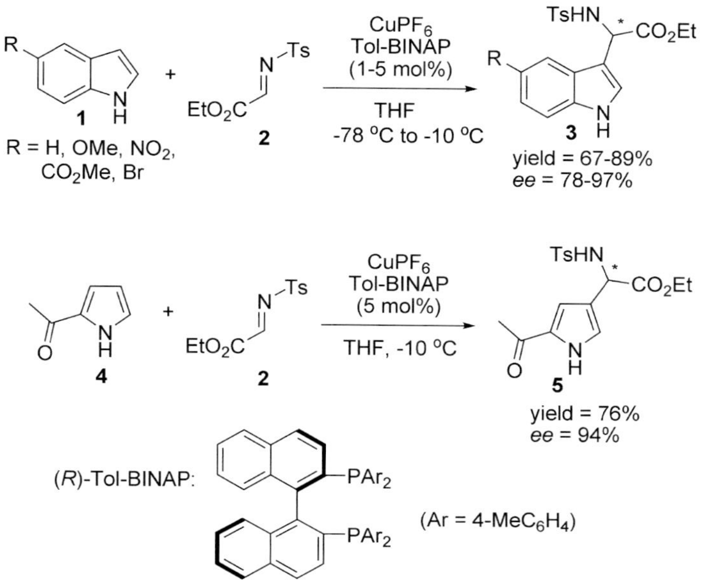

Scheme 3.22 Cu(I)-catalyzed asymmetric FC reaction of N-Ts α-imino esters.

same catalytic system (CuX-Tol-BINAP). In 2000, Jørgensen and coworkers realized the highly enantioselective FC reactions of aromatic compounds with α-imino esters [42]. In the presence of 5 mol% of CuPF$_6$ and (R)-Tol-BINAP, the reaction of N,N-dimethylaniline with α-imino esters was carried out in THF at −78 °C to give the protected aromatic amino esters in 75% yield with 93% ee (Table 3.5). A wide range of structurally diverse arenes were examined to afford the desired products **8** with up to 98% ee.

Further studies showed that this catalytic system could be applied to many other electron-rich heteroaromatic compounds such as substituted furans, thiophenes, and pyrroles to give N-alkoxycarbonyl-α-amino esters in good yields with enantioselectivities up to 96% ee for the furans, 94% ee for the thiophenes, and 59% ee for the N-methyl pyrrole (Table 3.6) [43]. It should be noted that the imines used in the reaction were prepared by an aza-Wittig reaction and used directly without further purification, which is certainly a great advantage for practical applications. Interestingly, the aza-ylide showed impact on the yields and ees of the FC reaction products. In particular, the aza-ylide **9**, in which one phenyl group was replaced with the more electron-donating benzyl group, gave the optimal results here.

Using benzoylhydrazones as the alkylating reagent, enantioselective FC alkylation of arenes and heteroarenes was realized by Leighton et al. in 2005 [44]. The reaction provided the alkylated aryl and heteroaryl glycine derivatives **15** in 54–92% yield with 87–95% ee (Table 3.7). Although 1.5 equiv. of the chiral silane reagent was required to obtain decent results, the silane reagent could be prepared on a large scale in a single step from the optically pure pseudoephedrine and PhSiCl$_3$. Moreover, the former was reported to be easily recovered in nearly quantitative yield.

Table 3.5 Cu(I)-catalyzed asymmetric FC reaction of α-imino esters with aromatic compounds.

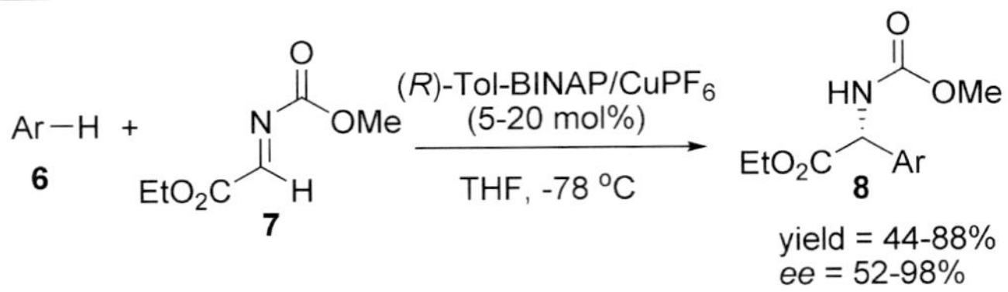

Ar–H + EtO₂C–CH=N–C(O)OMe → (with (R)-Tol-BINAP/CuPF₆ (5–20 mol%), THF, −78 °C) → EtO₂C–CH(Ar)–NH–C(O)OMe

6 + **7** → **8**

yield = 44–88%
ee = 52–98%

Entry	Substrate (6)	Catalyst loading (mol%)	Yield (%)	ee (%)
1	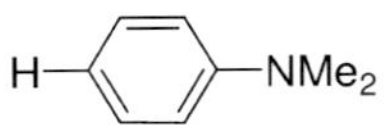 H–C₆H₄–NMe₂	5	75	93
2	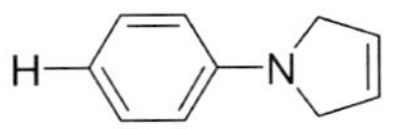 H–C₆H₄–(2,5-dihydropyrrol-1-yl)	5	88	86
3	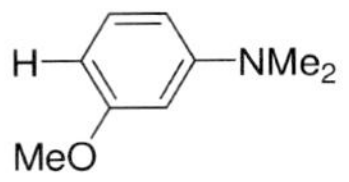 H–C₆H₃(OMe)–NMe₂	10	79	76
4	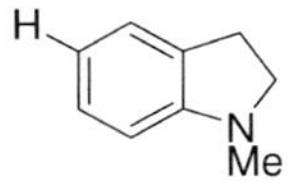 N-Me indoline	20	79	98
5	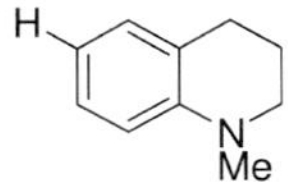 N-Me tetrahydroquinoline	5	82	87
6	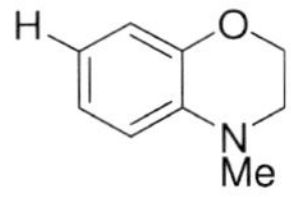 N-Me benzomorpholine	10	44	52
7	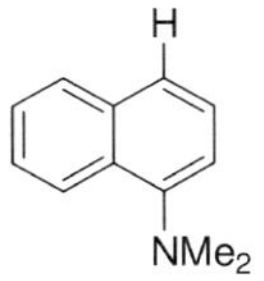 4-NMe₂-naphthalene	5	75	92

3-Indolyl methanamine is a popular structural unit found in numerous indole alkaloids and synthetic indole derivatives. Efficient catalytic asymmetric FC reactions of indoles with imines will provide direct, convergent and versatile methods for the enantioselective construction of this class of compounds. In 2006, Zhou *et al.* realized the chiral Lewis acid-catalyzed Friedel–Crafts alkylation of indoles by using aryl aldimines instead of the α-imino esters [45].

In the presence of a catalytic amount of Cu(OTf)₂/(S)-Bn-bisoxazoline complex, the reaction of indoles with N-Ts or N-Ns imines went smoothly to afford the 3-indolyl

Table 3.6 Cu(I)-catalyzed asymmetric FC reaction of α-imino esters with heteroaromatic compounds.

[Reaction scheme: MeO₂C-N=PBnPh₂ (9) + MeO₂C-C(O)H (10) → imine 11 (MeO₂C-N=CH-C(O)OMe), with (R)-Tol-BINAP/CuPF₆ (10 mol%), toluene/CH₂Cl₂ (19/1), −78 °C, Het−H → product 12 (HN(P(O)...)-CH(Het)-CO₂Me, OMe); yield = 24–82%, ee = 59–96%]

Entry	Substrate	Substrate amount (equiv.)	Yield (%)	ee (%)
1	furan-2-carbaldehyde (furan-H)	2	59	89
2	5-Et-furan-H	2	64	88
3	5-Bu^t-furan-H	10	33	96
4	5-methoxy(OMe)-furan-H	1.2	63	96
5	2-Me-furan-H	2	55	72
6	2,5-diMe-furan-H	2	24	84
7	5-OMe-thiophene-H	1.2	75	94
8	N-Me-pyrrole-H	1.2	55	59

methanamine derivatives with up to 96% *ee* (Scheme 3.23). In this reaction, the 1,3-binding mode of the nitrogen atom of imine and the oxygen atom of the sulfonyl coordinated to a Cu center was proposed.

3.2.1.1.2 Organocatalysis

In analogy to Lewis acid activation of imine compounds, a variety of chiral organic molecules such as ureas, thioureas and acids undergo hydrogen bond interaction with C=N double bond giving rise to LUMO-lowering activation (Scheme 3.24) [46]. This concept has been applied to the FC reaction, providing a large number of

Table 3.7 Chiral silane Lewis acid promoted asymmetric FC reaction with benzoyl hydrazones.

Ar—H + BzHN-N=CH-CO₂ⁱPr → (S,S)-**13** (1.5 equiv), toluene, −20 °C, 22–48 h → Ar-CH(NHNHBz)-CO₂ⁱPr
6 (3 equiv)   **14**   **15**

yield = 54–92%
ee = 87–95%

Entry	Substrate (6)	Yield (%)	ee (%)
1	H–C₆H₄–NMe₂ (para)	65	95
2	H–C₆H₄–pyrrolidinyl (para)	72	93
3	H–C₆H₃(OMe)–NMe₂	92	90
4	N-Me indoline (H at 5-position)	84	90
5	N-Me tetrahydroquinoline (H at 6)	86	87
6	5-O₂N-N-Me-indole (H at 3)	74	88
7	2-methoxythiophene (H at 5)	91	89
8	N-Bn pyrrole	76	88

enantioselective processes. Several chiral Brønsted acids that will be discussed in this chapter are listed in Figure 3.1.

In 2004, Terada and coworkers reported the first organocatalytic asymmetric aza-FC alkylation where 2-methoxyfuran was used as the nucleophilic donor [47]. In the presence of 2 mol% of chiral phosphoric acid (CPA) (*R*)-**19a** as the catalyst, the 1,2-aza-FC reaction of 2-methoxyfuran with *N*-Boc aldimine proceeded smoothly at −35 °C to afford furan-2-ylamine derivatives in 80–96% yields with 86–97% *ee*s (Scheme 3.25). Most notably, the reaction could be performed on a gram scale even using 0.5 mol% of the catalyst to afford the desired product without losing

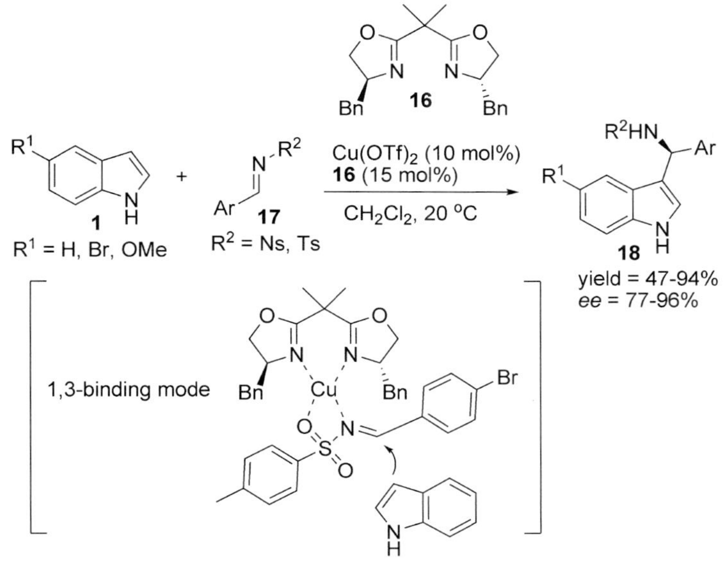

Scheme 3.23 Cu(II)-catalyzed asymmetric FC reaction of indoles with N-sulfonyl aldimines.

any enantioselectivity (95% yield, 97% ee). Furthermore, the synthetic utility of the product was demonstrated by oxidative cleavage of the furan ring to form a 1,4-dicarbonyl compound that could be further derivatized to a γ-butenolide **25**, which is a common building block in the synthesis of various natural products. There was no loss of optical purity during the transformation.

In 2006, Deng and coworkers applied bifunctional cinchona alkaloid-derived thioureas as efficient catalysts for the asymmetric FC reaction of indoles with imines [48]. With 10 mol% of the catalyst **26**, the 3-indolyl methanamine derivatives were obtained in high yields with up to 97% ee (Scheme 3.26). Interest-

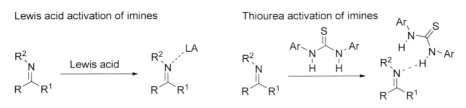

Scheme 3.24 Different imine activation modes.

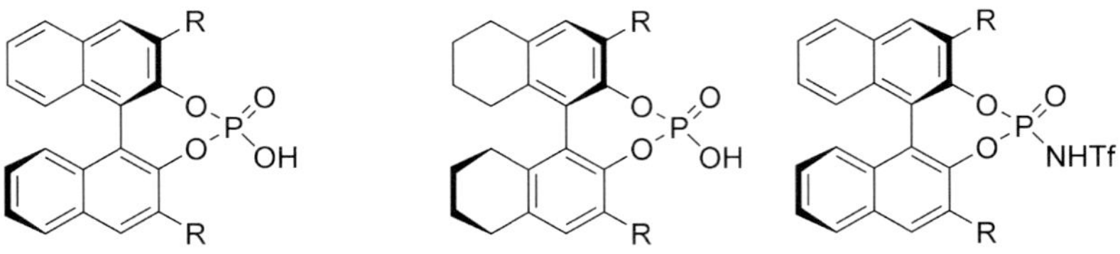

(R)-19a R = 3,5-dimesitylphenyl
(R)-19b R = 1-naphthyl
(R)-19c R = 3,5-diphenylphenyl
(R)-19d R = SiPh$_3$
(R)-19e R = 2,4,6-(^{i}Pr)$_3$C$_6$H$_2$

(R)-19f R = CHPh$_2$

(R)-20 R = 4-NO$_2$C$_6$H$_4$

Figure 3.1 Several chiral Brønsted acids.

ingly, both enantiomers of the products in high *ee*s could be obtained by catalysts derived from different chiral sources. Notably, this catalytic system is suitable not only for indoles with various electronic properties but also for both aryl and alkyl imines.

Very recently, He and coworkers described a readily recycled and regenerated heterogeneous catalyst **27**, synthesized in two steps from **26** and SBA-15 that is one kind of widely recognized mesoporous silica with large and accessible channels as well as good thermal and hydrothermal stability [49]. Interestingly, an enhancement of enantioselectivity was observed for the asymmetric FC reaction of indoles with imines when using this supported catalyst. The recycling ability was tested, and the yield and *ee* of the reaction were retained at the same level for up to 6 runs (Scheme 3.26).

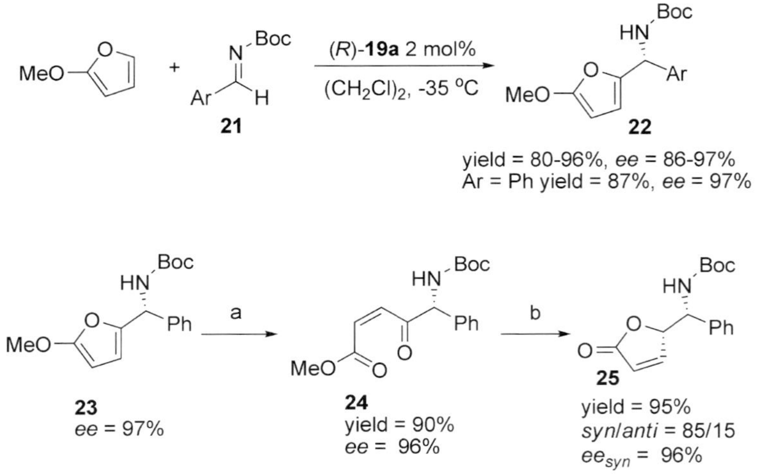

21

yield = 80–96%, *ee* = 86–97%
Ar = Ph yield = 87%, *ee* = 97%

23
ee = 97%

24
yield = 90%
ee = 96%

25
yield = 95%
syn/anti = 85/15
ee$_{syn}$ = 96%

Conditions: (a) NBS, NaHCO$_3$, Et$_2$O/H$_2$O, 0 °C, 30 min; (b) CeCl$_3$ · 7H$_2$O, NaBH$_4$, MeOH, −78 °C to room temperature, 5h

Scheme 3.25 CPA-catalyzed asymmetric FC reaction of 2-methoxyfuran with N-Boc imines.

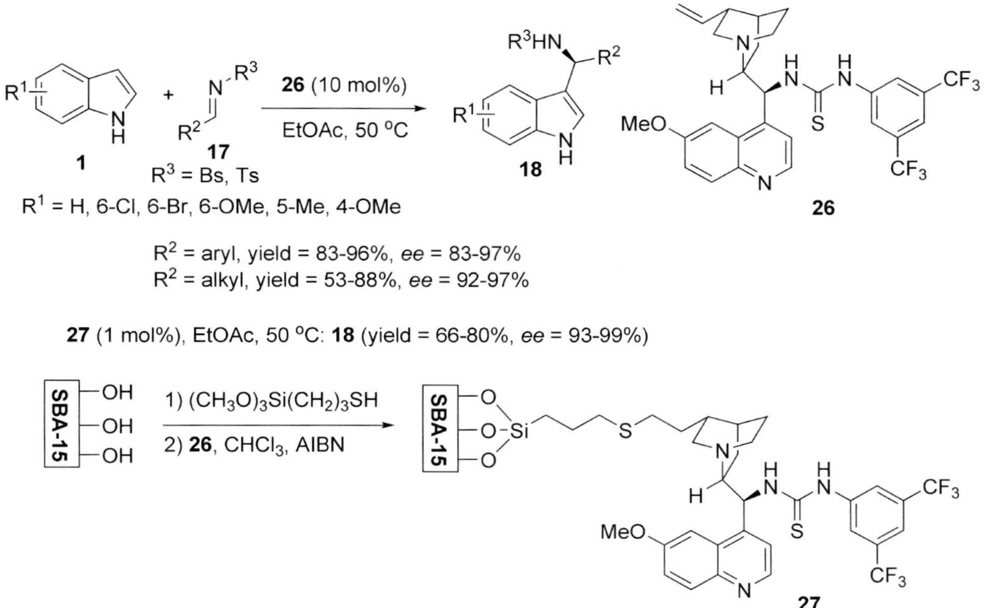

Scheme 3.26 Chiral thiourea-catalyzed asymmetric FC reaction of indoles with N-sulfonyl imines.

In 2007, You and coworkers also realized this enantioselective FC reaction by using a chiral phosphoric acid as the catalyst [50]. Utilizing 10 mol% of chiral phosphoric acid (S)-**19b**, the reaction of indoles with N-sulfonyl imines went smoothly and led to the products in good to excellent yields with up to >99% ee (Table 3.8). It should be noted that the reaction proceeded to completion within minutes to hours for most of the substrates. For a large scale application, the reaction of indole with N-benzylidenebenzenesulfonamide (R^2 = Ph, R^3 = Bs in **17**) was carried out on a 10 mmol scale (for imine), in the presence of 5 mol% of the catalyst. Here, the desired product (R^1 = H, R^2 = Ph, R^3 = Bs in **18**) was obtained in 94% yield with >99% ee. The chemistry tends to work better for imines derived from aromatic aldehydes than those derived from aliphatic aldehydes.

At the same time, Terada and coworkers reported the asymmetric FC reaction of N-TBS-protected indoles with N-Boc imines catalyzed by the chiral phosphoric acid (R)-**19c** [51]. With 2–10 mol% of the catalyst in 1,1,2,2-tetrachloroethane at −40 °C, the desired 3-indolyl methanamines were obtained in good yields for most of the substrates with excellent enantioselectivities up to 98% ee (Scheme 3.27).

Meanwhile, Antilla and coworkers also realized the similar reaction with chiral phosphoric acid (S)-**19d** as the catalyst. Highly enantioselective addition of N-benzyl indoles to N-acyl imines was achieved [52]. The addition products **32** were obtained in excellent yields (89–99%) and enantioselectivities (90–97% ee) (Scheme 3.28).

In addition to the FC reaction of indole, Antilla et al. recently realized the enantioselective FC reaction of pyrrole derivatives with N-acyl imines catalyzed by

Table 3.8 CPA-catalyzed asymmetric FC reaction of indoles with N-sulfonyl imines.

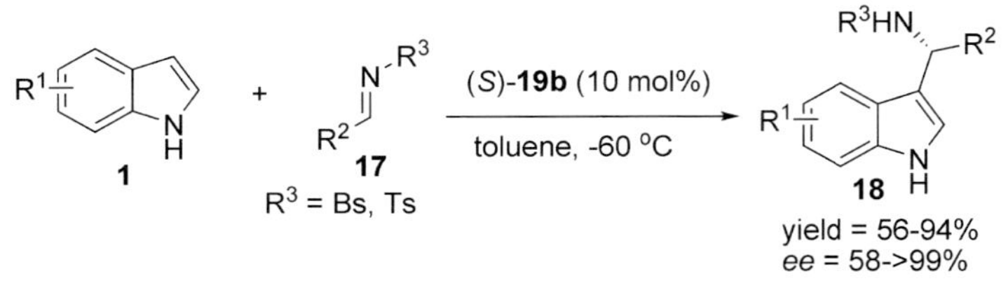
R1
1
N H
+
N R3
R2
17
R3 = Bs, Ts
(S)-19b (10 mol%)
toluene, -60 °C
R3HN
R2
R1
18
N H
yield = 56-94%
ee = 58->99%

Entry	R[1]	R[2]	R[3]	Yield (%)	*ee* (%)
1	H	C_6H_5	Ts	83	98
2	H	C_6H_5	Bs	88	99
3	5-OMe	C_6H_5	Bs	84	99
4	5-Me	C_6H_5	Bs	83	99
5	5-Br	C_6H_5	Bs	89	>99
6	6-Cl	C_6H_5	Bs	87	>99
7	H	4-MeC_6H_4	Ts	93	>99
8	H	3-$NO_2C_6H_4$	Ts	85	89
9	H	3-$MeOC_6H_4$	Bs	90	97
10	H	*c*-C_6H_{11}	Ts	56	58

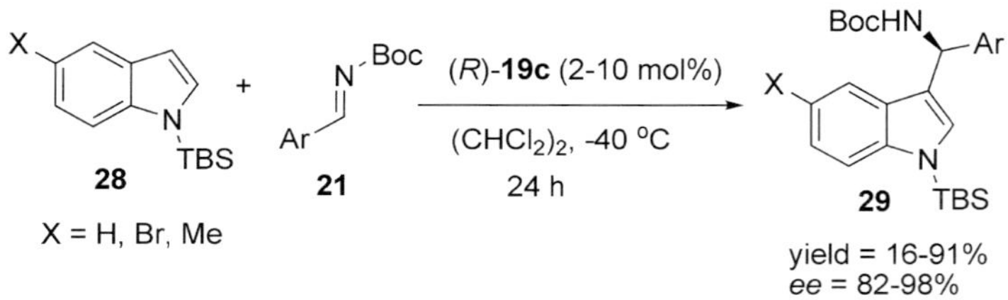
X
N
TBS
28
X = H, Br, Me
+
N Boc
Ar
21
(R)-19c (2-10 mol%)
(CHCl2)2, -40 °C
24 h
BocHN
Ar
X
N
29 TBS
yield = 16-91%
ee = 82-98%

Scheme 3.27 CPA-catalyzed asymmetric FC reaction of *N*-TBS indoles with *N*-Boc imines.

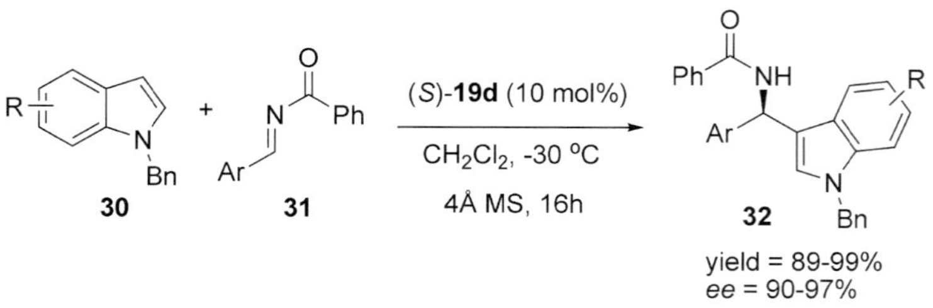
R
N
Bn
30
+
O
N Ph
Ar
31
(S)-19d (10 mol%)
CH2Cl2, -30 °C
4Å MS, 16h
O
Ph NH
Ar
R
N
32 Bn
yield = 89-99%
ee = 90-97%

Scheme 3.28 CPA-catalyzed asymmetric FC reaction of *N*-Bn indoles with *N*-acyl imines.

a chiral phosphoric acid [53]. In the presence of 5 mol% of (*S*)-**19d**, the reaction afforded the alkylation products in high yields and with up to 99% *ee* (Table 3.9).

Terada and Sorimachi reported the catalytic asymmetric FC reaction of indoles with enecarbamates [54]. In the presence of 5 mol% of chiral phosphoric acid (*R*)-**19e**,

Table 3.9 CPA-catalyzed asymmetric FC reaction of pyrroles with N-acyl imines.

Entry	R^1	R^2	R^3	R^4	Yield (%)	ee (%)
1	H	Me	H	H	86	90
2	4-Me	Me	H	H	95	89
3	4-OMe	Me	H	H	91	96
4	3-OMe	Me	H	H	97	>99
5	4-F	Me	H	H	96	85
6	4-Cl	Me	H	H	90	81
7	4-CF$_3$	Me	H	H	89	58
8	4-OMe	CH$_2$CH$_2$Br	H	H	87	92
9	4-OMe	allyl	H	H	66	91
10	4-OMe	Me	n-Bu	H	97	42
11	4-OMe	Me	H	Et	89	76

the reaction of indoles with enecarbamates led to enantioenriched 3-indolyl methanamine derivatives (with up to 96% ee). Because the imine was completely isomerized to enecarbamate during preparation, it could be considered that enecarbamate serves as a stable and useful precursor of aliphatic imine in the reaction. As shown in Scheme 3.29b, the geometric isomers (E)-**35a** and (Z)-**35a** gave the product **36a** with comparable enantioselectivity. These results suggested that both reactions proceeded through the common intermediate (**A**), composed of (R)-**19e** and an imine, generated by the protonation of enecarbamate **35a**.

Subsequently, Zhou and coworkers developed an asymmetric FC reaction of indoles with α-aryl enamides catalyzed by chiral phosphoric acid (S)-**19e** [55]. Using α-aryl enamides as the electrophilic partners, asymmetric FC reaction led to chiral amines with a quaternary carbon atom in high enantioselectivities (up to 97% ee) (Table 3.10). Possibly, the enamide form an equilibrium with the corresponding ketimine, which is protonated and activated by the chiral phosphoric acid to accept the nucleophilic attack of the indole.

Due to the dramatic difference in reactivity between the 2- and the 3-position of indole, most of the current studies on enantioselective FC reaction of indoles lead to 3-substituted indole derivatives. Developing highly enantioselective synthesis of 2-substituted indole derivatives represents a challenging task, especially for indoles without substituents on the 1- and 3-positions. Recently, Çavdar and Saraçoğlu found reaction of 4,7-dihydroindole with α,β-unsaturated compounds led to 2-substituted

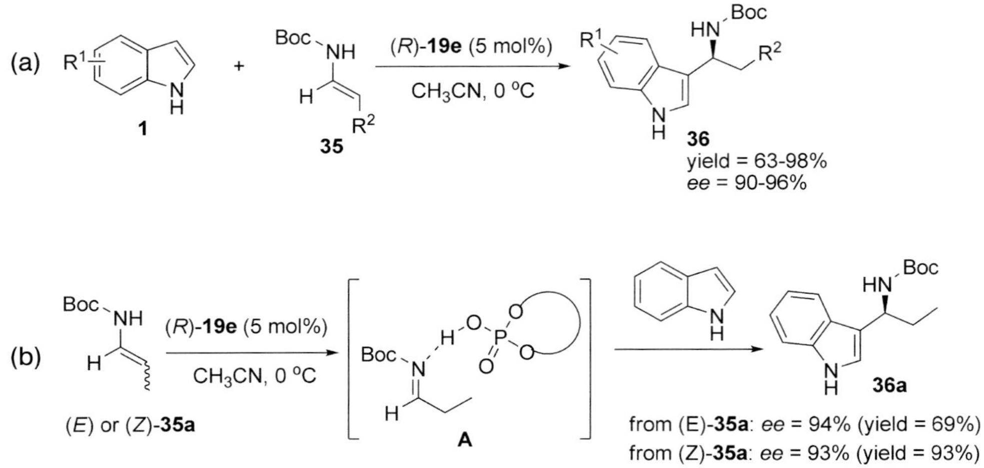

Scheme 3.29 CPA-catalyzed asymmetric FC reaction of indoles with enecarbamates.

indole derivatives after oxidation of the alkylation products [56]. Subsequently, Evans et al. [57a, b] and Pedro et al. [57c] realized the asymmetric version of this reaction utilizing chiral Lewis acid catalysts, providing easy access to enantioenriched 2-substituted indole derivatives. Utilizing this strategy, You and coworkers recently realized the asymmetric FC reaction of 4,7-dihydroindoles with imines catalyzed by chiral phosphoric acid (S)-**19d** [58]. The reaction provided the 2-substituted products

Table 3.10 CPA-catalyzed asymmetric FC reaction of indoles with α-aryl enamides.

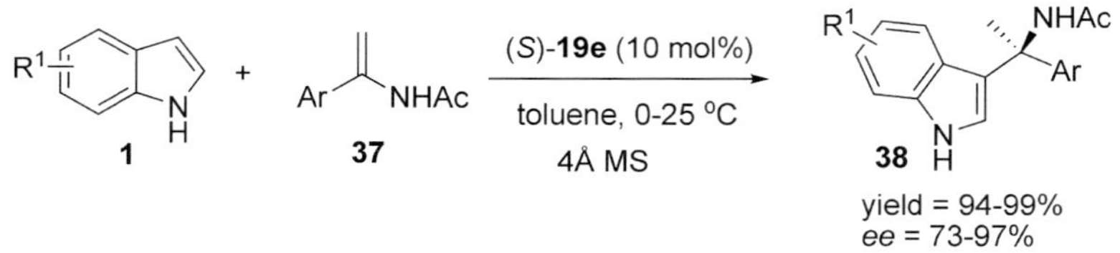

Entry	R^1	Ar	Yield (%)	ee (%)
1	H	C_6H_5	98	94
2	H	4-MeC$_6$H$_4$	99	92
3	H	4-MeOC$_6$H$_4$	94	90
4	H	4-BrC$_6$H$_4$	98	92
5	H	4-CF$_3$C$_6$H$_4$	98	93
6	H	3-MeOC$_6$H$_4$	99	97
7	H	2-MeOC$_6$H$_4$	95	73
8	4-OH	C_6H_5	95	86
9	5-Br	C_6H_5	98	90
10	5-MeO	C_6H_5	99	92

3.2 Imines

Table 3.11 CPA-catalyzed asymmetric FC reaction of 4,7-dihydroindoles with N-sulfonyl imines.

Entry	R^1, R^2	Ar	41, Yield (%)	41, ee (%)
1	H, H	C_6H_5	88	99
2	H, H	4-MeC_6H_4	81	98
3	H, H	4-BrC_6H_4	79	99
4	H, OMe	C_6H_5	74	>99
5	Me, H	C_6H_5	83	98

in good yields with excellent *ees* (up to >99% *ee*) (Table 3.11). In addition, 2-indolyl methanamine derivatives, a popular structure core in many biologically active natural and unnatural products, can be synthesized via a one-pot procedure including the FC reaction of 4,7-dihydroindoles with imines and oxidation of the products with *p*-benzoquinone.

Recently, Enders and coworkers realized the enantioselective FC reaction of indoles with *N*-Ts imines catalyzed by chiral *N*-triflyl phosphoramide (*R*)-**20** [59]. The 1,3-disubstituted isoindolines were obtained in high yields with excellent diastereo- and enantiomeric ratios by a subsequent intramolecular aza-Michael addition of the FC reaction products upon treatment with DBU (Table 3.12).

Hiemstra and coworkers developed catalytic asymmetric FC reaction of indole with tritylsulfenyl (Trs)- and 2-nitrophenylsulfenyl (Nps)-substituted glyoxyl imines to synthesize indolylglycine derivatives [60]. High yields and up to 86% *ee* for Nps-protected indolylglycine and 88% *ee* for Trs-indolylglycine were obtained (Scheme 3.30). The enantiopure product (99.5% *ee*) was obtained by a further crystallization. The advantage of the 2-nitrophenylsulfenyl substituent is its facile removal under acidic conditions allowing the ready access of amino ester **47** and unprotected (*S*)-(3)-indolyl glycine **48** after hydrolysis of the methyl ester.

The incorporation of a CF_3 group into organic molecules often leads to significant changes in the physical, chemical, and biological properties of the

Table 3.12 Chiral N-triflyl phosphoramide-catalyzed asymmetric synthesis of isoindolines.

Entry	R^1	R^2	R^3	Yield (%)	e.r. (%)
1	H	Me	H	94	95:5
2	H	Me	Br	99	94:6
3	H	Me	OMe	93	76:24
4	H	Me	H	86	61:39
5	H	Me	CO_2Me	71	86:14
6	F	Me	H	75	91:9
7	F	Me	OMe	95	88:12
8	H	tBu	H	75	88:12
9	H	tBu	Br	85	91:9
10	H	tBu	OMe	97	90:10

parent compounds. Consequently, trifluoromethylated compounds, structures in which the CF_3 group is attached to a stereogenic center, have attracted significant attention in organic synthesis, medicinal and agrochemical chemistry, and materials sciences. Very recently, Ma and coworkers reported a chiral phosphoric acid catalyzed three-component FC reaction [61]. The reaction was carried out between indoles with imines generated *in situ* from trifluoroacetaldehyde methyl hemiacetal and aniline. Novel chiral trifluoromethyl-containing compounds were synthesized in high yields with excellent enantioselectivities (up to 98% *ee*) (Scheme 3.31). Moreover, this methodology could be further extended to difluoro-acetaldehyde methyl hemiacetal providing a facile synthesis of enantiopure difluoromethyl compounds.

3.2.1.2 Pictet–Spengler Reaction

Tetrahydroisoquinoline and tetrahydro-β-carboline ring systems are popular subunits in numerous biologically active natural and unnatural products [62]. The Pictet–Spengler reaction, cyclization of electron-rich aryl or heteroaryl groups onto imine or iminium electrophiles, represents one of the most efficient methods for the construction of these structural motifs, Equation (3.2) [63].

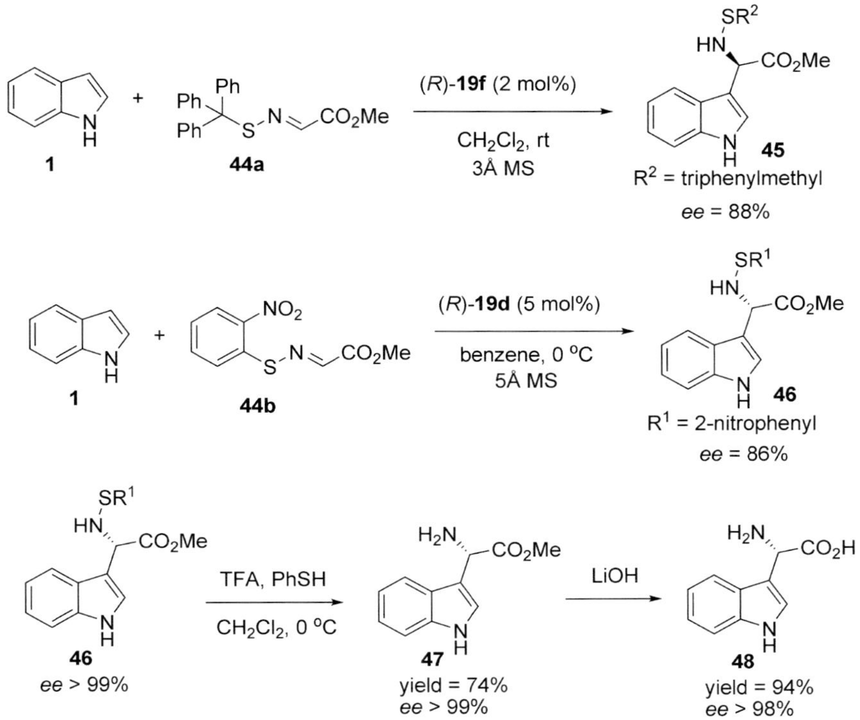

Scheme 3.30 CPA-catalyzed asymmetric synthesis of (S)-(3)-indolyl glycine.

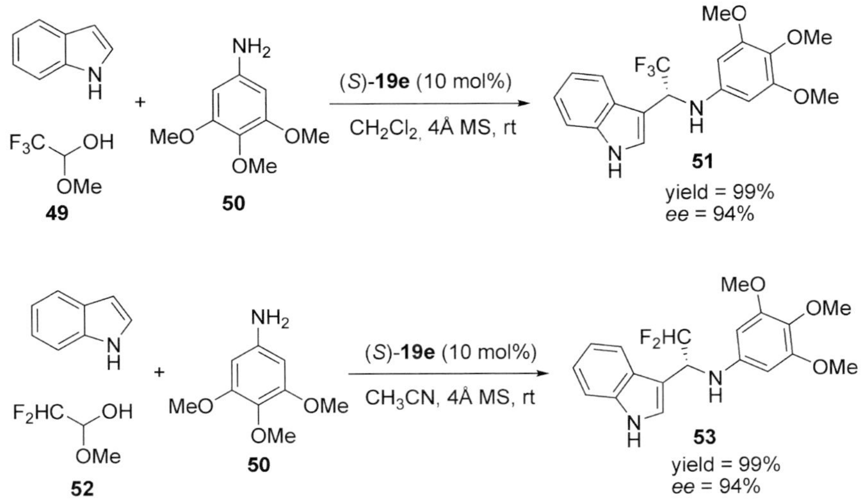

Scheme 3.31 CPA-catalyzed asymmetric synthesis of trifluoromethyl or difluoromethyl-containing compounds.

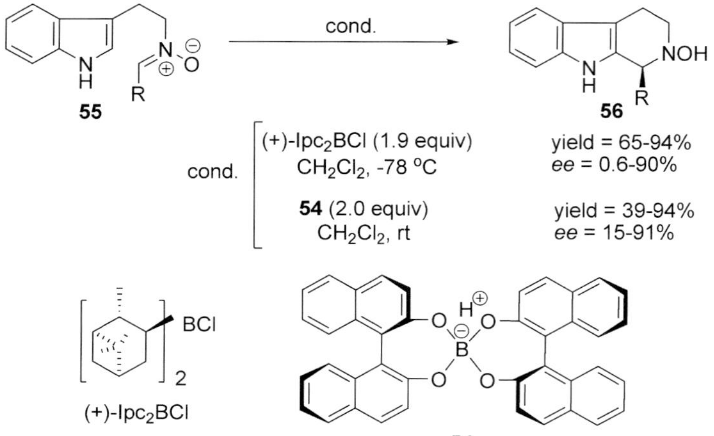

Scheme 3.32 Chiral boron reagents mediated Pictet–Spengler reaction.

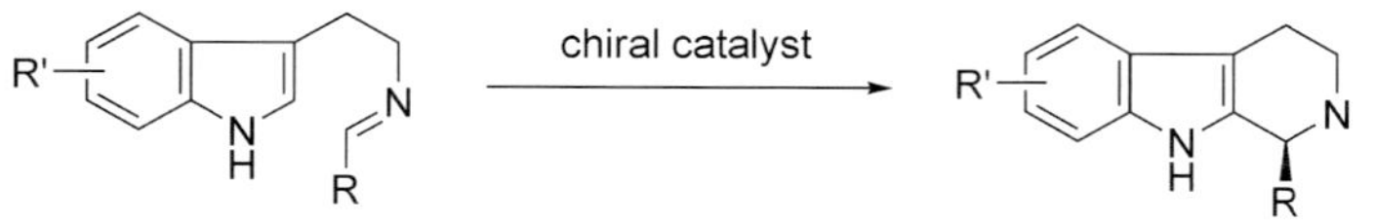

Their asymmetric variants are particularly attractive for accessing useful chiral building blocks and in complex alkaloid synthesis [64]. A number of elegant diastereoselective methods, including substrate-controlled Pictet–Spengler cyclizations, have been developed to access this important class of compounds. The successful diastereoselective approaches include the utilization of an enantiomerically pure chiral tryptamine such as a tryptophan derivative, chiral aldehyde, or a chiral auxiliary attached to the nitrogen atom [65]. The first example of an enantioselective Pictet–Spengler reaction, reported by Nakagawa and coworkers, requires the use of superstoichiometric quantities of an enantioenriched boron reagent. With 1.9 equiv. of (+)-Ipc$_2$BCl or 2.0 equiv. of **54**, the optically active N_β-hydroxytryptamine-derived nitrones **56** were obtained with up to 90% *ee* and 91% *ee*, respectively (Scheme 3.32) [66].

In 2004, Jacobsen and coworkers reported the enantioselective acyl-Pictet–Spengler reaction catalyzed by chiral thiourea derivatives [67]. The organocatalyst **57a** (10 mol%) was applied to activate the *N*-acyliminium ion intermediate, and the reaction provided access to a range of *N*-acetyl β-carbolines in high enantioselectivities (up to 95% *ee*) (Table 3.13). Subsequently, this method was applied to the total synthesis of (+)-yohimbine **63** (Scheme 3.33) [68].

This class of organocatalysts was further applied to enantioselective Pictet–Spengler-type cyclizations of hydroxylactams. Utilizing 10 mol% of chiral thiourea **57b**, good to excellent yields and high enantioselectivities were obtained in the cyclization of hydroxylactams **64** derived from a variety of succinimide and glutarimide precursors (Scheme 3.34) [69]. In a straightforward demonstration of the applicability of this new methodology, the enantioselective hydroxylactam cyclization product was

Table 3.13 Chiral thiourea-catalyzed enantioselective acyl-Pictet–Spengler reaction.

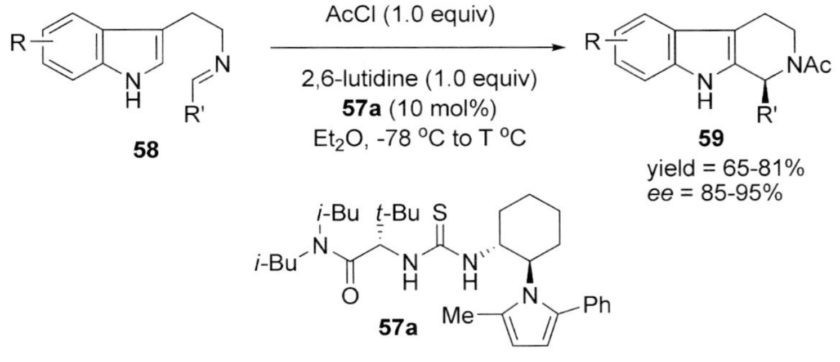

Entry	R	R'	T (°C)	Yield (%)	ee (%)
1	H	$CH(CH_2CH_3)_2$	−30	65	93
2	H	$CH(CH_3)_2$	−40	67	85
3	H	n-C_5H_{11}	−60	65	95
4	H	$CH_2CH(CH_3)_2$	−60	75	93
5	H	$CH_2CH_2OTBDPS$	−60	77	90
6	5-MeO	$CH(CH_2CH_3)_2$	−40	81	93
7	6-MeO	$CH(CH_2CH_3)_2$	−50	76	86

converted to (+)-harmicine **66** by subsequent LiAlH$_4$ reduction in 95% yield without loss of stereochemical purity. Interestingly, the enantioselective control was proposed through the dissociation of the chloride counter ion and formation of a chiral N-acyl iminium chloride–thiourea complex.

The regio- and enantioselective cyclization of pyrroles onto N-acyliminium ions generated *in situ* from hydroxylactams was also reported by Jacobsen and coworkers [70]. Excellent enantioselectivities up to 98% *ee* and high yields were obtained in these Pictet–Spengler-type reactions by a chiral thiourea catalyst **57b** (Scheme 3.35). Without a protecting group on the nitrogen of pyrrole, the cyclization took place selectively at the C-2 position. Interestingly, by introducing a sterically demanding

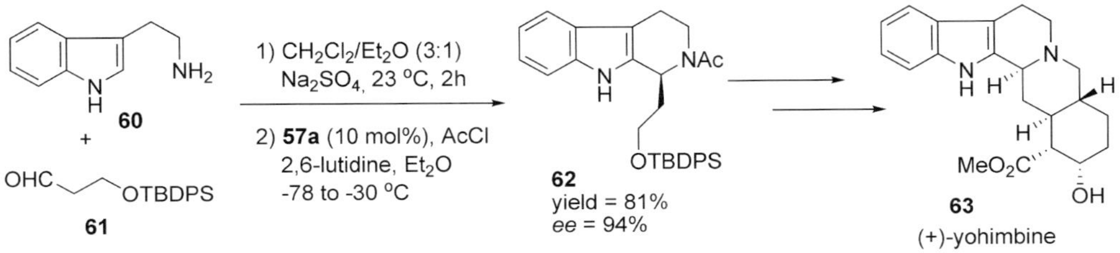

Scheme 3.33 Total synthesis of (+)-yohimbine via enantioselective acyl-Pictet-Spengler reaction.

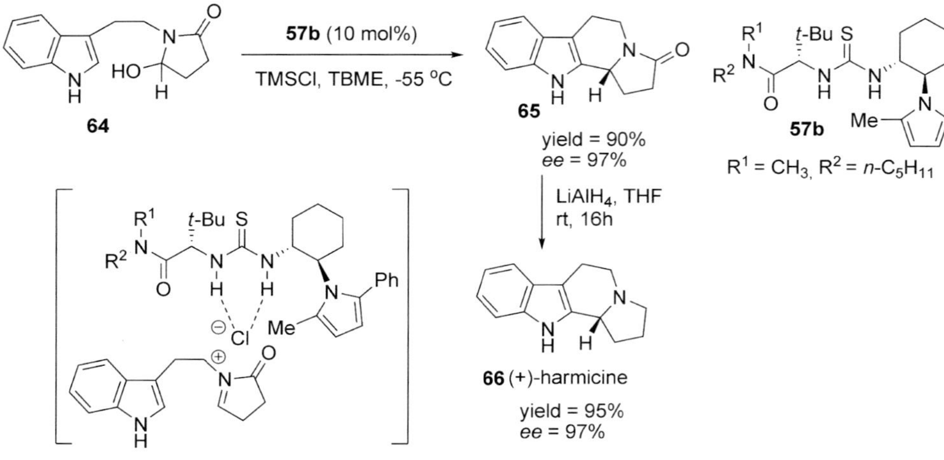

Scheme 3.34 Enantioselective Pictet–Spengler-type cyclization of hydroxylactams.

triisoproylsilyl group on the pyrrole nitrogen atom, the enantioselective cyclization could occur at the C-4 position of the pyrrole with high selectivity.

List *et al.* realized the Pictet–Spengler reaction by using the chiral phosphoric acid to activate the imines [71]. In the presence of 20 mol% of chiral phosphoric acid (S)-**19e**, the catalytic asymmetric Pictet–Spenlger reaction of various tryptamines with both aromatic and aliphatic aldehydes went smoothly to afford the tetrahydro-β-carbolines with up to 96% *ee* (Table 3.14).

In 2007, Hiemstra and coworkers also realized the Pictet–Spengler reaction by using the chiral phosphoric acid as catalyst [72]. Different from the report by Jacobsen and coworkers, where the necessary *N*-acetyl group was difficult to remove from the final product, Hiemstra and coworkers used *N*-sulfenyliminium ions as intermedi-

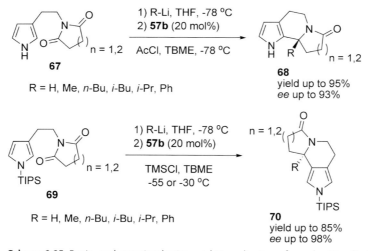

Scheme 3.35 Regio- and enantioselective catalytic cyclization of pyrroles *N*-acyliminium ions.

Table 3.14 CPA-catalyzed enantioselective Pictet–Spengler reaction.

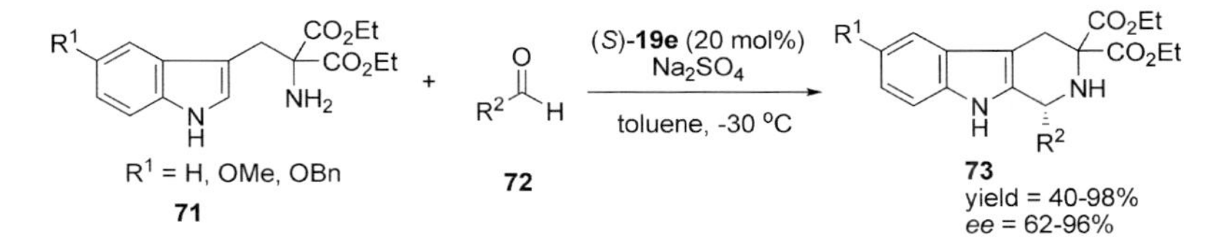

Entry	R¹	R²	Yield (%)	ee (%)
1	OBn	Et	98	90
2	OMe	Et	96	90
3	H	Et	76	88
4	OMe	n-Pr	98	88
5	OMe	n-Bu	90	87
6	OMe	i-Bu	96	80
7	OMe	i-Pr	85	81
8	OMe	2-pentyl	50	84
9[a]	OMe	4-NO$_2$C$_6$H$_4$	98	96
10[a]	OMe	C$_6$H$_5$	82	62

[a] At $-10\,°C$ in CH_2Cl_2.

ates, with the consequent easy removal of the N-protecting group from the desired product upon the treatment with HCl (Scheme 3.36a). Since the N-tritylsulfenyl product appeared to be rather labile, during the Pictet–Spenlger reaction, the addition of a trace amount of 3,5-di(*tert*-butyl)-4-hydroxytoluene (BHT), as a radical scavenger to the reaction mixture, would prevent the problems associated with the decomposi-

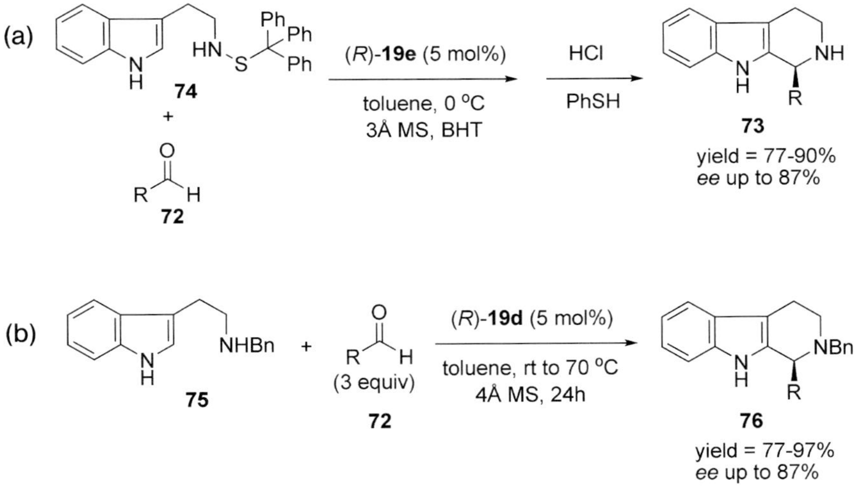

Scheme 3.36 (a) CPA-catalyzed enantioselective Pictet–Spengler reaction via sulfenyliminium ions or (b) from N-benzyltryptamine.

tion of the cyclization product. Very recently, the same group developed the asymmetric Pictet–Spengler reaction of N-benzyltryptamine with a variety of aldehydes [73]. With 2 mol% of (R)-**19d**, both aromatic and aliphatic aldehydes could be used and the corresponding adducts were obtained in 77–97% yields with up to 87% *ee* (Scheme 3.36b).

3.2.2
Selected Procedures

Chiral phosphoric acid-catalyzed asymmetric FC reaction of indoles with N-sulfonyl imines (Table 3.8, R^1 = H, R^2 = Ph, R^3 = Bs) [50]

In a dry Schlenk tube, N-sulfonyl imine **17** (81 mg, 0.25 mmol) and phosphoric acid (S)-**19b** (15 mg, 0.025 mmol) were dissolved in toluene (1 mL) under argon. The solution was stirred for 10 min at room temperature and then for another 5 min at −60 °C. Subsequently, indole (146 mg, 1.25 mmol) was added in one portion at −60 °C. After the reaction was complete (monitored by TLC), 10% $NaHCO_3$ (3 mL) was added to quench the reaction. The mixture was extracted with ethyl acetate (10 mL). The organic layer was washed with H_2O (5 mL) and brine (5 mL), separated, and dried over anhydrous Na_2SO_4. The solvents were removed under reduced pressure and the residue was purified by flash chromatography (ethyl acetate/petroleum ether = 1/3) to afford the product (97 mg, 88% yield). R_f = 0.40 (ethyl acetate/petroleum ether = 1/2, v/v); colorless solid, 99% *ee* [Daicel Chiralcel OD-H, Hexanes/IPA = 70/30, 0.8 ml min^{-1}, λ = 254 nm, *t* (major) = 11.96 min, *t* (minor) = 21.69 min]; $[\alpha]_D^{20}$ = +18.1° (*c* = 0.77, Acetone). ^{1}H NMR (300 MHz, CDCl$_3$) δ 5.22 (d, *J* = 7.2 Hz, 1H), 5.88 (d, *J* = 7.5 Hz, 1H), 6.62 (d, *J* = 2.4 Hz, 1H), 7.00 (t, *J* = 7.5 Hz, 1H), 7.13–7.31 (m, 10H), 7.43 (t, *J* = 7.5 Hz, 1H), 7.65 (d, *J* = 7.2 Hz, 2H), 8.02 (br, 1H); ^{13}C NMR (75 MHz, CDCl$_3$) δ 55.0, 111.3, 116.2, 119.2, 120.0, 122.5, 123.8, 125.3, 127.0, 127.1, 127.4, 128.3, 128.6, 132.2, 136.5, 140.0, 140.4.

Chiral thiourea-catalyzed enantioselective acyl-Pictet–Spengler reaction (Table 3.13, R = H, R′ = n-pentyl) [67]

In a flame-dried round-bottomed flask, tryptamine (40 mg, 0.25 mmol, 1.0 equiv.) was dissolved in dichloromethane/diethyl ether (3 : 1 v/v, 12.5 mL). Hexanal (32 mL, 0.275 mmol, 1.05 equiv.) was added dropwise by syringe at 23 °C, and the mixture was stirred at 23 °C for 90 min. Sodium sulfate (500 mg) was added and the mixture was stirred for an additional 30 min. The resulting solution was filtered by cannula transfer to a flame-dried round-bottomed flask. The desiccant was rinsed twice with dichloromethane (5 mL) and the rinses combined with the filtrate by cannula transfer. The solution was concentrated *in vacuo*, yielding the imine as a pale brown oil which was immediately dissolved in diethyl ether (5.0 mL). Catalyst **57a** (13.5 mg, 0.025 mmol, 10 mol%) was added, and the solution cooled to −78 °C in a dry ice/acetone bath. 2,6-Lutidine (29 mL, 0.25 mmol, 1.0 equiv.), then acetyl chloride (18 mL, 0.25 mmol, 1.0 equiv.), were added dropwise by syringe. The mixture was stirred at −78 °C for 5 min, then warmed to −60 °C and stirred for 23 h. The resulting heterogeneous mixture was allowed to warm to 23 °C and concentrated *in vacuo*. The residue was purified by chromatography on silica (33% ethyl acetate in hexanes), yielding the title compound as a white solid (47 mg, 65% yield, contaminated with 2% enamide byproduct). The enantiomeric excess was determined to be 95% by chiral HPLC (Pirkle (L)-leucine, 10% ethanol/hexanes, 1.2 mL min^{-1}, *t* (minor) = 10.4 min, *t* (major) = 12.3 min. $[\alpha]_D^{20} = +92°$ (*c* = 0.5, methanol). mp = 171–173 °C. ^{1}H NMR (500 MHz, CDCl$_3$): the compound exists as a 4 : 1 mixture of amide rotamers. Signals corresponding to the major rotamer: δ 8.23 (1H, br s), 7.45 (1H, d, *J* = 7.5 Hz), 7.32 (1H, d, *J* = 8.0 Hz), 7.15 (1H, ddd, *J* = 7.5, 7.5, 1.0 Hz), 7.09 (1H, ddd, *J* = 7.5, 7.5, 1.0 Hz), 5.78 (1H, dd, *J* = 9.0, 5.5 Hz), 4.01 (1H, dd, *J* = 13.5, 3.5 Hz), 3.55–3.49 (1H, m), 2.84–2.77 (2H, m), 2.25 (3H, s), 1.92–1.85 (1H, m), 1.82–1.75 (1H, m), 1.55–1.39 (2H, m), 1.36–1.27 (4H, m), 0.89 (3H, t, *J* = 6.5 Hz). Representative signals corresponding to the minor rotamer: δ 8.01 (1H, br s), 7.50 (1H, d, *J* = 8.0 Hz), 4.96 (1H, dd, *J* = 13.0, 5.0 Hz), 4.88–4.86 (1H, m), 3.03–2.97 (1H, m), 2.71 (1H, dd, *J* = 16.0, 3.5 Hz), 2.20 (3H, s). ^{13}C NMR (100 MHz, CDCl$_3$), signals corresponding to the major rotamer: d 169.9, 136.2, 135.1, 126.9, 121.9, 119.7, 118.1, 111.2, 107.6, 49.4, 41.3, 34.9, 32.2, 26.2, 22.8, 22.3, 22.2, 14.3. IR (CH$_2$Cl$_2$ film): 3271 (m), 2930 (m), 2857 (m), 1632 (s), 1451 (s). LRMS (CI): 284.6 (100%) [M]$^+$, 285.1 (30%) [M + H]$^+$.

Acknowledgments

Financial support was provided by the Chinese Academy of Sciences and the National Natural Science Foundation of China. I thank Dr. Qiang Kang for the help with the manuscript preparation.

Abbreviations

Ac	acetyl
AIBN	azobisisobutyronitrile
BHT	3,5-di(tert-butyl)-4-hydroxytoluene
Bn	benzyl
Boc	*tert*-butyloxycarbonyl
Bs	benzenesulfonyl
Bz	benzoyl
CPA	chiral phosphoric acid
DBU	1,8-diazabicyclo[5,4,0]undec-7-ene
Mes	mesityl
Nps	2-nitrophenylsulfenyl
Ns	4-nitrobenzenesulfonyl
TBS	*tert*-butyldimethylsilyl
TBME	*tert*-butyl methyl ether
TMSCl	trimethylsilyl chloride
Tf	trifluoromethanesulfonyl

TFA   trifluoroacetic acid
TIPS   triisopropylsilyl
Trs   tritylsulfenyl
Ts   p-toluenesulfonyl

References

1 (a) Olah, G.A. (1973) *Friedel–Crafts Chemistry*, Wiley-Interscience, New York; (b) Roberts, R.M. and Khalaf, A.A. (1984) *Friedel–Crafts Alkylation Chemistry. A Century of Discovery*, Marcel Dekker, New York; (c) Heaney, H. (1991) *Comprehensive Organic Synthesis*, Vol. 2 (eds B.M. Trost and I. Fleming), Pergamon, New York, p. 733; (d) Smith, M.B. (1994) *Organic Synthesis*, McGraw-Hill, New York, p. 1313; (e) Olah, G.A. (1964) *Friedel-Crafts and Related Reactions*, part 1, Vol. II, Wiley-Interscience, New York; (f) Olah, G.A., Krisnamurthy, R. and Prakash, G.K.S. (1991) *Friedel-Crafts Alkylation in Comprehensive Organic Synthesis*, Vol. III, 1st edn (eds B.M. Trost and I. Fleming), Pergamon, Oxford, p. 293.

2 Bigi, F., Casiraghi, G., Casnati, G. and Sartori, G. (1985) *The Journal of Organic Chemistry*, **50**, 5018–5022.

3 Erker, G. and van der Zeijden, A.A.H. (1990) *Angewandte Chemie – International Edition in English*, **29**, 512–514.

4 Wang, Y., Ding, K. and Dai, L. (2001) *Chemtracts*, **14**, 610–615; (b) Wang, Y. and Ding, K. (2001) *Chinese Journal of Organic Chemistry*, **21**, 763–767; (c) Bandini, M., Melloni, A. and Umani-Ronchi, A. (2004) *Angewandte Chemie – International Edition*, **43**, 550–556; (d) Poulsen, T.B. and Jørgensen, K.A. (2008) *Chemical Reviews*, **108**, 2903–2915.

5 For reactions with chloral, see e.g.: (a) Casiraghi, G., Casnati, G., Sartori, G. and Catellani, M. (1979) *Synthesis*, 824–825; (b) Menegheli, P., Rezende, M.C. and Zucco, C. (1987) *Synthetic Communications*, **17**, 457–467.

6 For reactions with benzaldehyde, see e.g.: (a) Sasakura, K., Terui, Y. and Sugasawa, T. (1985) *Chemical & Pharmaceutical Bulletin*, **33**, 1836–1842; (b) Adams, S.R., Kao, J.P.Y., Grynkeiwicz, G., Minta, A. and Tsien, R.Y. (1988) *Journal of the American Chemical Society*, **110**, 3212–3220.

7 (a) For reactions using phenylglyoxal derivatives, see: (a) Coan, S.B., Trucker, D.E. and Becher, E.I. (1955) *Journal of the American Chemical Society*, **77**, 60–66; (b) Gualieri, F. and Riccieri, F.M. (1965) *Bollettino Chimico Farmaceutico*, **104**, 149–154; (c) Christy, M.E., Colton, C.D., MacKay, M., Staas, W.H., Wong, J.B., Engelhardt, E.L., Torchiana, M.L. and Stone, C.A. (1977) *Journal of Medicinal Chemistry*, **20**, 421–430; (d) Sohda, D.J., Mizuno, K., Imamiya, E., Tawada, H., Meguro, K., Kawamatsu, Y. and Yamamoto, Y. (1982) *Chemical & Pharmaceutical Bulletin*, **30**, 3601–3616; (e) Bridge, A.W., Fenton, F., Halley, F., Hursthouse, M.B., Lehmann, C.W. and Lythgoe, D.J. (1993) *Journal of the Chemical Society-Perkin Transactions 1*, 2761–2772.

8 For reaction using *tert*-butyl glyoxal, see: Hahn, B., Köphke, B. and Voß, J. (1981) *Liebigs Annalen der Chemie*, **1**, 10–19.

9 (a) Bigi, F., Bocelli, G., Maggi, R. and Satori, G. (1999) *The Journal of Organic Chemistry*, **64**, 5004–5009; (b) Bigi, F., Sartori, G., Maggi, R., Cantarelli, E. and Galaverna, G. (1993) *Tetrahedron: Asymmetry*, **4**, 2411–2414; (c) Bigi, F., Casnati, G., Sartori, G., Dalprato, C. and Bortolini, R. (1990) *Tetrahedron: Asymmetry*, **1**, 857–860.

10 Casiraghi, G., Bigi, F., Casnati, G., Sartori, G., Soncini, P., Gasparri Fava, G. and Ferrari Belicchi, M. (1988) *The Journal of Organic Chemistry*, **53**, 1779–1785.

11 Bigi, F., Casnati, G., Sartori, G., Araldi, G. and Bocelli, G. (1989) *Tetrahedron Letters*, **30**, 1121–1124.

12 (a) El Kaim, L., Guyoton, S. and Meyer, C. (1996) *Tetrahedron Letters*, **37**, 375–378; (b) Costa, P.R.R., Cabral, L.M., Alencar, K.G., Schmidt, L.L. and Vasconcellos, M.L.A.A. (1997) *Tetrahedron Letters*, **38**, 7021–7024.

13 For recent reviews, see: (a) (2001) *Organofluorine Compounds* (ed. T. Hiyama), Springer, Heidelberg; (b) Kirsch, P. (2004) *Modern Fluoroorganic Chemistry: Synthesis, Reactivity, and Applications*, Wiley-VCH, New York; (c) (1999) *Enantiocontrolled Synthesis of Fluoroorganic Compounds: Stereochemical Challenges and Biomedicinal Targets* (ed. V.A. Soloshonok), Wiley, New York.

14 For reviews on the versatility of BINOL ligands in asymmetric synthesis, see: (a) Chen, Y., Yekta, S. and Yudin, A.K. (2003) *Chemical Reviews*, **103**, 3155–3211; (b) Brunel, J.M. (2005) *Chemical Reviews*, **105**, 857–897. For representative reviews on titanium compounds in organic synthesis, see: (c) Mikami, K., Matsumoto, Y. and Shiono, T. (2003) *Science of Synthesis*, Vol. 2 (ed. T. Imamoto), Georg Thieme Verlag, Stuttgart, p. 457; (d) Ramon, D.J. and Yus, M. (2006) *Chemical Reviews*, **106**, 2126–2208.

15 Ishii, A., Soloshonok, V.A. and Mikami, K. (2000) *The Journal of Organic Chemistry*, **65**, 1597–1599.

16 Due to the intrinsic instability of many aminomethyl and hydroxymethyl aromatic systems, polysubstitution reactions leading to bisaryl compounds are commonly encountered. For examples, see: (a) Hao, J., Taktak, S., Aikawa, K., Yusa, Y., Hatano, M. and Mikami, K. (2001) *Synlett*, 1443–1445; (b) Yadav, J.S., Subba Reddy, B.V., Murthy, Ch. V.S.R., Mahesh Kumar, G. and Madan, Ch. (2001) *Synthesis*, 783–787; (c) Ramesh, C., Banerjee, J., Pal, R. and Das, B. (2003) *Advanced Synthesis and Catalysis*, **345**, 557–559.

17 Mikami, K. and Matsukawa, S. (1997) *Nature*, **385**, 613–615.

18 (a) Long, J., Hu, J., Shen, X. and Ding, K. (2002) *Journal of the American Chemical Society*, **124**, 10–11; (b) Yuan, Y., Zhang, X. and Ding, K. (2003) *Angewandte Chemie – International Edition*, **42**, 5478–5480.

19 Yuan, Y., Wang, X., Li, X. and Ding, K. (2004) *The Journal of Organic Chemistry*, **69**, 146–149.

20 Corey, E.J., Barnes-Seeman, D., Lee, T.W. and Goodman, S.N. (1997) *Tetrahedron Letters*, **38**, 6513–6516.

21 (a) Li, D.-P., Guo, Y.-C., Ding, Y. and Xiao, W.-J. (2006) *Chemical Communications*, 799–801; (b) Li, C.-F., Liu, H., Liao, J., Cao, Y.-J., Liu, X.-P. and Xiao, W.-J. (2007) *Organic Letters*, **9**, 1847–1851; (c) Chen, J.-R., Li, C.-F., An, X.-L., Zhang, J.-J., Zhu, X.-Y. and Xiao, W.-J. (2008) *Angewandte Chemie – International Edition*, **47**, 2489–2492.

22 Dong, H.-M., Lu, H.-H., Lu, L.-Q., Chen, C.-B. and Xiao, W.-J. (2007) *Advanced Synthesis and Catalysis*, **349**, 1597–1603.

23 Gathergood, N., Zhuang, W. and Jørgensen, K.A. (2000) *Journal of the American Chemical Society*, **122**, 12517–12522.

24 (a) Zhuang, W., Gathergood, N., Hazell, R.G. and Jørgensen, K.A. (2001) *The Journal of Organic Chemistry*, **66**, 1009–1013; (b) Jørgensen, K.A. (2003) *Synthesis*, 1117–1125.

25 Gothelf, A.S., Hansen, T. and Jørgensen, K.A. (2001) *Journal of the Chemical Society – Perkin Transactions 1*, 854–860.

26 For the synthesis of chiral ligand (+)-**35** and its application in asymmetric copper (II)-catalyzed cyclopropanation reactions of alkenes and diazoesters, see: Lyle, M.P. and Wilson, P.D. (2004) *Organic Letters*, **6**, 855–857.

27 Lyle, M.P., Draper, N.D. and Wilson, P.D. (2005) *Organic Letters*, **7**, 901–904.

28 Zhao, J.-L., Liu, L., Sui, Y., Liu, Y.-L., Wang, D. and Chen, Y.-J. (2006) *Organic Letters*, **8**, 6127–6130.

29 (a) Berkessel, A. and Gröger, H. (2005) *Asymmetric Organocatalysis; From Biomimetic Concepts to Applications in Asymmetric Synthesis*, Wiley-VCH, Weinheim, Germany; (b) (2007) *Enantioselective Organocatalysis* (ed. P.I. Dalko), Wiley-VCH, Weinheim, Germany.

30 For general reviews on asymmetric organocatalysis, see: (a) Dalko, P.I. and Moisan, L. (2004) *Angewandte Chemie – International Edition*, **43**, 5138–5175; (b) See also special issues on asymmetric organocatalysis: (c) (2004) *Accounts of Chemical Research*, **37** (8); (d) (2004) *Advanced Synthesis and Catalysis*, **346**, (9–10); (e) (2007) *Chemical Reviews*, **107**, (12).

31 For recent reviews on asymmetric catalysis by chiral hydrogen-bond donors, see: (a) Akiyama, T., Itoh, J. and Fuchibe, K. (2006) *Advanced Synthesis and Catalysis*, **348**, 999–1010; (b) Taylor, M.S. and Jacobsen, E.N. (2006) *Angewandte Chemie-International Edition*, **45**, 1520–1543; (c) Bolm, C., Rantanen, T., Schiffers, I. and Zani, L. (2005) *Angewandte Chemie – International Edition*, **44**, 1758–1763.

32 Zhuang, W., Poulsen, T.B. and Jørgensen, K.A. (2005) *Organic and Biomolecular Chemistry*, **3**, 3284–3289.

33 Török, B., Abid, M., London, G., Esquibel, J. and Török, M. (2005) *Angewandte Chemie – International Edition*, **44**, 3086–3089.

34 Li, H.-M., Wang, Y.-Q. and Deng, L. (2006) *Organic Letters*, **8**, 4063–4065.

35 Corma, A., García, H., Moussaif, A., Sabater, M.J., Zniber, R. and Redouane, A. (2002) *Chemical Communications*, 1058–1059.

36 For reviews: (a) Olah, G.A. (1963) *Friedel – Crafts and Related Reactions*, Wiley, New York; (b) Olah, G.A. (1973) *Friedel-Crafts Chemistry*, Wiley, New York; (c) Olah, G.A., Krishnamurti, R. and Prakash, G.K.S. (1991) *Comprehensive Organic Synthesis*, Vol. 3 (eds B.M. Trost and I. Fleming), Pergamon, Oxford, p. 293.

37 For reviews on asymmetric Friedel–Crafts reaction: (a) Bandini, M., Melloni, A. and Umani-Ronchi, A. (2004) *Angewandte Chemie*, **116**, 560–566; (2004) *Angewandte Chemie – International Edition*, **43**, 550–556; (b) Bandini, M., Melloni, A., Tommasi, S. and Umani-Ronchi, A. (2005) *Synlett*, 1199–1222; (c) Sheng, Y.-F., Zhang, A.J., Zheng, X.-J. and You, S.-L. (2008) *Chinese Journal of Organic Chemistry*, **28**, 605–614; (d) Poulsen, T. and Jørgensen, K.A. *Chemical Reviews*, **108**, 2903–2915.

38 Paras, N.A. and MacMillan, D.W.C. (2002) *Journal of the American Chemical Society*, **124**, 7894–7895, and references cited therein.

39 Selected examples: (a) Gong, Y. and Kato, K. (2001) *Tetrahedron: Asymmetry*, **12**, 2121–2127; (b) Chen, Y.-J., Lei, F., Sui, Y., Liu, L. and Wang, D. (2003) *Synlett*, 1160–1164; (c) Lei, F., Chen, Y.-J., Sui, Y., Liu, L. and Wang, D. (2003) *Tetrahedron*, **59**, 7609–7614; (d) Jiang, B. and Huang, Z.-G. (2005) *Synthesis*, 2198–2204; (e) Cheng, L., Liu, L., Sui, Y., Wang, D. and Chen, Y.-J. (2007) *Tetrahedron: Asymmetry*, **18**, 1833–1843; (f) Abid, M., Teixeira, L. and Török, B. (2008) *Organic Letters*, **10**, 933–935.

40 (a) Barrett, G.C. (1985) *Chemistry and Biochemistry of the Amino Acids*, Chapman and Hall, London; (b) Williams, R.M. and Hendrix, J.A. (1992) *Chemical Reviews*, **92**, 889–917.

41 Johannsen, M. (1999) *Chemical Communications*, 2233–2234.

42 Saaby, S., Fang, X., Gathergood, N. and Jørgensen, K.A. (2000) *Angewandte Chemie*, **112**, 4280–4282; (2000) *Angewandte Chemie – International Edition*, **39**, 4114–4116.

43 (a) Saaby, S., Bayón, P., Aburel, P.S. and Jørgensen, K.A. (2002) *The Journal of*

Organic Chemistry, **67**, 4352–4361; (b) Jørgensen, K.A. (2003) *Synthesis*, 1117–1125.

44. Shirakawa, S., Berger, R. and Leighton, J.L. (2005) *Journal of the American Chemical Society*, **127**, 2858–2859.

45. Jia, Y.-X., Xie, J.-H., Duan, H.-F., Wang, L.-X. and Zhou, Q.-L. (2006) *Organic Letters*, **8**, 1621–1624.

46. Reviews on chiral Brønsted acid catalysis: (a) Taylor, M.S. and Jacobsen, E.N. (2006) *Angewandte Chemie*, **118**, 1550–1573; (2006) *Angewandte Chemie – International Edition*, **45**, 1520–1543; (b) Akiyama, T., Itoh, J. and Fuchibe, K. (2006) *Advanced Synthesis and Catalysis*, **348**, 999–1010; (c) Connon, S.J. (2006) *Angewandte Chemie*, **118**, 4013–4016; (2006) *Angewandte Chemie – International Edition*, **45**, 3909–3912; (d) Akiyama, T. (2007) *Chemical Reviews*, **107**, 5744–5758.

47. Uraguchi, D., Sorimachi, K. and Terada, M. (2004) *Journal of the American Chemical Society*, **126**, 11804–11805.

48. Wang, Y.-Q., Song, J., Hong, R., Li, H. and Deng, L. (2006) *Journal of the American Chemical Society*, **128**, 8156–8157.

49. Yu, P., He, J. and Guo, C. (2008) *Chemical Communications*, 2355–2357.

50. Kang, Q., Zhao, Z.-A. and You, S.-L. (2007) *Journal of the American Chemical Society*, **129**, 1484–1485.

51. Terada, M., Yokoyama, S., Sorimachi, K. and Uraguchi, D. (2007) *Advanced Synthesis and Catalysis*, **349**, 1863–1867.

52. Rowland, G.B., Rowland, E.B., Liang, Y., Perman, J.A. and Antilla, J.C. (2007) *Organic Letters*, **9**, 2609–2611.

53. Rowland, G.B., Rowland, E.B., Liang, Y., Perman, J.A. and Antilla, J.C. (2007) *Organic Letters*, **9**, 4065–4068.

54. Terada, M. and Sorimachi, K. (2007) *Journal of the American Chemical Society*, **129**, 292–293.

55. Jia, Y.-X., Zhong, J., Zhu, S.-F., Zhang, C.-M. and Zhou, Q.-L. (2007) *Angewandte Chemie*, **119**, 5661–5663; (2007) *Angewandte Chemie – International Edition*, **46**, 5565–5567.

56. (a) Çavdar, H. and Saraçoğlu, N. (2005) *Tetrahedron*, **61**, 2401–2405; (b) Çavdar, H. and Saraçoğlu, N. (2006) *The Journal of Organic Chemistry*, **71**, 7793–7799.

57. (a) Evans, D.A. and Fandrick, K.R. (2006) *Organic Letters*, **8**, 2249–2252; (b) Evans, D.A., Fandrick, K.R., Song, H.-J., Scheidt, K.A. and Xu, R. (2007) *Journal of the American Chemical Society*, **129**, 10029–10041; (c) Blay, G., Fernández, I., Pedro, J.R. and Vila, C. (2007) *Tetrahedron Letters*, **48**, 6731–6734.

58. Kang, Q., Zheng, X.-J. and You, S.-L. (2008) *Chemistry – A European Journal*, **14**, 3539–3542.

59. Enders, D., Narine, A.A., Toulgoat, F. and Bisschops, T. (2008) *Angewandte Chemie*, **120**, 5744–5748; (2008) *Angewandte Chemie – International Edition*, **47**, 5661–5665.

60. Wanner, M.J., Hauwert, P., Schoemaker, H.E., Gelder, R., van Maarseveen, J.H. and Hiemstra, H. (2008) *European Journal of Organic Chemistry*, **180**, 180–185.

61. Zhang, G.-W., Wang, L., Nie, J. and Ma, J.-A. (2008) *Advanced Synthesis and Catalysis*, **350**, 1457–1463.

62. For reviews: (a) Brown, R.T. (1983) *Indoles* (ed. J.E. Saxton), Wiley-Interscience, New York, Part 4; (b) Bentley, K.W. (2004) *Natural Product Reports*, **21**, 395–424, and references cited therein.

63. (a) Pictet, A. and Spengler, T. (1911) *Berichte der Deutschen Chemischen Gesellschaft*, **44**, 2030–2036; For a comprehensive review: (b) Cox, E.D. and Cook, J.M. (1995) *Chemical Reviews*, **95**, 1797–1842.

64. Chrazanowska, M. and Rozwadowska, M.D. (2004) *Chemical Reviews*, **104**, 3341–3370.

65. Selected examples: (a) Cox, E.D., Hamaker, L.K., Li, J., Yu, P., Czerwinski, K.M., Deng, L., Bennett, D.W. and Cook, J.M. (1997) *The Journal of Organic Chemistry*, **62**, 44–61; (b) Czarnocki, Z., Suh, D., MacLean, D.B., Hultin, P.G. and Szarek, W.A. (1992) *Canadian Journal of Chemistry*, **70**,

1555–1561; (c) Czarnocki, Z., Mieckzkowsi, J.B., Kiegiel, J. and Araźny, Z. (1995) *Tetrahedron: Asymmetry*, **6**, 2899–2902; (d) Schmidt, G., Waldmann, H., Henke, H. and Burkard, M. (1996) *Chemistry – A European Journal*, **2**, 1566–1571; (e) Gremmen, C., Willemse, B., Wanner, M.J. and Koomen, G.-J. (2000) *Organic Letters*, **2**, 1955–1958; (f) Gremmen, C., Wanner, M.J. and Koomen, G.-J. (2001) *Tetrahedron Letters*, **42**, 8885–8888; (g) Tsuji, R., Nakagawa, M. and Nishida, A. (2003) *Tetrahedron: Asymmetry*, **14**, 177–180; (h) Suzuki, K. and Takayama, H. (2006) *Organic Letters*, **8**, 4605–4608; (i) Shi, X.-X., Liu, S.-L., Xu, W. and Xu, Y.-L. (2008) *Tetrahedron: Asymmetry*, **19**, 435–442.

66 Yamada, H., Kawate, T., Matsumizu, M., Nishida, A., Yamaguchi, K. and Nakagawa, M. (1998) *The Journal of Organic Chemistry*, **63**, 6348–6354.

67 Taylor, M.S. and Jacobsen, E.N. (2004) *Journal of the American Chemical Society*, **126**, 10558–10559.

68 Mergott, D.J., Zuend, S.J. and Jacobsen, E.N. (2008) *Organic Letters*, **10**, 745–748.

69 Raheem, I.T., Thiara, P.S., Peterson, E.A. and Jacobsen, E.N. (2007) *Journal of the American Chemical Society*, **129**, 13404–13405.

70 Raheem, I.T., Thiara, P.S. and Jacobsen, E.N. (2008) *Organic Letters*, **10**, 1577–1580.

71 Seayad, J., Seayad, A.M. and List, B. (2006) *Journal of the American Chemical Society*, **128**, 1086–1087.

72 Wanner, M.J., vander Haas, R.N.S., deCuba, K.R., van Maarseveen, J.H. and Hiemstra, H. (2007) *Angewandte Chemie*, **119**, 7629–7631; (2007) *Angewandte Chemie – International Edition*, **46**, 7485–7487.

73 Sewgobind, N.V., Wanner, M.J., Ingemann, S., de Gelder, R., van Maarseveen, J.H. and Hiemstra, H. (2008) *The Journal of Organic Chemistry*, **73**, 6405–6408.

4
Nucleophilic Allylic Alkylation and Hydroarylation of Allenes

Marco Bandini and Achille Umani-Ronchi

Summary

The synthesis of benzylic stereocenters containing vinyl units is a stimulating and challenging target for organic chemists due to the wide distribution of this key structural motif in biologically active natural products and pharmaceutical compounds. Aromatic electrophilic allylic substitution and intramolecular hydroarylation of allenes are the most promising synthetic routes to the framework. In this chapter, a collection of inter- and intramolecular approaches, working under a catalytic regime, is presented, with particular emphasis on their scope and limitations. Recent developments in the field of asymmetric Friedel–Crafts-type allylic alkylation (*ee* constantly higher than 90%) are also presented. The known chiral catalysts employed, for enantioselective FC-allylation of arenes, are Ir, Pd, Pt and Au-organometallic species. Interestingly, no examples of organocatalyzed allylation of arenes have been described to date.

4.1
Introduction

The design and development of new catalytic and stereoselective protocols for the alkylation of aromatic and heteroaromatic compounds (Friedel–Crafts (FC) reactions) represent an important challenge for organic chemists [1]. These strategies, associated with excellent regio- and chemoselectivity, allow easy access to several classes of polyfunctionalized enantiomerically enriched compounds which are valuable building blocks in organic synthesis [2].

Substitution of C–H bonds is a convenient, efficient and economical synthetic strategy because C–H bonds are the most ubiquitous and inexpensive chemical linkage in nature [3]. Moreover, particular attention has been paid to the synthetic application of the C–H functionalization of arenes via inter- and intramolecular processes, since new methods for the formation of useful polycyclic products are anticipated.

Numerous aromatic compounds, including benzenes carrying electron-donating substituents, furans, pyrroles and indoles, have been successfully applied in a number of FC reactions with diverse electrophiles, and their enantioselective variants have also been deeply investigated in the presence of chiral metal complexes or organic molecules. Nevertheless, most reports in this area still focus on the relatively more reactive indole or pyrrole compounds; a few exceptions are the asymmetric FC reactions of benzene derivatives bearing powerful electron-donating groups.

The introduction of vinyl units in benzylic stereocenters is a goal of great importance due to the wide distribution of this key structural motif in biologically active natural products and pharmaceutical compounds. Metal-catalyzed aromatic electrophilic allylic substitution [4] and intramolecular hydroarylation of allenes are the most promising synthetic routes to the construction of vinyl-benzylic stereocenters (Figure 4.1).

Nucleophilic allylic alkylation processes have been deeply investigated in the presence of chiral and achiral Pd-, Ir-, Mo-, W- and Rh-based catalysts; they are attracting considerable interest and significant progress has been made. Important intermediates in these reactions are π- and σ-allyl complexes, generally prepared from halides, acetates, or carbonates. A modest number of papers report the use of allyl alcohols as alkylating agents, which would be economically and environmentally more desirable.

Functionalization of allenes is a highly atom economical approach complementary to allylic alkylations of arenes for the construction of vinyl-containing stereocenters. Late transition metals, featuring marked π-acidic character, are generally optimal catalysts for the process, displaying high functional group tolerance. Cationic Au(I)/(III), Pt(II), Ag(I) and Ru(III) salts/complexes are among the most used systems for the manipulation of unactivated carbon–carbon double bonds with nucleophilic species. Interestingly, over the last two years (2006–2008) great emphasis has been given to the topic of catalytic cycloarylation reactions of allenes, as a valuable synthetic methodology for the construction of complex polycyclic aromatic compounds. Preliminary efforts toward the development of asymmetric variants of these processes are also worthy of note.

This chapter is organized according to these types of reactions and the topics covered here include intermolecular and intramolecular approaches. Other catalyzed

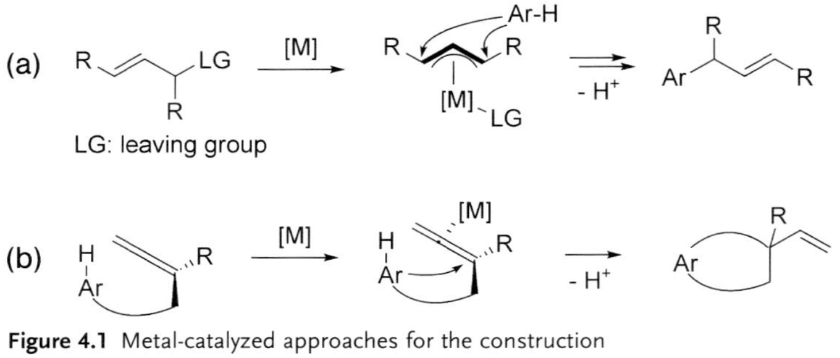

Figure 4.1 Metal-catalyzed approaches for the construction of vinyl-benzyl stereocenters: (a) allylic substitution, (b) hydroarylation of allenes.

processes such as Claisen rearrangement and Heck-type direct metal-catalyzed oxidative cycloarylations are, however, excluded, since these have been already extensively reviewed and in deference to space limitations [5].

4.2 Allylic Alkylations

4.2.1 Intermolecular Approach

The widespread natural occurrence and potential for further functionalization of allyl aromatic compounds justifies the continued interest in synthetic methods for the introduction of allyl groups onto the aromatic nucleus. Moreover, there is an urgent requirement to develop novel enantioselective FC reactions, especially for benzene-like arenes. In the following sections, a summary of the most representative synthetic methodologies addressing the catalytic allylic alkylation of arene is presented.

4.2.1.1 Benzene-Like Compounds

The main topic of the book concerns the catalytic alkylations of *unactivated* arenes through selective C—H bond functionalization. However, the recent finding by Lee and colleagues on the bimetallic-catalyzed cross-coupling of iodo-arenes, with allyl acetates, can be selected as an introductory study to the catalytic allylation of simple arenes due to its impact on the synthetic organic chemistry community. In particular, it was discovered that allylindium species, generated *in situ* by reductive transmetallation of allylpalladium(II) complexes with an indium/indium(III) chloride mixture, are effective nucleophilic cross-coupling partners in Pd-catalyzed allylic alkylations of arenes (Scheme 4.1a) [6]. A variety of electrophilic cross-coupling

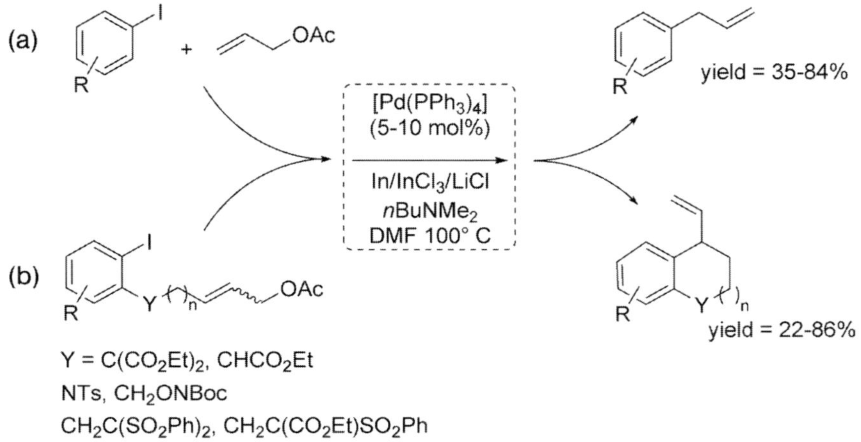
(a)
OAc
R
[Pd(PPh3)4]
(5-10 mol%)
In/InCl3/LiCl
nBuNMe2
DMF 100° C
yield = 35-84%
(b)
Y
n
yield = 22-86%
Y = C(CO2Et)2, CHCO2Et
NTs, CH2ONBoc
CH2C(SO2Ph)2, CH2C(CO2Et)SO2Ph

Scheme 4.1 Intra- and intermolecular Pd-In-mediated cross-coupling reaction of iodobenzenes with allyl acetates.

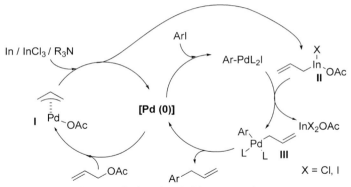

Scheme 4.2 Mechanism of Pd-catalyzed allyl cross-coupling reaction.

reagents, such as aryl iodides, vinyl bromide and triflates, are reacted smoothly under optimal reaction conditions. The encouraging results prompted the authors to validate the scope of the process also with intramolecular transformations by using tailored arene-allyl acetates (Scheme 4.1b) [7]. In all the cases, moderate to good yields (22–86%) of the cyclized products were obtained.

The reaction mechanism for the Pd–In mediated allyl cross-coupling reaction of aryl iodides with allyl acetate is shown in Scheme 4.2. Here, the initially formed [allylPd(II)] intermediate **I** undergoes transmetallation with In/InCl$_3$ to give the *in situ* [allylIn(III)XOAc] reagents **II**. The subsequent cross-coupling event with the haloarene ArX was mediated by the presence of [Pd(0)] species in solution.

The functionalization of aromatic compounds through carbon fragment preactivation has revolutionized the way to synthesize molecules over the past three decades. However, incorporation of halides or other leaving groups on the aromatic systems can be laborious, with the concomitant production of conspicuous amounts of hazardous wastes. For these reasons much effort has been addressed toward the development of directed alkylation (allylation) of unactivated arenes under a catalytic regime.

Transition-metal-catalyzed alkylation of electron-rich aromatics and heteroaromatics, with allyl acetates, was initially introduced by Billups (Pd-catalysis) [8a] and subsequently Kocòvsky [8b]. In the latter case, [Mo(II)] salts were synthesized and tested in the FC allylic alkylations and, among them, [Mo(CO)$_4$Br$_2$]$_2$ **1** (5 mol%) worked smoothly in the intermolecular functionalization of phenols, anisoles, indoles and furans Scheme 4.3.

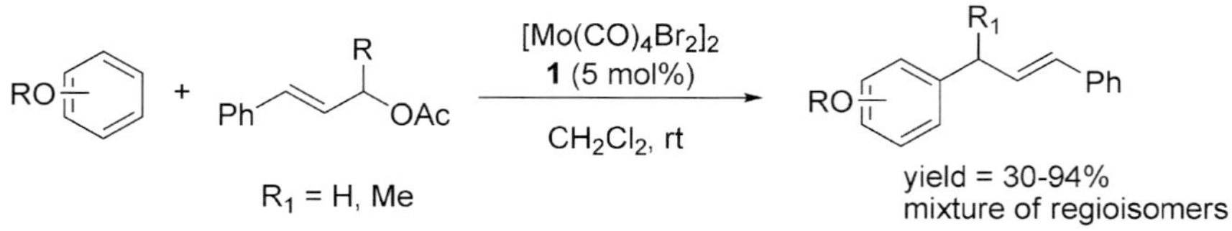

Scheme 4.3 Mo-catalyzed allylation of electron-rich arenes, only the cases for acyclic acetates are shown.

Several [C$_5$Me$_5$Ru] complexes (C$_5$Me$_5$ = Cp*) have been used effectively as catalysts in the allylation chemistry of aromatic compounds. Among them, Trost's complex [Ru(Cp*)-(CH$_3$CN)$_3$]PF$_6$ [9] and analogs deserve mention for their catalytic performances and selectivity.

Based on mechanistic considerations on allylic alkylation, amination and phenolation reactions, a new highly active dicationic [Ru(IV)] complex [Ru(Cp*)(CH$_3$CN)$_2$(η^3-PhCH=CHCH$_2$)](PF$_6$)$_2$ **2** (3 mol%), was successfully prepared and employed by Pregosin and coworkers in FC-type allylations of various electron-rich arenes with allyl carbonates (Scheme 4.3) [10]. Specifically, phenols, 2-naphthols, anisoles and *o*-xylol can be selectively allylated without the use of Lewis acid main group compounds or other additives and the alkylation reaction proceeds under relatively mild conditions. However, a mixture of *ortho*, *meta*, and *para* isomers was generally isolated for several arene compounds. In many cases the reactions are regioselective with the arene attack occurring at the less-substituted allyl carbon atom exclusively.

Very recently, alcohols started to emerge as a new class of electrophilic agents for the catalytic allylic alkylation of arenes. Environmental and economic aspects support the use of alcohols in FC reactions (H$_2$O as the only by-product), however, large amounts of LAs are generally required due to deactivation phenomena of the catalysts. This approach was pioneered by Fukuzawa and coworkers who disclosed the efficiency and recoverability of Sc(OTf)$_3$ (10 mol%) in promoting the allylation and benzylation of benzene by using the arene as solvent (at 115–120 °C) [11].

Further developments in the field are also ascribable to Pregosin who described the use of allyl alcohols in the alkylation of phenols and naphthols with **2a**, featuring comparable performances to those of carbonates [12] (Scheme 4.4).

A tentative mechanistic cycle for the ruthenium(IV)-catalyzed C–C coupling alkylation reaction using allyl alcohols as the alkylating agent is shown in Scheme 4.5.

An alternative synthetic route for carbon–carbon bond forming reactions between electron-rich arenes and primary and secondary allyl alcohols has been reported recently by Chan and coworkers using gold(III) catalysis under mild conditions [13]. The FC allylic alkylation of a wide variety of aromatic compounds was performed with asymmetrically substituted allylic alcohols in the presence of AuCl$_3$ (5 mol%)

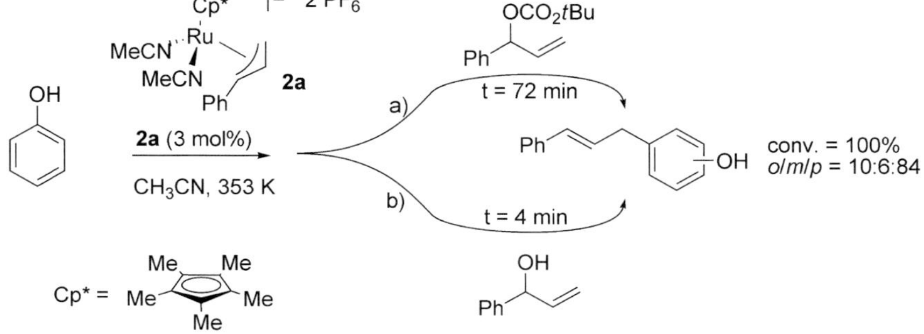

Scheme 4.4 Ru-catalyzed phenol allylic alkylation with branched allyl carbonates.

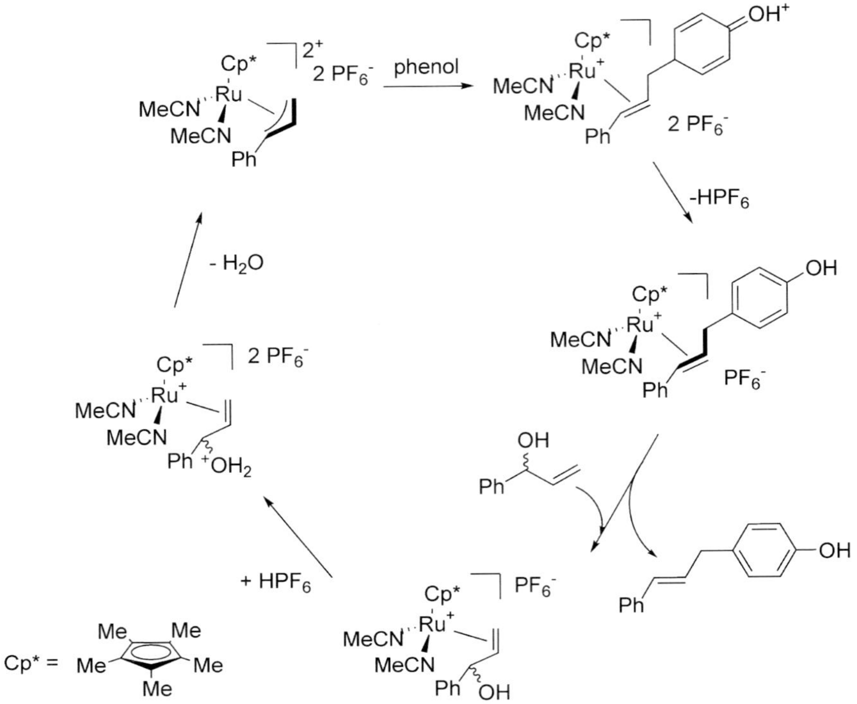

Scheme 4.5 Proposed mechanism for Ru-catalyzed C—C bond formation.

as the catalyst (CH_2Cl_2, rt). The chemical yields were good to high (50–99%) and the corresponding alkylated species isolated also in a highly regioselective manner. In fact the carbon–carbon bond formation occurs at the less sterically hindered carbon center of the allylic moiety Scheme 4.6. A few examples describing the regioselective alkylation of indole, pyrrole and furan were also included in this report.

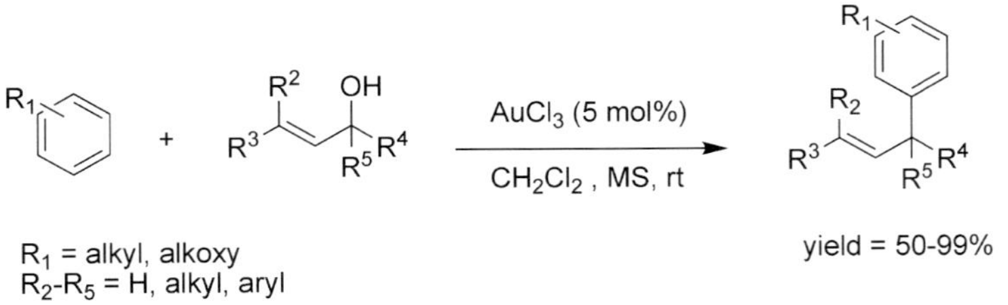

R_1 = alkyl, alkoxy
R_2-R_5 = H, alkyl, aryl

yield = 50-99%

Scheme 4.6 Gold(III)chloride-catalyzed allylic alkylation of aromatic and heteroaromatic compounds with allylic alcohols.

4.2.1.2 Indoles

The C(3)-allylated indolyl core is a very important structural motif for the preparation of many naturally occurring indolyl-based alkaloids [14], such as auxins [15a], and of unnatural potent HIV inhibitors, BMS-378806 [15b].

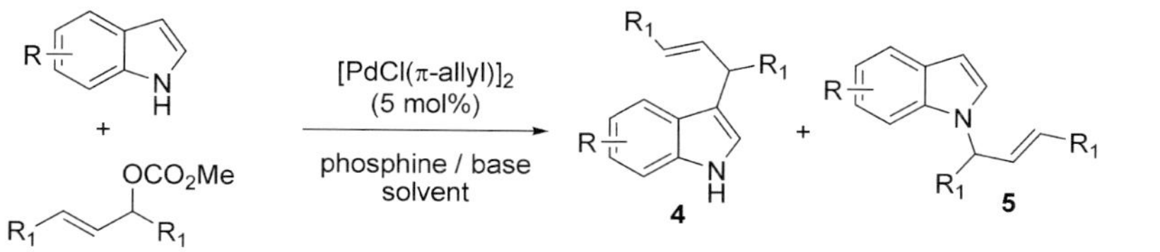

3: R_1 = Me, Ph, -$(CH_2)_3$-

C(3)-alkylation: CH_2Cl_2, Li_2CO_3
N(1)-alkylation, THF/DMF, K_2CO_3/Cs_2CO_3

Scheme 4.7 Pd-catalyzed allylic alkylation of indoles.

After Kocòvsky's seminal paper [8], Bandini and Umani-Ronchi described inter- and intramolecular catalytic regioselective allylic alkylation of indoles with cyclic and acyclic carbonates in the presence of [(PdCl(π–allyl)]$_2$/phosphine complexes. Interestingly, the combined use of low coordinating solvents (DCM) and Li_2CO_3, as the base, drives the reaction course toward the exclusive formation of the thermodynamic C(3)-attack, while the strongest bases (i.e., Cs_2CO_3, K_2CO_3) and more coordinating solvents (DMF, THF) promoted the *N*(1)-allylation Scheme 4.7 [16].

The strategy has also proven to be successful in the regiochemical alkylation of indoles with asymmetric 4-phenylbut-3-enyl methyl carbonate **3a**. In fact, only attack at the less-hindered position of the asymmetric carbonate was detected Scheme 4.8.

Subsequently, Chen and coworkers developed a mild and efficient I_2-catalyzed (5 mol%) allylic alkylation of indole derivatives with allylic and propargylic acetates in high regioselectivity and excellent yields [17].

A breakthrough in the field of metal-catalyzed allylation of indoles with allyl alcohols was the discovery by Tamaru *et al.* [18]. The use of a substoichiometric amount of Et_3B (30 mol%) combined with $Pd(PPh_3)_4$ (5 mol%) worked nicely in the C (3) selective allylation starting from α-,γ-methyl and α-,γ-phenyl-substituted allyl alcohols. Remarkably, no N(1)-allylation was detected. Unsymmetrical allyl alcohols showed almost the same regioselectivities, indicating [π-allylpalladium] species as common reaction intermediates Scheme 4.9 [19].

As previously mentioned, cationic [Ru(IV)] complexes have also found application in the catalytic allylic alkylation of heteroaromatics. Here, the former [Ru(IV)] catalyst **2a** was slightly modified to the more easily prepared [Ru(η^3-C_3H_5(Cp*)

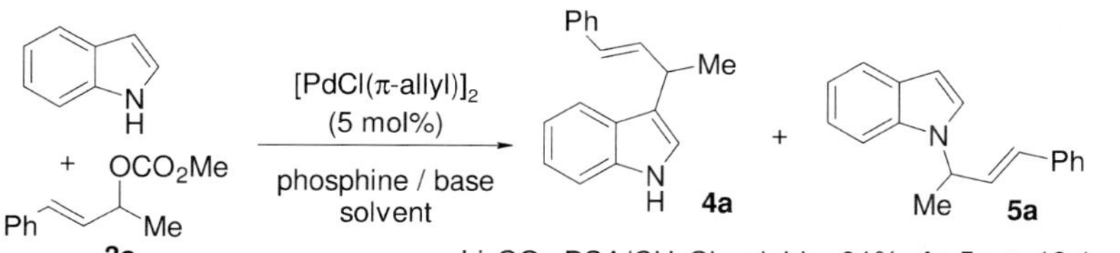

3a

Li_2CO_3-BSA/CH_2Cl_2: yield = 91%, **4a**:**5a** -> 10:1
Cs_2CO_3/DMF: yield = 84%, **4a**:**5a** -> 1:50

Scheme 4.8 Indole alkylation with asymmetric 4-phenyl-3-enyl methyl carbonate **3a**.

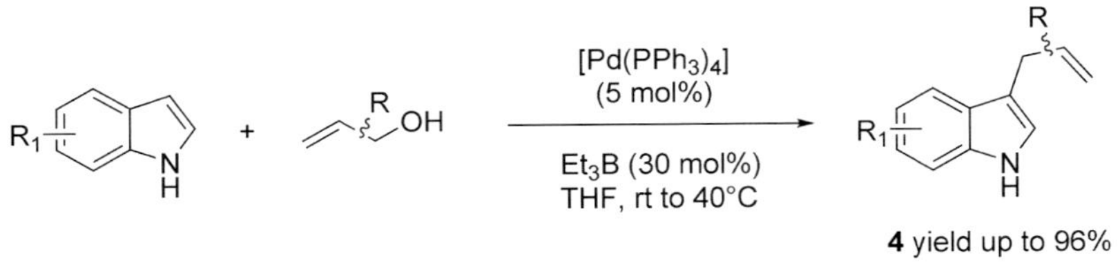

Scheme 4.9 Pd-catalyzed allylation of indoles with alcohols.

(CH$_3$CN)$_2$](PF$_6$)$_2$**2b** (5 mol%), that promoted the C—C bond making reaction of indoles with allyl alcohol at room temperature [20]. The major product was the C(3)-substituted allyl compound (**4**) with the double allylation product (**6**) as a minor component Scheme 4.10.

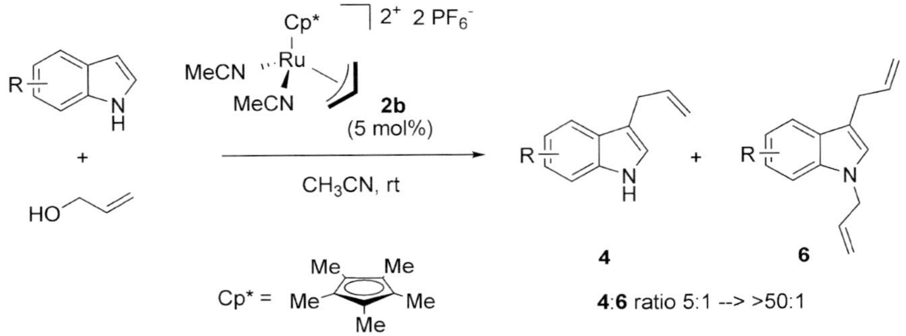

Scheme 4.10 Ru-catalyzed allylation of indoles with allyl alcohol.

Moreover, the same team has very recently disclosed the effectiveness of the Trost-sulfonate catalyst [Ru(Cp*)-(CH$_2$CH=CHC$_6$H$_5$)(CH$_3$CN)(CH$_3$C$_6$H$_4$SO$_3$)]$^+$ in the highly regioselective alkylation of pyrroles [21].

Breit and coworkers discussed a tailored self-assembed palladium phosphane catalyst, made from [(η^3-allyl)Pd(cod)]BF$_3$ and DPPon(CF$_3$)$_2$ ligand **7**, for the direct C(3)-allylation of indoles with allylic alcohols [22]. In accord with experimental observations, the authors suggest that the residual water contained in the catalytic system, can be incorporated into the ligand through a hydrogen bonding network Scheme 4.11.

The use of highly reactive secondary allylic alcohols in the regioselective allylation of indoles in the presence of InCl$_3$ [23], FeCl$_3$ [24] and calix[n]arene sulfonic acids has also been discussed [25].

4.2.1.3 Asymmetric Allylations

The synthesis of enantiopure indole derivatives is of significant importance because of the wide distribution of indole moieties in biologically active natural products and pharmaceutical compounds. In 2002 MacMillan stated "...*the indole framework has become widely identified as a "privileged" structure or pharmacaphore, with representation in over 3000 natural isolates and 40 medicinal agents of diverse therapeutic*

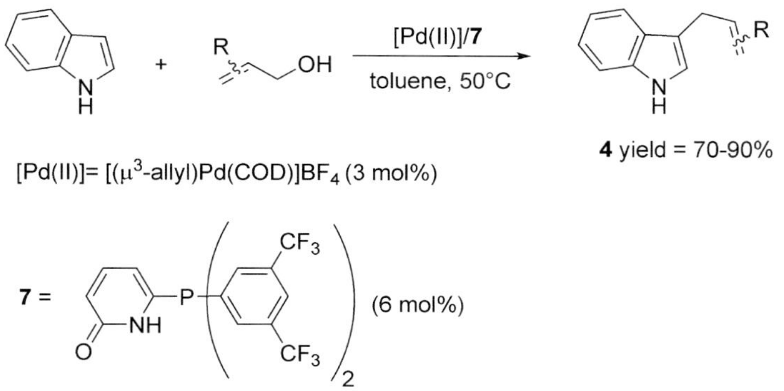

4.2 Allylic Alkylations | 153

Scheme 4.11 Pd-catalyzed indole alkylation.

action..." [26]. Since then, enantioselective Friedel–Crafts reactions targeting indoles have attracted considerable interest and there has been significant progress in the area [2, 27].

Focusing on allylic alkylation processes, very recently, Chan et al. reported an effective Pd-catalyzed asymmetric allylic alkylation of indoles with (±)-1,3-diphenyl-2-propenyl acetate **3b** under mild reaction conditions Scheme 4.12 [28].

Such a protocol utilized a new class of air-stable chiral P/S ligands based on ferrocene and heterocyclic scaffolds that are readily available from Ugi's amine. Generally, substituted indoles **4** were isolated in good yields and enantiomeric excess up to 96%. A collection of results is summarized in Table 4.1.

Subsequent studies in this area were performed by You et al. who reported on the use of a well-developed Ir-catalytic system in the alkylation of indoles with 1,3-unsymmetrical allylic substrate [29]. High enantioselectivities up to 92% were achieved in the presence of 2 mol% of [Ir(COD)Cl]$_2$, 4 mol% of **9** and 1 equiv of Cs$_2$CO$_3$. To test the generality of the reaction, different indoles were reacted with

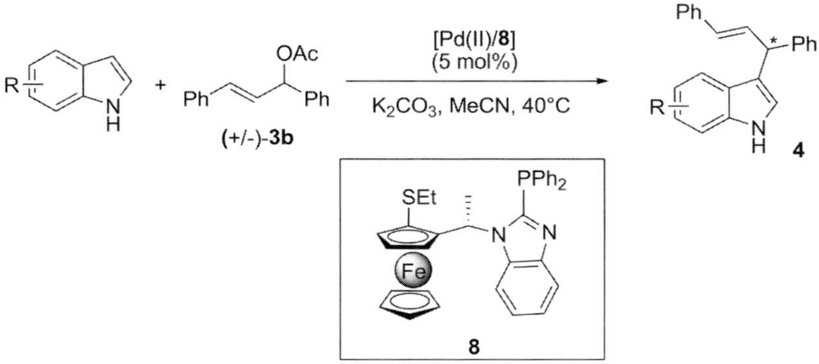

Scheme 4.12 Catalytic enantioselective allylic alkylation of indoles through the use of Pd-**8** chiral catalyst.

Table 4.1 Scope of indoles in the Pd-catalyzed AA Friedel–Crafts-type alkylation.[a]

Indole	4	Yield (%)	ee (%)
indole	4ba	74	95
2-Ph indole	4bb	77	92
2-Me indole	4bc	78	95
7-Me indole	4bd	77	96
4-Me indole	4be	75	96
7-Br indole	4bf	66	94
7-Cl indole	4bg	61	96
5-MeO indole	4bh	85	94
5-BnO indole	4bi	56	94

[a]All the reactions were carried out in CH_3CN at 40 °C, with K_2CO_3 for 18 h.

variously substituted allyl carbonates, leading to branched products with up to >97/3 branched-linear ratio (Scheme 4.13).

Conceptually analogous to the Tamaru's report [18], is the enantioselective process developed by Trost and coworker that, taking advantage of the intrinsic nucleophilicity of indoles at the C(3) position, synthesized quaternary stereocenters in a highly enantiomerically enriched form directly from 3-substituted indoles [30a]. High enantioselectivities were achieved for the synthesis of 3,3-disubstituted indolenines and indolines (**11**) starting from allyl alcohol as the electrophile, 3-substituted indoles, 9-BBN-C_6H_{13} as the promoter of the reaction, $Pd_2(dba)_3CHCl_3$ and ligand (*S,S*)-**10** (Table 4.2). The high C(3):N(1) selectivity depends on the nature of the borane,

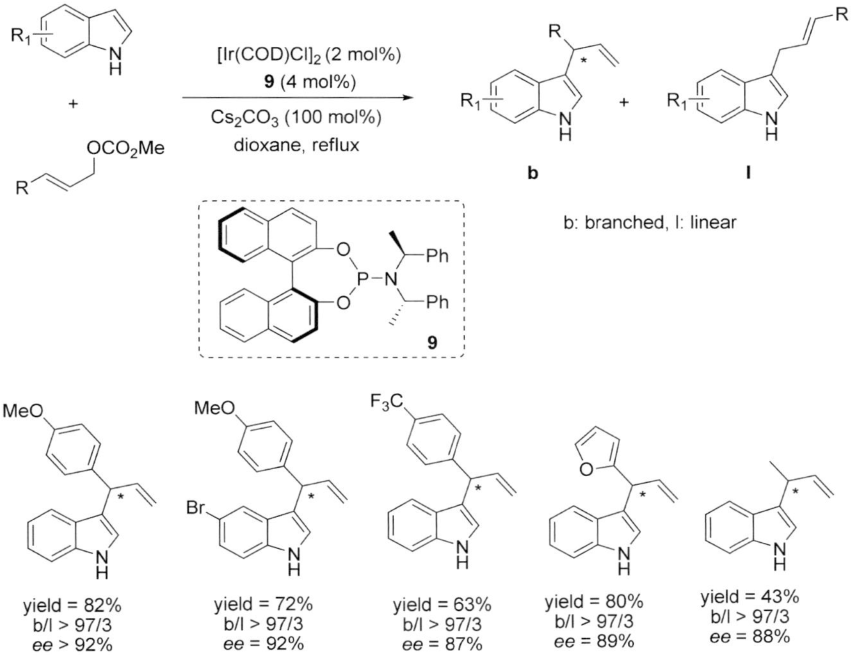

4.2 Allylic Alkylations

Scheme 4.13 Ir-catalyzed enantioselective functionalization of indoles with 1,3-unsymmetrical allylic substrates.

suggesting that, in addition to promoting the ionization of allyl alcohol, the boron is directly involved in the enantiodiscriminating step.

The synthetic versatility of the protocol was finally demonstrated with the synthesis of indolines with a *cis*-5,5-or 5,6-fused ring.

4.2.1.4 Experimental: Selected Procedures

Representative procedure for the Ir-catalyzed enantioselective allylation of indoles (Scheme 4.13) [29]

In a solution of [Ir(COD)Cl]$_2$ (0.004 mmol) in anhydrous THF (0.5 mL) the phosphoramidite ligand **9** (0.008 mmol) was dissolved, followed by propylamine (0.3 mL). The reaction mixture was heated at 50 °C (0.5 h) and then the volatiles removed under vacuum. To the resulting yellow, allylic carbonate (0.20 mmol), the desired indole (0.40 mmol), Cs$_2$CO$_3$ (0.20 mmol) and 1,4-dioxane (1.0 mL) were added in sequence. The mixture was refluxed until complete consumption (checked by TLC). After the usual work-up (filtration on celite, evaporation), the ratio of regioisomers was determined by ^{1}H-NMR and the crude residue was purified by flash column chromatography. Analytical data for the reaction between methyl carbonate (R = 4-methoxyphenyl) and indole (R$_1$ = H): Reaction time 4 h, yield = 82% as a yellow oil, regioisomers ratio (b/l) > 99/1, petroleum ether/ethyl acetate = 30/1. *ee* = 92%.

Table 4.2 Scope of the Pd-catalyzed allylation of C(3)-substituted indoles.

R	Product (11)	Yield (%)	ee (%)
5-NO$_2$	11a	nr[a]	—
5-Br	11b	89	60
H	11c	92	74
5-MeO	11d	95	84
5-BnO	11e	87	83
5-OH	11f	88	86
5-Bn$_2$N	11g	94	90
4-MeO	11h	92	83
6-MeO	11i	89	78
7-MeO	11j	nr[a]	—

[a]No reaction.

CHIRALCEL OD-H, nHex/IPA = 90/10; flow rate = 1.0 mL min^{-1}; tR = 13.5 (minor), 13.9 (major) min. [α]$_D$ 20 = −2.8° (c 1.0, CHCl$_3$). ^{1}H NMR (300 MHz, CDCl$_3$) b 7.97 (br s, 1H), 7.43 (d, J = 7.8 Hz, 1H), 7.35 (d, J = 8.1 Hz, 1H), 7.26–7.16 (m, 3H), 7.05 (t, J = 7.5 Hz, 1H), 6.88–6.85 (m, 3H), 6.36 (ddd, J = 6.9, 9.9, 17.1 Hz, 1H), 5.20 (d, J = 10.2 Hz, 1H), 5.08 (d, J = 17.1 Hz, 1H), 4.94 (d, J = 7.2 Hz, 1H), 3.81(s, 3H). ^{13}C NMR (75 MHz, CDCl$_3$) b 140.7, 136.6, 135.3, 129.3, 126.7, 122.4, 121.9, 119.8, 119.2, 118.7, 115.1, 113.6, 111.0, 55.2, 46.1. IR (cm^{-1}) 3420, 3058, 2957, 2836, 1664, 1637, 1608, 1510, 1457, 1247, 1178, 1034, 823, 743.

4.2.2
Intramolecular Approaches

4.2.2.1 Introduction and Fundamental Examples
The search for efficient, mild and economical catalytic *intramolecular* allylic alkylations of aromatic and heteroaromatic compounds is a goal for the whole chemical community.

4.2 Allylic Alkylations

A large number of catalytic systems for allylation reactions have been developed, however, intramolecular variants have still not achieved their full potential. An efficient catalytic intramolecular allylic alkylation of arenes has been reported by Bandini and Umani-Ronchi, with Kocòvsky's catalyst [Mo(CO)$_4$Br$_2$]$_2$ in combination with AgOTf [31].

The potential of the methodology relied also on the easy accessibility (gram-scale) of the starting allyl-carbonates (**12**) via alkylation of substituted diethyl benzylmalonate and (Z)-bromocarbonate. The air-sensitive [Mo(II)] salt catalyzed the cyclization of a wide range of functionalized and unfunctionalized arenes ([Mo(CO)$_4$Br$_2$]$_2$ 2.5 mol%, ClCH$_2$CH$_2$Cl, 80 °C, 16 h) with high yields (70–97%) even in the absence of moisture restrictions (Scheme 4.14a). A representative collection of 1-vinyl-tetrahydronaphthalenes (**13**) is shown in Scheme 4.14b.

Previously, Cook et al. described the use of indium(III) salts as highly effective catalysts for the intramolecular and intermolecular Friedel–Crafts reaction of simple allylic bromides and arenes [32]. In particular, the cyclization of allyl bromides **14** was promoted to a great extent by InCl$_3$ in dichloromethane at room temperature in the presence of molecular sieves. Rigorous anhydrous conditions were necessary in order to guarantee high conversions and to prevent the decomposition

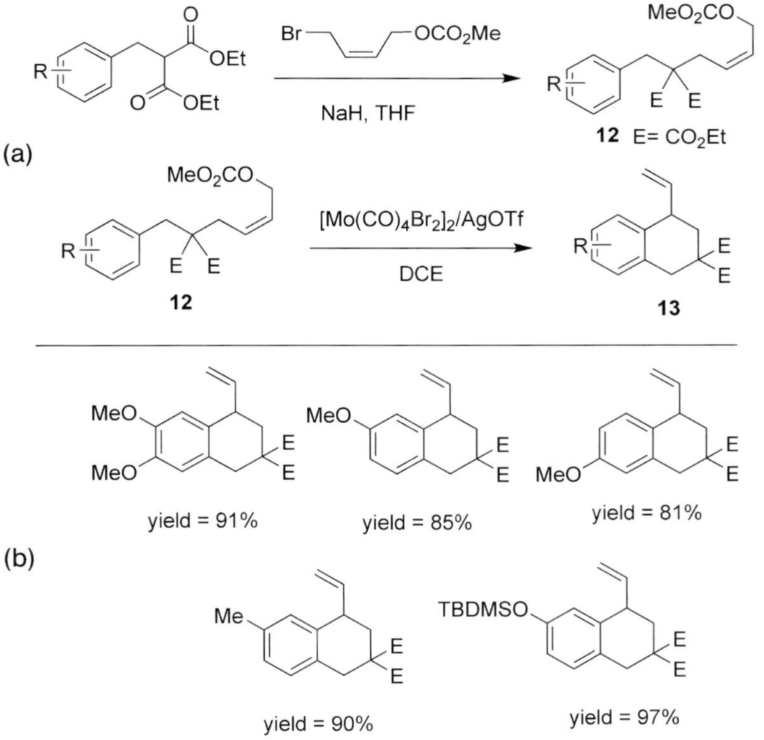

Scheme 4.14 Synthesis of 1-vinyl-tetrahydronaphthalenes **13** via intramolecular Mo-catalyzed allylic alkylation of arenes.

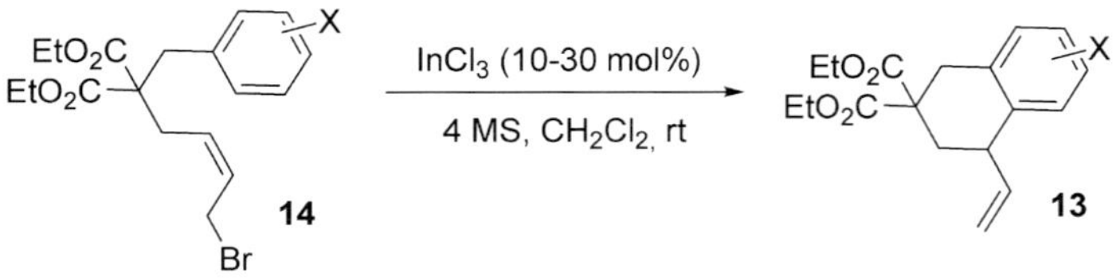

Scheme 4.15 Intramolecular allylic alkylation of electron-deficient arenes via In(III)-catalysis.

of water-sensitive starting materials. A wide variety of aromatic rings cyclized to benzocarbocycles under the optimal reaction parameters. This protocol emerged as one of the few examples in which arenes, carrying electron-withdrawing groups (i.e., 3-F(Cl), 4-F(Cl)...), underwent FC alkylation smoothly under catalytic regimes (Scheme 4.15).

An alternative route to functionalized tetrahydronaphthalenes based on the Hg(OTf)$_2$ catalyzed (0.5 mol%) cyclizations of (*E*)-6-(alkoxyphenyl)-2-hexen-1-ol (**15**), (Scheme 4.16) [33]. The reaction takes place via protonation of the hydroxy group of the organomercuric intermediate **16** by TfOH to generate an oxonium cation **17**. The demercuration of **17** restores the catalyst Hg(OTf)$_2$ affording the desired product **13**. Note that the configuration of the double bond of the allylic alcohol does not affect the reaction outcome. In fact, (*Z*)-6-(3,5-dimethoxyphenyl)-2-hexen-1-ol affords the cyclization product in comparable 95% yield. Only electron-rich and electron-neutral compounds proved to be suitable substrates under optimal reaction conditions.

In this context, Bandini and Umani-Ronchi have recently expanded the synthetic utility of such an approach by demonstrating a mild and environmentally friendly silver(I) catalysis in the intramolecular alkylation of arenes with tethered (*Z*)-allyl alcohols. A range of acyclic aromatic compounds was tested in the presence of AgOTf (10 mol%) in DCE under reflux. Silver (AgOTf, 10 mol%) and gold ([PP$_3$AuCl]/AgOTf, 10 mol%) catalysts were assessed through comparative experiments, demonstrating the generally higher efficiency of the ligand-free silver catalyst (Scheme 4.17) [34].

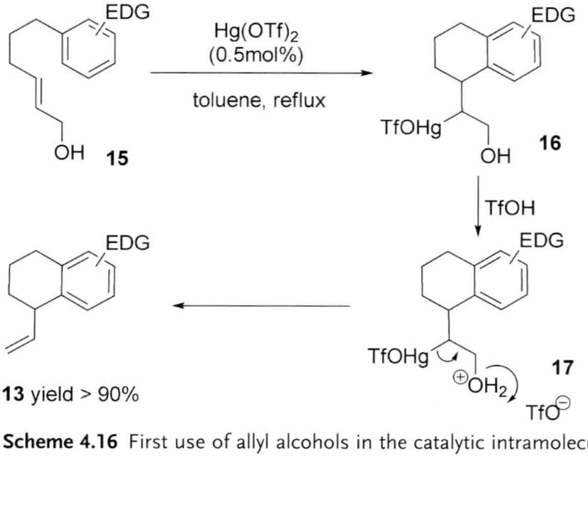

Scheme 4.16 First use of allyl alcohols in the catalytic intramolecular alkylation of arenes.

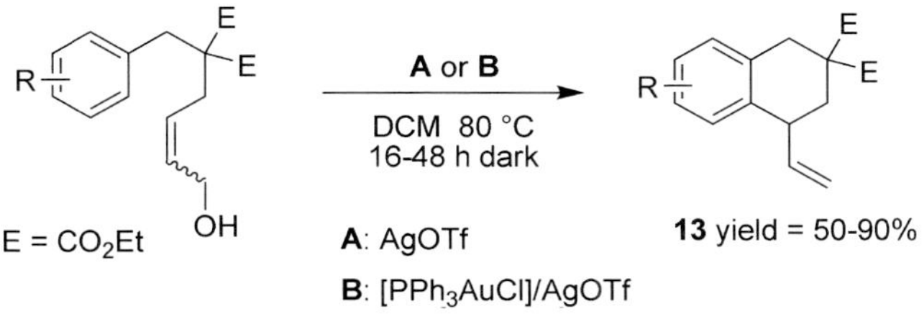

4.2 Allylic Alkylations

Scheme 4.17 Friedel–Crafts allylic cycloalkylation of aryl-alcohols.

The afore-mentioned Ag-catalysis is among the few examples of silver-based LA-catalysis in Friedel–Crafts alkylation reactions [35].

Up to now, only one example of intramolecular catalytic enantioselective allylic alkylation of arenes (indoles) has been described [36]. It represents a synthetic alternative to the well known Pictect–Spengler (PS) condensation [37] for the preparation of highly functionalized tetra-hydro-β-carbolines in enantiomerically pure form.

This study is an extension of the regioselective intramolecular Pd-catalyzed allylic alkylation of indoles [16]. In particular, by replacing PPh₃ with Trost's chiral ligands **19,22**, 4-substituted-THBCs **20** were isolated with high regio- and stereoselectivity ($ee = 82$–97%). Remarkably, the protocol proved tolerance toward to substitutions on both the indole ring and the amino side-chain, allowing even quaternary stereocenters (**20**-R_1 = Me) to be obtained in 90% ee (Scheme 4.18a). Moreover, the strategy was successfully employed for the stereoselective synthesis of tetrahydro-γ-carbolines **23** (ee up to 93%) simply by building up the carbonate chain at the C(3) position of the indole ring (**21**, Scheme 4.18b).

4.2.2.2 Experimental: Selected Procedures

Representative procedure for the enantioselective synthesis of tetrahydro-β-carbonlines 20 of arenes (Scheme 4.18) [36]

A solution of [Pd₂dba₃]·CHCl₃ (5 mol%) and **19** (11 mol%) in anhydrous CH₂Cl₂ (1.0 mL) was stirred until a deep orange color appeared. Then, **18** (0.07 mmol) dissolved in 0.5 mL of CH₂Cl₂ and Li₂CO₃ (2 equiv) were added. The reaction was stirred overnight and then quenched with water (4 mL) and extracted with AcOEt. The crude product was purified by passage through a pad of silica gel and the enantiomeric excess determined by HPLC analysis with a chiral column. The C/N regiochemistry of the cyclization was determined on the reaction crude by HPLC and confirmed after flash chromatography. Analytical data for (S)-**20** with R = 5-OMe, $R_1 = R_2 =$ H: Yellow solid. (CH₂Cl₂:AcOEt 8:2), mp = 138–141 °C, yield = 98% (C/N > 50 : 1), ee = 90%, $[\alpha]^D$ = 45.2 (c 0.9, CHCl₃). HPLC analysis: AD column (225 nm), method: n-Hex:IPA 90 : 10 flow 0.6 mL min⁻¹, t_S = 19.9 min, t_R = 27.5 min. ¹H-NMR (200 MHz, CDCl₃): 2.66 (dd, J = 7.0, 11.8 Hz, 1H), 2.97 (dd, J = 2.6, 11.8 Hz, 1H), 3.63 (s, 3H), 3.77 (d, J = 3.0 Hz, 2H), 3.83 (s, 3H), 5.14–5.31 (m, 2H), 5.85–6.08 (m, 1H), 6.77 (dd, J = 2.2, 8.8 Hz, 1H), 7.01 (d, J = 1.8 Hz, 1H), 7.17

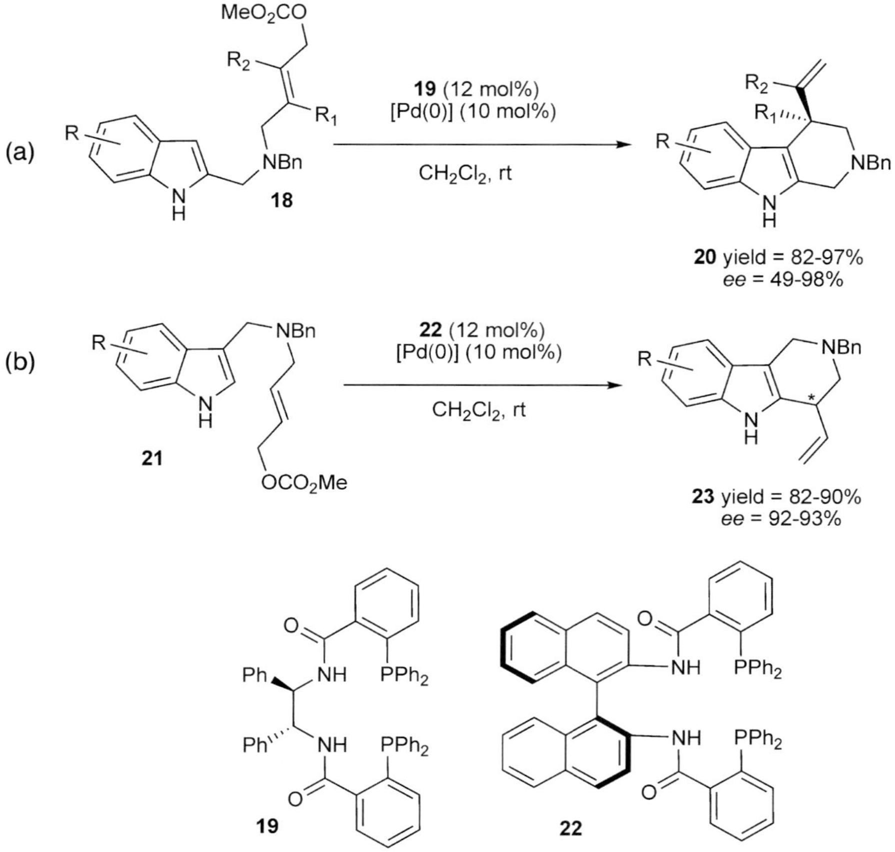

Scheme 4.18 Stereoselective synthesis of tetrahydro-β- and γ-carbolines (**20** and **23**) via asymmetric allylic alkylation.

(d, $J = 8.8$ Hz, 1H), 7.27–7.38 (m, 5H), 7.60 (br, 1H); ^{13}C-NMR (75 MHz, CDCl$_3$): 38.8, 50.5, 56.1, 57.5, 62.2, 101.9, 110.0, 110.9, 111.5, 115.7, 127.5, 127.8(2C), 128.6 (2C), 129.3, 131.4, 133.3, 138.6, 140.5, 154.0; LC-ESI-MS: 319 (M + 1); IR (nujol): ν 3395(w), 3188(m), 2726(s), 1731(m), 1460(s), 1373(s) cm^{-1}.

4.3
Metallo-Catalyzed Hydroarylation of Allenes

4.3.1
Introduction and Fundamental Examples

The metal-catalyzed hydroarylation of allenes [38a] is an atom economical complementary approach to nucleophilic allylic alkylation, for the introduction of vinyl units onto benzylic positions of aromatic compounds [38b]. At the present, gold catalysis has arisen to prominence in this field even if only intramolecular approaches have been described [39].

The catalytic hydrofunctionalization of allenes with indoles was pioneered by Widenhoefer and coworkers and underlined the high efficiency of cationic Au(I) salts in the synthesis of 4-substituted tetrahydro-carbazoles **26** via intramolecular hydroarylation of 2-allenyl indoles **24** [40]. Among the plethora of metal salts tested, Au[P(t-Bu)$_2$(o-biphenyl)Cl] **25**, combined with an equimolar amount of AgOTf, emerged as the catalyst of choice, promoting the ring-closing event in excellent yield (37–95%) and short reaction time (0.5–1 h). Finally, under optimal reaction parameters the presence of either an electron-withdrawing or electron-releasing group at the C(5)-position of the indolyl core was well tolerated (Scheme 4.19a).

Subsequently, the same team reported the enantioselective variant of the hydroarylation of allenes with indoles by means of chiral bis(gold) complexes of general formula (P-P)Au$_2$X$_2$ with (P-P) = bidentate phosphine [41]. After a survey of reaction conditions the use of preformed (S)-tBu-4-MeO-MeBIPHEP-(AuCl)$_2$ complex (2.5 mol%), combined with AgBF$_4$ (5 mol%, toluene −10 °C), led to the desired carbazoles **26** in high yields and *ee* up to 92% (Scheme 4.19b).

The tremendous potential of cationic gold(I) complexes for the hydroarylation of allenes was subsequently exploited by Nelson and coworkers as a key step for the total synthesis of (−)-rhazinilam (**30**) [42]. Here, the authors succeeded in discovering an effective catalyst for annulation processes involving pyrroles and enantiomerically pure allenes. With the aim to exploit a substrate-directed catalyst association the model substrate **28** was synthesized and allowed to cyclize under several reaction conditions. In the best conditions, the use of PPh$_3$-AuOTf (5 mol%) provided the desired bicyclic pyrrole **29** in 92% yield and 94% *de* (Scheme 4.20).

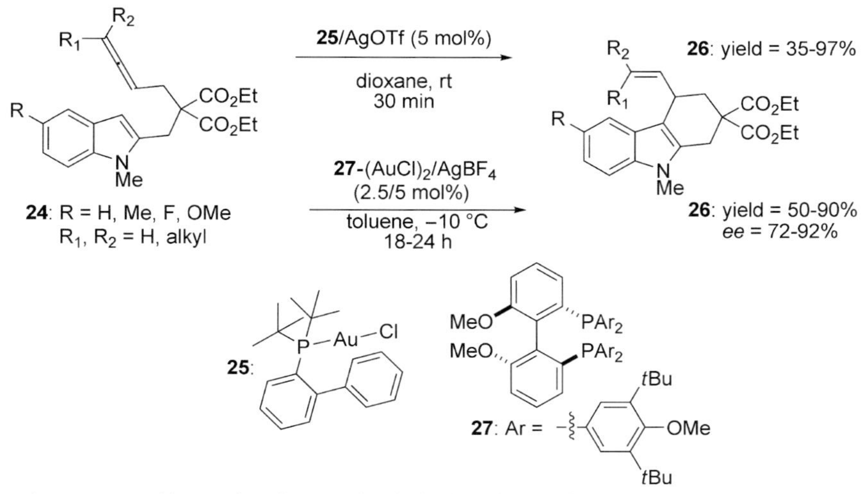

Scheme 4.19 Gold(I)-mediated intramolecular hydroarylation of allenes with indoles: catalysis and stereoselection.

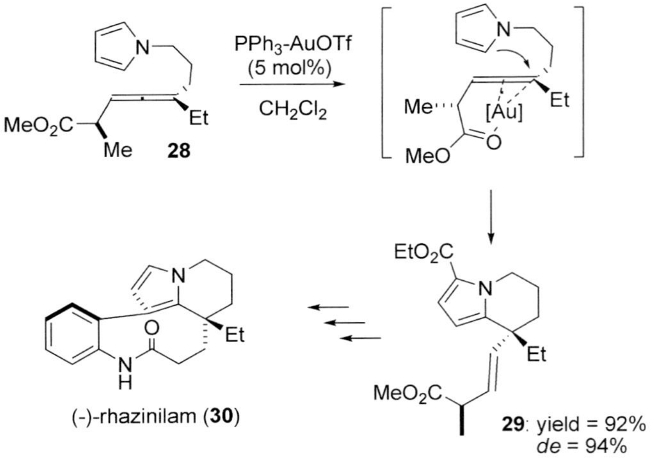

Scheme 4.20 Au(I)-catalyzed annulation of enantioenriched allenes with pyrroles in total synthesis.

Finally, very recently Gagné and coworker realized the efficient gold(I) catalyzed intramolecular *exo*-hydroarylation of substituted 4-allenyl electron-rich arenes to furnish vinyl-substituted benzocycles **13** (Scheme 4.21) [43]. The bench stable triphenylphosphite gold(I) catalyst $(PhO_3)_3PAuCl$ (**32**), in combination with $AgSbF_6$, produced the cyclized products in moderate to high yields.

Most likely, all these examples of gold-catalyzed hydroarylation of arenes proceed via an activation of the allene unit by the π-acid gold(I) monocation to allow the intramolecular *anti*-attack of the aromatic ring (outer-sphere) and consequent generation of a gold-alkenyl intermediate **33**. Finally, protonolysis of the C–Au bond leads to the cyclic product **13** (retention of C–C double bond configuration for the poly-substituted allenes) and the release of the catalytically active gold species. A pictorial representation of such an hypothesis for the model allenyl arene **31** is shown in Scheme 4.22.

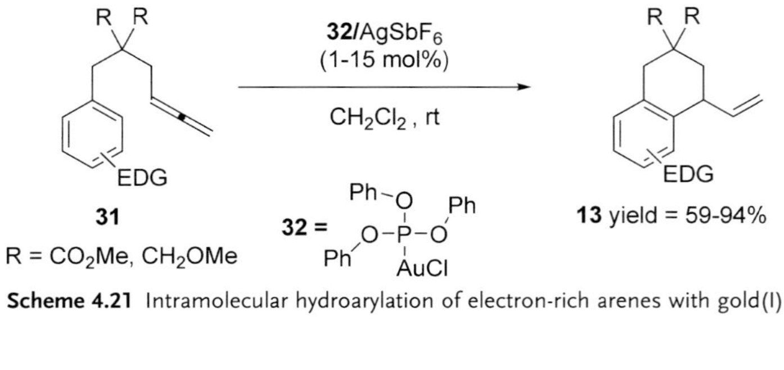

Scheme 4.21 Intramolecular hydroarylation of electron-rich arenes with gold(I)-catalysis.

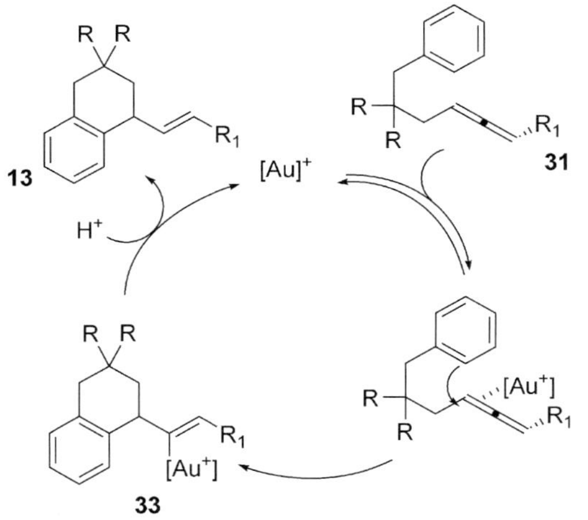

Scheme 4.22 Sketch of the mechanistic cycle for the Au-catalyzed hydroarylation of allenes.

4.3.2
Experimental: Selected Procedures

Representative procedure for the hydroarylation of arenes (Scheme 4.19) [41]

A mixture of (S)-**27**-(AuCl)$_2$ (5.1 mg) and AgBF$_4$ (1.2 mg) in toluene (0.2 mL) was stirred at $-20\,°C$ for 10 min. Then 0.13 mmol of **24** in toluene (0.3 mL) was added via syringe. The resulting mixture was stirred at $-10\,°C$ for 17 h. Analytical data for **26** (R = F; R$_1$ = R$_2$ = H): white solid, 90%, ee = 75%, mp = 42–46 °C. Rf = 0.48 (hexanes–EtOAc = 2 : 1). ^{1}H-NMR: 7.17 (dd, J = 2.4, 9.6 Hz, 1 H), 7.12 (d, J = 4.4, 8.8 Hz, 1 H), 6.86 (dt, J = 2.7, 9.2 Hz, 1 H), 5.82–5.73 (m, 1 H), 5.29–5.16 (m, 2 H), 3.76 (s, 3 H), 3.70 (s, 3 H), 3.68–3.65 (m, 1 H), 3.63 (s, 3 H), 3.43 (d, J = 8.0 Hz, 1 H), 3.12 (dd, J = 2.2, 8.2 Hz, 1 H), 2.66 (ddd, J = 0.8, 6.0, 13.2 Hz, 1 H), 2.03 (dd, J = 10.0, 13.2 Hz, 1 H). ^{13}C NMR 172.0, 171.1, 157.7 (d, J = 232 Hz), 141.3, 134.6 (d, J = 34 Hz), 127.0, 116.2, 109.4 (d, J = 10 Hz), 109.1, 104.8 (d, J = 33 Hz), 54.3, 53.3, 36.6, 36.2, 29.8, 28.4. The enantiomers eluted at 13.2 and 14.1 min (nHex–IPA = 90 : 10 at 0.5 mL min^{-1} (Chiralpak AD-H column).

Representative procedure for the hydroarylation of arenes (Scheme 4.21) [43]

A 5 mL vial was charged with **32** (0.05 mmol, 10 mol%), AgSbF$_6$ (0.07 mmol) and CH$_2$Cl$_2$ (1.0 mL). After 2 min, allene **31** (0.5 mmol) was added. The suspension turned deep green within 20 min. After completion, checked by TLC, (6–24 h), the reaction crude was directly purified by flash chromatography (ethyl acetate/hexanes). Analytical data for 3,5-diMeO-substitued arene: Clear oil, yield = 85%, ^{1}H NMR (400 MHz, CDCl$_3$): 6.23 (s, 2H), 5.74 (m, 1H), 4.90 (d, 1H, J = 10.4 Hz), 4.68 (d, 1H, J = 17.2 Hz), 3.77 (s, 3H), 3.70 (s, 3H), 3.66 (s, 3H), 3.64 (s, 3H), 3.34 (d, 1H, J = 16 Hz), 2.99 (d, 1H, J = 16.4 Hz), 2.48 (m, 1H), 2.29 (m, 1H). ^{13}C (100 MHz CDCl$_3$): 171.8, 171.7, 159.1, 158.6, 141.3, 135.8, 113.3, 104.3, 97.1, 55.3, 55.2, 52.6, 52.4, 35.4, 35.0, 34.3.

Acknowledgments

Financial support was provided by MIUR, Rome, and Alma Mater Studiorum, University of Bologna.

Abbreviations

LG   leaving group
FC   Friedel–Crafts
PS   Pictet–Spengler
EDG   electron-donating group
EWG   electron-withdrawing group

References

1. (a) Friedel, C. and Crafts, J.M. (1877) *Comptes Rendus Hebdomadaires des Seances de l' Academie des Sciences*, **84**, 1392; (b) Friedel, C. and Crafts, J.M. (1877) *Comptes Rendus Hebdomadaires des Seances de l'Academie des Sciences*, **84**, 1450.
2. (a) Wang, Y., Ding, K. and Dai, L. (2001) *Chemtracts-Organic Chemistry*, **14**, 610–615; (b) Jørgensen, K.A. (2003) *Synthesis*, 1117–1126; (c) Bandini, M., Melloni, A. and Umani-Ronchi, A. (2004) *Angewandte Chemie*, **116**, 560–566; (2004) *Angewandte Chemie – International Edition*, **43**, 550–556; (d) Poulsen, T.B. and Jørgensen, K.A. (2008) *Chemical Reviews*, **108**, 2903–2912.
3. Trost, B.M. (1991) *Science*, **254**, 1471–1477.
4. (a) Trost, B.M., Fraisse, P.L. and Ball, Z.T. (2004) *Angewandte Chemie*, **114**, 1101–1103; (2002) *Angewandte Chemie – International Edition*, **41**, 1059–1061; (b) Trost, B.M. and Crawley, M.L. (2003) *Chemical Reviews*, **115**, 2921–2944.
5. For representative examples see: (a) Ferreira, E.M. and Stoltz, B.M. (2003) *Journal of the American Chemical Society*, **125**, 9578–9579; (b) Zhang, H., Ferreira, E.M. and Stoltz, B.M. (2004) *Angewandte Chemie – International Edition*, **43**, 6144–6148; (2004) *Angewandte Chemie*, **116**, 6270–6274.
6. Lee, P.H., Seomon, D., Lee, K., Kim, S., Kim, H., Kim, H., Shim, E., Lee, M., Lee, S. and Kim, M. (2004) *Advanced Synthesis and Catalysis*, **346**, 1641.
7. Seomoon, D., Lee, K., Kim, H. and Lee, P.H. (2007) *Chemistry – A European Journal*, **13**, 5197–5206.
8. (a) Billups, M.E., Erkes, R.S. and Reed, L.E., (1980) *Synthetic Communications*, **10**, 147–154; (b) Malkov, A.V., Davis, S.L., Baxendale, I.R., Mitchell, W.L. and Kočovsky, P. (1999) *The Journal of Organic Chemistry*, **62**, 2751–2764.
9. (a) Trost, B.M., Fraisse, P.L. and Ball, Z.T. (2002) *Angewandte Chemie*, **114**, 1101–1103; (2002) *Angewandte Chemie – International Edition*, **41**, 1059–1061; (b) Mbaye, M.D., Demerseman, B., Renaud, J.L., Toupet, L. and Bruneau, C. (2003) *Angewandte Chemie*, **115**, 5220–5222; (2003) *Angewandte Chemie – International Edition*, **42**, 5066–5068.
10. (a) Hermatschweiler, R., Fernández, I., Breher, F., Pregosin, P.S., Veiros, L.F. and Calhorda, M.J. (2005) *Angewandte Chemie*, **117**, 4471–4474; (2005) *Angewandte Chemie – International Edition*, **44**, 4397–4400.

11 Tsuchimoto, T., Tobita, K., Hiyama, T. and Fukuzawa, S. (1997) *The Journal of Organic Chemistry*, **62**, 6997–7005.

12 (a) Nieves, I.F., Schott, D., Gruber, S. and Pregosin, P.S. (2007) *Helvetica Chimica Acta*, **90**, 271–276; (b) Fernández, I., Hermatschweiler, R., Breher, F., Pregosin, P.S., Veiros, L.F. and Calhorda, M.J. (2006) *Angewandte Chemie*, **118**, 6535–6540; (2006) *Angewandte Chemie – International Edition*, **45**, 6386–6391.

13 Rao, W. and Way Hong Chan, P. (2008) *Organic and Biomolecular Chemistry*, **6**, 2426–2433.

14 (a) Sundberg, R.J. (1996) *Indoles*, Academic Press, London, pp. 105–118; (b) Zhu, X. and Ganesan, A. (2002) *The Journal of Organic Chemistry*, **67**, 2705–2708.

15 (a) Brown, J.B., Henbest, H.B. and Jones, E.R. (1952) *Journal of the Chemical Society*, 3172–3176; (b) Wang, T., Zhang, Z., Wallace, O.B., Deshpande, M., Fang, H., Yang, Z., Zadjura, L.M., Tweedie, D.L., Huang, S., Zhao, F., Ranadive, S., Robinson, B.S., Gong, Y.-F., Ricarrdi, K., Spicer, T.P., Deminie, C., Rose, R., Wang, H.-G.H., Blair, W.S., Shi, P.-Y., Lin, P.-F., Colonno, R.J. and Meanwell, N.A. (2003) *Journal of Medicinal Chemistry*, **46**, 4236–4239.

16 Bandini, M., Melloni, A. and Umani-Ronchi, A. (2004) *Organic Letters*, **6**, 3199–3202.

17 Liu, Z., Liu, L., Shafic, Z., Wu, Y.C., Wang, D. and Chen, Y.J. (2007) *Tetrahedron Letters*, **48**, 3963–3967.

18 Kimura, M., Futamata, M., Mukai, R. and Tamaru, Y. (2005) *Journal of the American Chemical Society*, **127**, 4592–4593.

19 For preliminary findings in the molybdenum(IV)-catalyzed allylic alkylation of alkoxyarenes with cinnamyl-alcohols see: Malkov, A.V., Spoor, P., Vinader, V. and Kočovsky, P. (1999) *The Journal of Organic Chemistry*, **64**, 5308–5311.

20 Zaitsev, A.B., Gruber, S. and Pregosin, P.S. (2007) *Chemical Communications*, 4692–4693.

21 Zaitsev, A.B., Gruber, S., Plüss, P.A., Pregosin, P.S., Veiros, L.F. and Wörle, M. (2008) *Journal of the American Chemical Society*, **130**, 11604–11605.

22 Usui, I., Schmidt, S., Keller, M. and Breit, B. (2008) *Organic Letters*, **10**, 1207–1210.

23 Yadav, J.S., Reddy, B.V.S., Aravind, A., Narayana Kumar, G.G.K.S. and Reddy, A.S. (2007) *Tetrahedron Letters*, **48**, 6117–6120.

24 (a) Jana, U., Maiti, S. and Biswas, S. (2007) *Tetrahedron Letters*, **48**, 7160–7163; (b) Liu, Z., Liu, L., Shafic, Z., Wang, D. and Chen, Y.J. (2007) *Letters in Organic Chemistry*, **4**, 256–260.

25 Liu, Y.-L., Liu, L., Wang, Y.-L., Han, Y.-C., Wang, D. and Chen, Y.-J. (2008) *Green Chemistry*, **10**, 635–640.

26 Austin, J.F. and MacMillan, D.W.C. (2002) *Journal of the American Chemical Society*, **124**, 1172–1173.

27 (a) Bandini, M., Melloni, A., Tommasi, S. and Umani-Ronchi, A. (2005) *Synlett*, 1199–1222; (b) Bandini, M., Eichholzer, A. and Umani-Ronchi, A. (2007) *Mini-Reviews in Organic Chemistry*, **4**, 115–124.

28 Cheung, H.Y., You, W.Y., Lam, F.L., Au-Yeung, T.T.-L., Zhou, Z., Chan, T.H. and Chan, A.S.C. (2007) *Organic Letters*, **9**, 4295–4298.

29 Liu, W-Bo., He, H., Dai, Li-Xin. and You, Shu-Li. (2008) *Organic Letters*, **10**, 1815–1818.

30 (a) Trost, B.M. and Quancard, J. (2006) *Journal of the American Chemical Society*, **128**, 6314–6315; (b) For different asymmetric Pd-catalyzed alkylations of indoles/oxindoles with allyl acetates see: Trost, B.M. and Brennan, M.K. (2006) *Organic Letters*, **8**, 2027–2030; Trost, B.M., Krische, M.J., Bert, V. and Grenzer, E.M. (2002)*Organic Letters*, **4**, 2005–2008.

31 Bandini, M., Eichholzer, A., Kotrusz, P. and Umani-Ronchi, A. (2008) *Advanced Synthesis and Catalysis*, **350**, 531–536.

32 (a) Hayashi, R. and Cook, G.R. (2007) *Organic Letters*, **9**, 1311–1314; (b) Kaneko,

M., Hayashi, R. and Cook, G.R. (2007) *Tetrahedron Letters*, **48**, 7085–7087.

33 Namba, K., Yamamoto, H., Sasaki, I., Mori, K., Imagawa, H. and Nishizawa, M. (2008) *Organic Letters*, **10**, 1767–1770.

34 Bandini, M., Eichholzer, A., Kotrusz, P., Tragni, M., Troisi, S. and Umani-Ronchi, A. (2009) *Organic Biomolecular Chemistry*, **7**, 1501–1507.

35 (a) Youn, S.W. and Eom, J.I. (2006) *The Journal of Organic Chemistry*, **71**, 6705–6707; (b) Shafiq, Z., Liu, L., Liu, Z., Wang, D. and Chen, Y.-J. (2007) *Organic Letters*, **9**, 2525–2528.

36 Bandini, M., Melloni, A., Piccinelli, F., Sinisi, R., Tommasi, S. and Umani-Ronchi, A. (2006) *Journal of the American Chemical Society*, **128**, 1424–1425.

37 (a) Pictet, A. and Spengler, T. (1911) *Berichte der Deutschen Chemischen Gesellschaft*, **44**, 2030–2036; for a comprehensive review: Cox, E.D. and Cook, J.M. (1995) *Chemical Reviews*, **95**, 1797–1842.

38 (a) Ma, S. (2005) *Chemical Review*, **105**, 2829–2872; (b) Bandini, M., Emer, E., Tommasi, S. and Umani-Ronchi, A. (2006) *European Journal of Organic Chemistry*, 3527–3544.

39 For recent reviews on Au-catalyzed hydroarylation of unactivated olefins see: (a) Widenhoefer, R.A. (2008) *Chemistry – A European Journal*, **14**, 5382–5391; (b) Arcadi, A. (2008) *Chemical Reviews*, **108**, 3266–3325; (c) Chianese, A.R., Lee, S.J. and Gagné, M.R. (2007) *Angewandte Chemie*, **119**, 4118–4136; (2007) *Angewandte Chemie-International Edition*, **46**, 4042–4059; (d) Fürstner, A. and Davies, P.W. (2007) *Angewandte Chemie*, **119**, 3478–3519; (2007) *Angewandte Chemie – International Edition*, **46**, 3410–3449.

40 Zhang, Z., Liu, C., Kinder, R.E., Han, X., Qian, H. and Widenhoefer, R.A. (2006) *Journal of the American Chemical Society*, **128**, 9066–9073.

41 Liu, C. and Widenhoefer, R.A. (2007) *Organic Letters*, **9**, 1935–1938.

42 Liu, Z., Wasmuth, A.S. and Nelson, S.G. (2006) *Journal of the American Chemical Society*, **128**, 10352–10353.

43 Tarselli, M.A. and Gagné, M.R. (2008) *The Journal of Organic Chemistry*, **73**, 2439–2441.

5
Nucleophilic Substitution on Csp³ Carbon Atoms

Marco Bandini and Pier Giorgio Cozzi

Summary

The functionalization of aromatic compounds through catalytic stereoselective reactions is predominant in modern organic chemistry, and constitutes the *leitmotiv* in approaching new complex chiral molecules. In this framework, asymmetric Friedel–Crafts alkylations, based on nucleophilic substitution of pro-stereogenic Csp³ carbon atoms are becoming of great importance.

In this chapter, we present an overview of the work published in this area, with particular emphasis on the selective ring-opening of enantiomerically pure epoxides via mild Lewis acids catalysis, kinetic resolution of chiral racemic *cis*- and *trans*-oxiranes and desymmetrization of meso molecules with indoles and chiral organometallic catalysis.

The emerging field of direct activation of alcohols in stereocontrolled FC alkylations is also discussed. Both diastereo- and enantioselective approaches are described. Here, particular emphasis has been devoted to the peculiar attitude of ferrocenyl alcohols in participating in catalytic alkylations of arenes "on water".

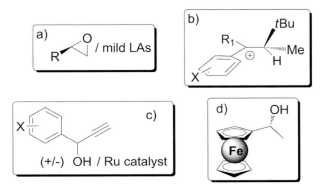

Catalytic Asymmetric Friedel–Crafts Alkylations. Edited by M. Bandini and A. Umani-Ronchi
Copyright © 2009 WILEY-VCH Verlag GmbH & Co. KGaA, Weinheim
ISBN: 978-3-527-32380-7

5.1
Ring-Opening of Epoxides

5.1.1
Introduction

Among the main alkylating agents employed in Friedel–Crafts (FC) alkylations, it is known that oxygenated compounds such as alcohols and ethers generally require higher catalyst loadings than alkyl halides and alkenes. This is mainly ascribable to the higher coordinating capability of the former compounds that interact strongly with the promoting agents preventing a favorable turnover of the catalytic processes. Despite this trend, cyclic ethers (mainly epoxides) have been investigated as alkylating agents for aromatics for a long time, with stimulating and frequently argumentative mechanistic discussions [1].

In this field, the seminal investigation by Milstein describes the use of racemic 1,2-epoxypropane **1a** as an alkylating agent for benzene, in the presence of stoichiometric amounts of $AlCl_3$. The regiochemical attack at benzene on the more substituted carbon atom led the authors to propose an S_N1-type mechanism with the formation of a carbocation intermediate Scheme 5.1a [2]. A few years later, a step toward the use of enantiomerically pure oxiranes in FC alkylations was achieved by Suga and coworkers. By rationalizing the stereochemical course (inversion of configuration) of the ring-opening reaction of (R)-1,2-epoxybutane **1b** with C_6H_6 ($AlCl_3$ or $SnCl_4$), the authors contradicted the previous paper, demonstrating that an S_N2-type mechanism was operating (Scheme 5.1b) [3].

Arene-mediated ring-opening of higher cyclic ethers such as THF and pyrans are generally characterized by significantly lower reaction rates with respect to epoxides and are accompanied by undesired side-reactions. For these reasons, examples of larger cyclic ethers in the alkylation of arenes are rare and only one stereoselective approach has been reported, using (S)-(+)-2-methyltetrahydrofuran for the alkylation of benzene with partial stereoinversion [4].

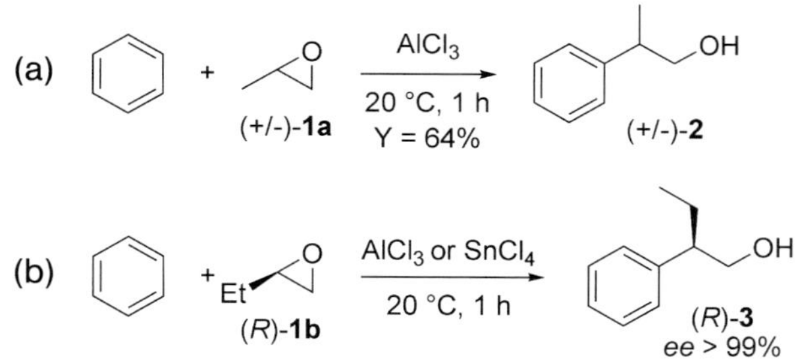

Scheme 5.1 Pioneering studies on the benzene-based ring-opening of racemic (a) and enantiopure (b) epoxides.

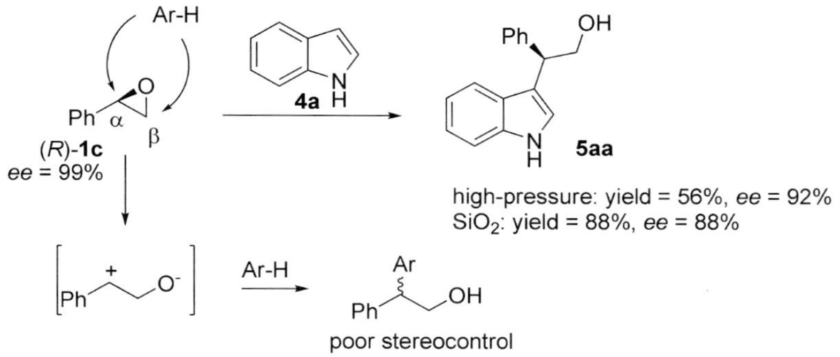

Scheme 5.2 Indole alkylation based on the ring-opening of styrene oxide: regio- and stereoselection.

5.1.2
Enantiomerically Pure Epoxides

5.1.2.1 Introduction

Epoxides are an ideal source of chemical diversity, because they can be easily opened with a plethora of N-, O- and S- nucleophiles furnishing functionally diverse compounds [5]. However, the use of enantiomerically pure oxiranes, as alkylating agents of aromatic compounds, remained poorly unexplored until Kotsuki and coworkers [6, 7] reported the regio- and stereocontrolled alkylation of indole **4a** with enantiomerically pure (R)-styrene oxide **1c** (Scheme 5.2).

Regiocontrol in nucleophilic ring-opening of oxiranes is often an issue. In this case, the use of an aromatic epoxide guaranteed the selective indole attack at the benzylic carbon (α). Moreover, mild Lewis acid/conditions were necessary in order to avoid undesired acid-catalyzed rearrangements and to prevent the formation of benzylic cation intermediates with consequent loss of stereocontrol [8].

The use of high-pressure conditions (10 kbar) partially absolved this requirement, giving the β-indolyl-alcohol **5aa** in moderate yield (56%, 24 h) and good stereoretention ($ee = 92\%$). In contrast, the use of silica as an acidic promoter slowed down the reaction rate remarkably (7 days, yield = 88%, $ee = 88\%$).

Aliphatic epoxides were also considered. Moderate yields of the ring-opening products were obtained by the simultaneous employment of high pressure (10 kbar) and mild Lewis acids.

Worth noting is the example reported by Fráter and coworkers, with the ring-opening of enantiopure methyloxirane **1a** by 1,1,2,3,3-pentamethylindane **6** as the key step, for the synthesis of perfumery synthetic Galaxolide **8** [9]. A complete stereoinversion of **1a** was observed although an almost quantitative amount of TiCl$_4$ (0.8 eq.) was required to obtain the (S)-β-arylethan-1-ol **7** in 57% yield (Scheme 5.3).

In analogy, the use of stoichiometric amounts of trifluoromethansulfonic acid TfOH enabled the stereospecific ring-opening of enantiomerically pure methyl (2R)-glycidate **1d** by electron-rich benzene-like arenes (Scheme 5.4) [10]. The presence of

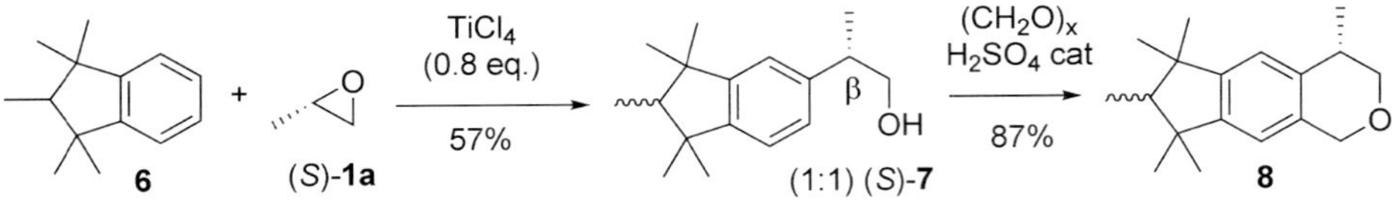

Scheme 5.3 Synthesis of Galaxolide intermediate (S)-**7** via intermolecular Ti-mediated alkylation of benzene derivative **6**.

electron-withdrawing substituents at the α-position of the epoxide drives the ring-opening process at the β-position exclusively, due to the formation of β-carbenium ion [11]. A highly reactive dioxonium dication **9** is proposed as a possible intermediate for the reaction, based on the formation of sulfonate **11** (main product) when the aromatic compound was not sufficiently reactive (i.e., benzene).

A different FC approach for the C-alkylation of substituted phenols via ring-opening of enantiomerically pure epoxides, was recently described by Pineschi and coworkers [12]. Triarylborates **12** displayed excellent reactivity toward internal as well as terminal three-membered heterocyclic rings high C- vs. O-alkylation ratio (10 : 1), even in the absence of external catalysts. Moreover, an unusual syn-stereoselection was observed during the ring-opening of disubstituted epoxides and triarylborates, suggesting that a retention of the configuration was operating during the ring-opening event. As an example, the reaction of (R)-(+)-styrene **1c** with electron-rich substituted borates led to a ratio of retention vs. inversion of configuration of 92 : 8 (Scheme 5.5).

The same authors exploited the reactivity of triarylborate for the ring-openig of activated enantioenriched aziridines [13]. Here, in contrast to the enormous impact of aziridines on the preparation of enantiomerically enriched building blocks for the pharmaceutical and fine-chemical industries [14], to date no example of catalytic ring-opening of enantiomerically enriched/pure aziridines based on aryl nucleophiles has

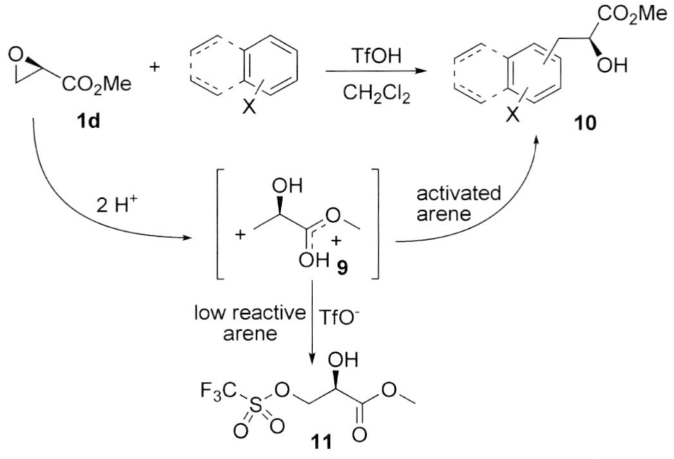

Scheme 5.4 Use of TfOH in the ring-opening of methyl (2R)-glycidate with electron-rich arenes.

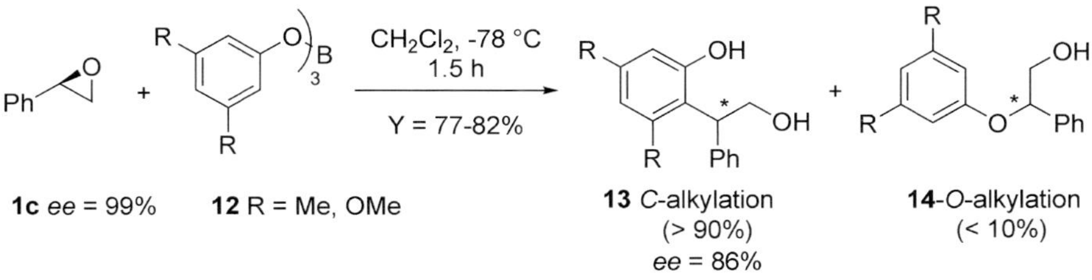

Scheme 5.5 Regio- and stereoselective FC alkylation of phenols via carbon–carbon coupling with enantiopure styrene oxide.

been reported. In fact, although a few examples of catalytic protocols involving racemic aziridines can be found [15], only Pineschi and Farr [16] have documented the nucleophilic ring-opening of enantiomerically pure aziridines N-X (X: Ts, Cbz, P(O)Ph$_2$) with aryl borates and BF$_3$·OEt$_2$/indoles, respectively, leading to arylated ethylamines with complete regioselectivity (benzylic position) and negligible epimerization.

In particular, in the latter example, the enantiospecific opening of the N-nosyl-aziridine **15** by **4b** was utilized for the gram scale preparation of GnRH antagonist, in the presence of a stoichiometric amount of BF$_3$·OEt$_2$ (Scheme 5.6).

The reaction medium can also play a decisive role in the FC-type alkylation of electron-rich arenes with enantiopure oxiranes. In particular, it has been demonstrated that aromatic and aliphatic epoxides can be attached to indoles/pyrroles in 2,2,2-trifluoroethanol without the need for catalysts or additives. The principle of electrophilic solvent assistance of S$_N$2-type reaction was demonstrated with excellent levels of enantioselectivity (ee > 99%) [17]. It is known, in fact, that the heterolysis of

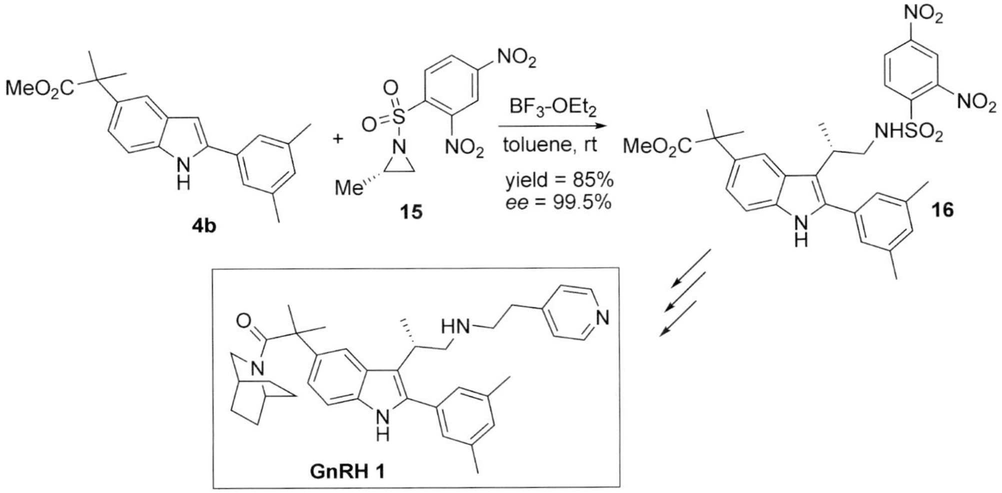

Scheme 5.6 Enantiospecific opening of N-nosyl-aziridine **15** by indole **4b** for the preparation of GnRH antagonist.

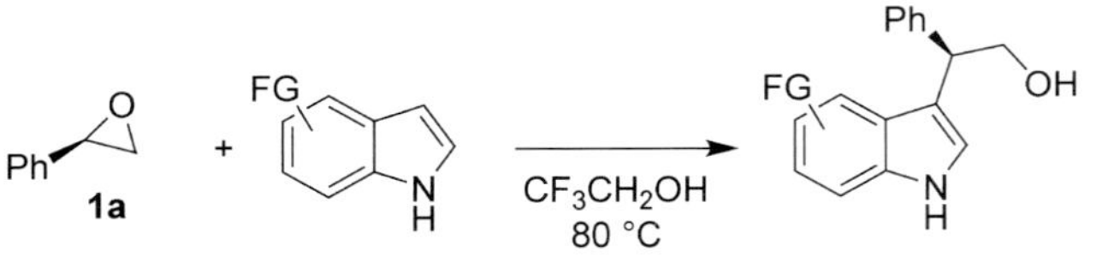

Scheme 5.7 Reactions of indoles with **1a** in protic solvents: FG = functional group.

C–O bonds can be facilitated by the employment of protic solvents characterized by high ionizing power [18].

Here, while **1a** did not react in MeOH, EtOH or aqueous CH_3CN at 70–90 °C, chemical yields (up to 79%) and high enantiomeric excesses were observed in 2,2,2-trifluoroethanol (Scheme 5.7).

The excellent chemical and stereochemical outcomes obtained in 2,2,2-trifluoroethanol are determined by its high ionizating power ($Y = 2.53$) and low solvent nucleophilicity [19]. The scope of the reaction with various substituted indoles has been investigated. The reaction takes place exclusively at position-3 of the indole ring and, accordingly to the nucleophilicity scale of indole [20], when indoles bearing electron-withdrawing groups were employed, inferior yields of the desired product were obtained. Interestingly, the reaction was extended to the more nucleophilic pyrrole, giving rise to a mixture of the two regioisomers (**17a** + **18a**) in a 2 : 1 ratio (Scheme 5.8).

In all the cases, the enantiomeric excess of the starting epoxide was retained in the product and no racemization occured. However, the chemistry of pyrrole was characterized by moderate yields, due to the tendency to produce tars and polymeric non-volatile adducts.

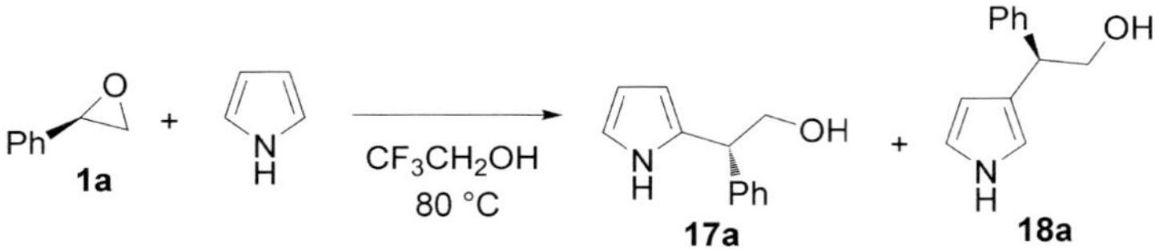

Scheme 5.8 Reaction of pyrrole with **1a** in 2,2,2-trifluoroethanol as the reaction medium.

5.1.2.2 Indium(III) Catalysis

The previously described examples clearly underlined the interest of the chemical community in the use of EP (enantiomerically pure) epoxides in FC alkylations. However, the first catalytic version of such a process came rather belatedly. In 2002 Cozzi and Umani-Ronchi first discovered the unique activity of In(III)Br as a highly regio- and stereoselective additive for the intermolecular alkylation of variously functionalized indoles with both terminal and internal enantiomerically pure aromatic epoxides [21]. No racemization of the starting oxiranes was observed, with the nucleophilic attack regiospecifically on the benzylic carbon. In Table 5.1 a collection of results is reported.

The loading of catalyst (1 mol% was best) played a crucial role in the stereochemical course of the alkylation. In particular, attempts to speed up the reaction rate by using a

Table 5.1 Addition of indoles to aromatic epoxides catalyzed by InBr$_3$.[a]

Entry	Epoxide	Indole	Product	Yield (%)[b]	ee (%)[c]
1	1c	4a	5ca	70	99
2	1c	4c	5cc	41[d]	70
3	1c	4d	5cd	54	99
4	1c	4e	5ce	54	99
5	1c	4f	5cf	24[e]	—[f]
6	1c	4g	5cg	79	99
7	1c	4h	5ch	68	99

Table 5.1 (Continued)

Entry	Epoxide	Indole	Product	Yield (%)[b]	ee (%)[c]
8	1e	4a	5ea	65	99
9	1e	4g	5eg	82	99
10	1f	4g	5fg	84[g]	83

[a] All the reactions were carried out in anhydrous CH_2Cl_2 at room temperature, employing 1 mol% of $InBr_3$ for 8–16 h unless otherwise specified.
[b] Isolated yields after chromatographic purification.
[c] The enantiomeric excesses were determined by HPLC analysis with a chiral column.
[d] The reaction was performed using 10 mol% of $InBr_3$ at room temperature for 16 h.
[e] The reaction was performed using 10 mol% of $InBr_3$ at room temperature for 96 h.
[f] The enantiomeric excess was not evaluated.
[g] The optically active epoxide (1R,2S)-1f was prepared using the asymmetric Jacobsen epoxidation in 83% ee.

larger amount of $InBr_3$ (i.e., 10 mol%) led to a significant drop in both chemical and optical yield (**3aa**, ee = 75%).

The growing demand for environmentally sustainable chemical transformations has prompted numerous research teams to develop new effective, easily removable recoverable and reusable chiral catalysts. The heterogenization of homogeneous promoting agents is one of the most promising strategies to address this issue [22]. Since the pioneering study of Neckers [23] numerous organic transformations involving metals anchored to acid ion-exchange resins have been described [24].

In this context, a new type of In(III) Lewis acid **19** supported on an ion-exchange resin (Amberlist-15) has been described by Bandini and Umani-Ronchi to be effective in promoting the ring-opening of enantiomerically pure aromatic epoxides with indoles (Scheme 5.9) [25].

Catalytic amounts of **19** (20 mol% [In] content) promoted the C(3)-regioselective alkylation of N-methyl indole **4h** with **1c** in reagent grade Et_2O with moderate isolated

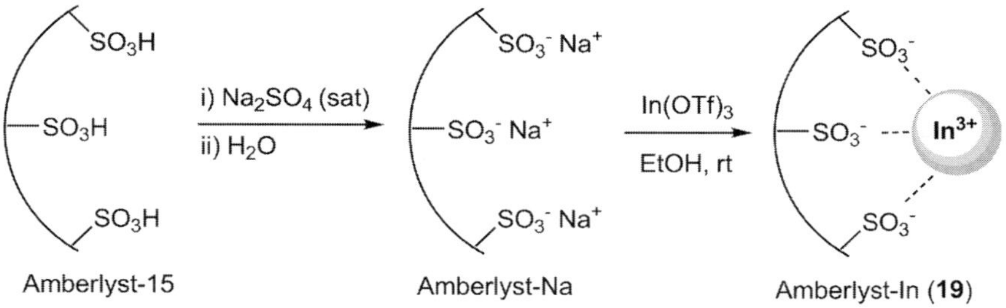

5.1 Ring-Opening of Epoxides

Scheme 5.9 Synthesis of the resin-supported indium Lewis acid **19** (estimated loading: 0.92 mmol g^{-1}, exchange percentage EP: 59%).

yield (64%) and high enantiomeric excess (98%). The active role of the indium was underlined by control experiments employing the precursors Amberlyst-15 (Amb-15) and Amberlyst-Na (Amb-Na) as additives.

Interestingly, while Amb-Na proved to be ineffective for the alkylation, Amb-15 led to **5ch** in poor yield and with a marked level of racemization (Scheme 5.10a). This model reaction was also adopted to investigate the recoverability and reusability of the active species. Here, five consecutive runs were carried out without recording significant loss of activity (yield = 52–67%; ee = 96–98%, Scheme 5.10b).

5.1.2.3 Mechanism of Indium(III) Catalysis

To preserve the stereochemical information of the starting oxirane a rigorous S_N2-type mechanism must be operating during the ring-opening reaction. This task

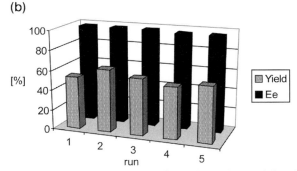

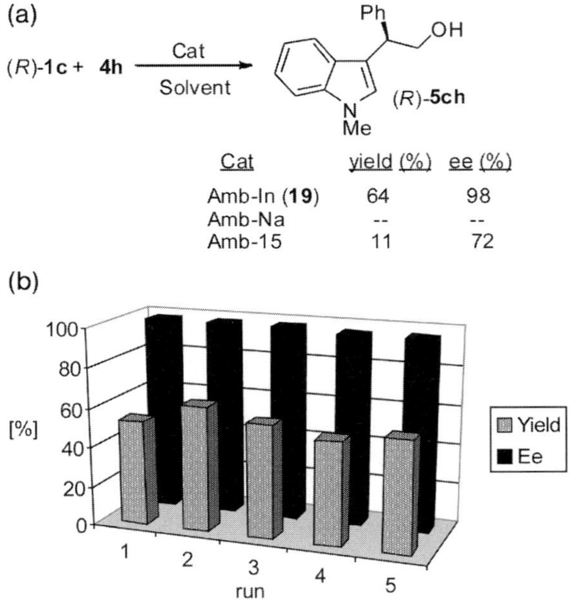

Cat	yield (%)	ee (%)
Amb-In (**19**)	64	98
Amb-Na	--	--
Amb-15	11	72

Scheme 5.10 Proof of reusability of **19** in the alkylation of **4h** with (R)-**1c**.

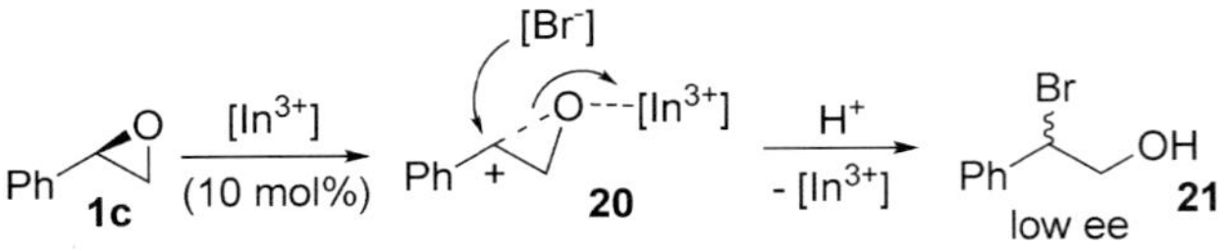

Scheme 5.11 Thiourea-gold-catalyzed intramolecular diastereoselective cycloalkylation of arenes.

is satisfied by a proper combination of reaction parameters such as, moderate heterophilicity of indium(III) salts [26], low loading of catalyst (1 mol%) and mild reaction conditions. In fact, as previously mentioned, the use of 10 mol% of catalyst speeded up the consumption of epoxides but the final isolated yield of β-indolyl-alcohol was lower (60%) with partial racemization (75%). The concomitant formation of the β-bromohydrine **21** in discrete amount accounts for the low selectivity of the reaction and suggests the activation of the oxiranes via In-epoxide interaction (**20**) as a crucial event of the process (Scheme 5.11).

These evidences suggested the use of In(OTf)$_3$, instead of InBr$_3$, as an ideal precursor for the synthesis of polymer-supported indium catalyst **19**. In fact, the negligible nucleophilic character of the triflate counteranion allowed the use of higher loading of catalyst (20 mol%) without appreciable loss in chemical and optical yields.

5.1.2.4 Gold(III) Catalysis

Benzene-type compounds have also been reacted with enantiomerically pure oxiranes for the synthesis of functionalized chromanols **22** [27]. In this scenario, He and coworkers reported on the effectiveness of the AuCl$_3$/AgOTf (2.5/7.5 mol%) catalytic system in promoting inter- and intramolecular ring-opening of terminal and internal epoxides by means of electron-rich arenes (Scheme 5.12).

The intramolecular attempts furnished exclusively the *endo*-product (six-membered ring) with rigorous *trans* diastereoselection when internal epoxides **1g,h** were employed.

Finally, a complete inversion of configuration of the benzylic carbon atom was recorded when the model cyclization was carried out on enantiomerically pure (1*S*,2*S*)-**1h**.

The gold-catalyzed cycloalkylation of aromatic epoxides was successfully exploited by Chen and coworkers for the synthesis of catechins in a highly regio- and

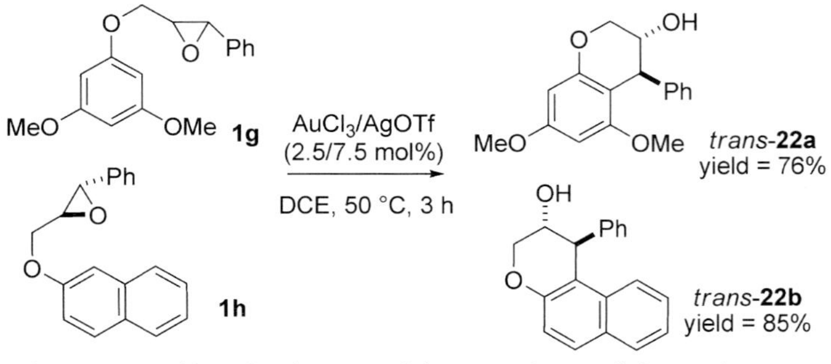

Scheme 5.12 Gold-catalyzed regio- and diastereoselective alkylation of arenes.

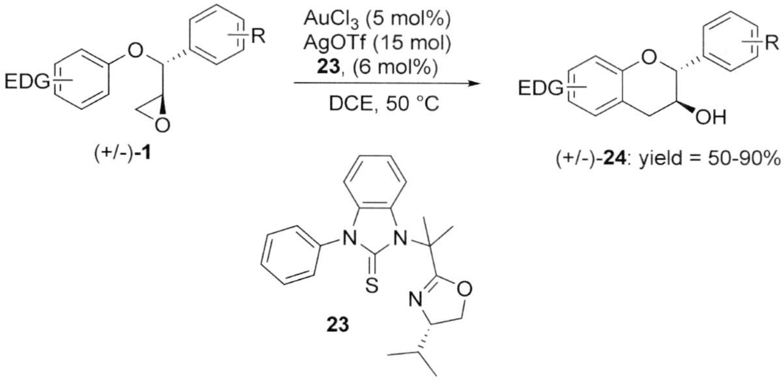

Scheme 5.13 Formation of bromohydrine **21** as the side-reaction of the In-catalyzed ring-opening of epoxides with a high loading of catalyst.

stereospecific manner [28]. The annulation precursors adopted are racemic monosubstituted aliphatic epoxides, readily obtainable from common starting materials.

The presence in the molecular skeleton of a labile benzylic ether moiety precluded the use of the original Au(III) reaction conditions. In fact, epoxide **1i** underwent decomposition when treated with a mixture of AuCl$_3$ and AgOTf (DCE, 50 °C, 4 h), with the desired catechin **24** isolated only in <20% yield. However, the introduction of the chiral thiourea **23**, as a gold ligand, adequately controlled the electronic properties of the metal-catalyst, minimizing the side-reaction and leading to **24** in 65% yield (Scheme 5.13). The protocol appeared limited to poly-alkoxy π-rich benzenes and a marked decrease in reaction rate occurred with less substituted arenes.

5.1.2.5 Mechanism of Gold(III) Catalysis

Two different reaction mechanisms are proposed for the gold-catalyzed alkylation of arenes: (1) a classic FC pathway involving Lewis acid activation of the epoxide by the Au(III) salt; (2) formation of an arylgold(III) intermediate [29] and subsequent intramolecular S$_N$2 attack on the epoxide (Scheme 5.14). Although more experimental details are needed, the authors provided evidence for an auration step during the analogous gold-catalyzed alkylation of arenes in combination with primary alcohol sulfonate esters [30a]. Again, although the role of the silver salt has not been completely elucidated [30b], the formation of stronger Au-based Lewis acids, via removal of the chlorine atoms, is proposed.

5.1.2.6 Experiments: Selected Procedures

Preparation of the Amb-In catalyst 19 (Scheme 5.9) [25]

A flamed two-necked flask was charged with 25 mL of EtOH, 2 g of Amb-Na and 1.12 g (2 mmol) of In(OTf)$_3$ under a nitrogen atmosphere and the resulting mixture was shaken overnight (16 h). The resin was collected by filtration, washed with 50 mL

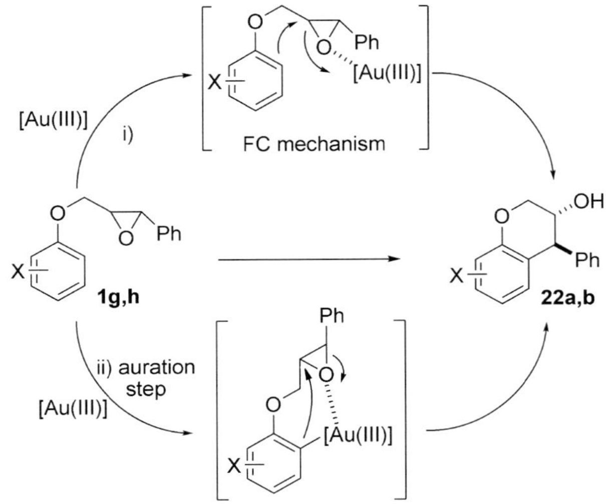

Scheme 5.14 Friedel–Crafts mechanism vs. auration-based pathway for the cycloalkylation of electron-rich benzenes.

of EtOH and dried under vacuum overnight. The content of indium in the resin (0.92 mmol g^{-1}) was determined by titration of the remaining amount of In^{3+} in the solution following the known procedure (EDTA 0.01 M, xylenol orange as the indicator, buffer: 20% hexamethylenetetraamine).

General procedure for the synthesis of 5ch catalyzed by 19 (Scheme 5.10) [25]

A sample vial containing 2 mL of reagent grade Et$_2$O was charged with 0.2 mmol of **1c**, 0.3 mmol of **4h** and 44 mg (20 mol% based on indium content) of Amb-In (**19**). The mixture was shaken for 24 h with a basic orbital mixer and then the catalyst was recovered by filtration. Evaporation of the solvent and subsequent purification by flash chromatography furnished the desired indolyl alcohols **5ch** in 64% yield. Yellow oil. $R_f = 0.32$ (c-Hex:AcOEt 8:2); $[\alpha]^{25}_D = -0.61$ (c 0.76, CHCl$_3$). ee = 98%, HPLC: isocratic 85/15 n-hex/IPA, $t_S = 22.3$ min, $t_R = 40.6$ min. ^{1}H NMR (200 MHz, CDCl$_3$): $\delta = 1.58$ (br, 1 H), 3.78 (s, 3 H), 4.16–4.25 (m, 2 H), 4.50 (t, $J = 6.8$ Hz, 1 H), 6.97 (s, 1 H), 7.01–7.49 (m, 9 H). ^{13}C NMR (50 MHz, CDCl$_3$): = δ 32.8, 45.6, 66.5, 109.2, 114.4, 119.0, 119.4, 121.8, 126.6 (2), 127.4, 128.2 (2), 128.5 (2), 137.2, 141.7; GC-MS m/z (relative intensity) 51 (5), 77 (10), 102 (15), 144 (10), 178 (20), 204 (25), 220 (100), 251 (30); IR (neat) 3549, 3400, 3060, 3030, 1610, 1600, 1470, 1360, 1335, 1020, 910, 735 cm^{-1}.

General procedure for ring-opening of enantiopure 1h (Scheme 5.12) [29]

To a suspension of anhydrous AuCl$_3$ (2.5 mol%) and AgOTf (7.5 mol%) in dry DCE (3.0 mL) 0.5 mmol of (1S,2S)-**1h** were added. The mixture was stirred for 3 h at 50 °C and after the usual work-up the reaction crude was purified by flash chromatography

(113 mg, 85% yield). $[\alpha]_D^{25} = 1.82$ (c 1.0, CHCl$_3$). ee > 98%. ^{1}H NMR (500 MHz, CDCl$_3$): δ = 2.39 (d, J = 8.6 Hz, 1 H), 4.13 (s, 2H), 4.19 (d, J = 7.2 Hz, 1 H), 4.65 (s, 1 H), 7.16 (m, 4 H), 7.27 (m, 4 H), 7.46 (d, J = 5.3 Hz, 1 H), 7.75 (m, 2 H). ^{13}C NMR (100 MHz, CDCl$_3$): δ = 45.9, 64.8, 69.7, 111.7, 118.5, 122.9, 123.5, 126.7, 126.9, 128.5, 128.6, 128.8, 129.3, 129.8, 133.4, 143.0, 151.7. MS of C$_{19}$H$_{16}$O$_2$ (m/z, ACPI): 259.1 (M$^+$ + 1-H$_2$O).

5.1.3
Asymmetric Ring-Opening of Racemic and *meso* Epoxides

5.1.3.1 Introduction

Metal Salen complexes are now considered privileged structures able to accomplish a large variety of useful synthetic transformations involving epoxides [31]. In particular, Salen-based Lewis acids were used to activate epoxides toward polymerization, or insertion reactions with CO$_2$ [32]. The most active catalysts for these transformations are [Cr(Salen)] complexes, in neutral or cationic form [33].

Although it is generally accepted that [Cr(Salen)] complexes are able to transmit chiral information via a double activation pathway [34], in which electrophile and nucleophile are arranged in close proximity to two molecules of catalyst, the peculiar Lewis acidic nature of [Cr(Salen)] can be used advantageously in single activation of electrophiles as well. In this context, Jacobsen discovered that chiral tetra- and tridentate Schiff base chromium(III) complexes catalyzed highly enantio- and diastereoselective hetero-Diels-Alder (HDA) reactions between aldehydes and mild nucleophilic dienes, through a simple catalyst activation of the aldehydes [35].

Activation of epoxides by [Cr(Salen)] complexes was reported by Jacobsen, and used in the highly enantioselective oxiranes ring-opening with Me$_3$SiN$_3$ [36]. However, a partial limitation in the scope of the reaction was encountered with highly substituted epoxides. In addition, although styrene oxide derivatives have been examined in kinetic resolution reactions promoted by [Cr(Salen)Cl] and [Co(Salen)OAc] [37], these strategies were not homogeneously effective with all types of aromatic epoxides (*cis, trans, meso*).

Because these metallo-salen catalyzed reactions did not proceed through free carbocation species, the electrophilic nature of the activation can be used to promote asymmetric FC transformation.

5.1.3.2 Salen-Chromium-Catalyzed Kinetic Resolution of Epoxides with Indoles

The high reactivity of indoles allows a new approach toward a general kinetic resolution of aromatic epoxides. As Lewis acids, like indium salts, were able to promote the addition of indoles to enantioenriched epoxides (Table 5.1), the same reaction was investigated in the context of a chiral environment. The chiral Lewis acid [Cr(Salen)] was chosen due to its ability to recognize chiral epoxide [38] and, through careful optimization of the reaction conditions, the employment of the easily handled, inexpensive and commercially available 2-methylindole as nucleophile was found to be better in terms of conversion, isolated yields, and enantiomeric excess. The reaction was run in non-coordinating solvents (CH$_2$Cl$_2$, TBME) and in highly

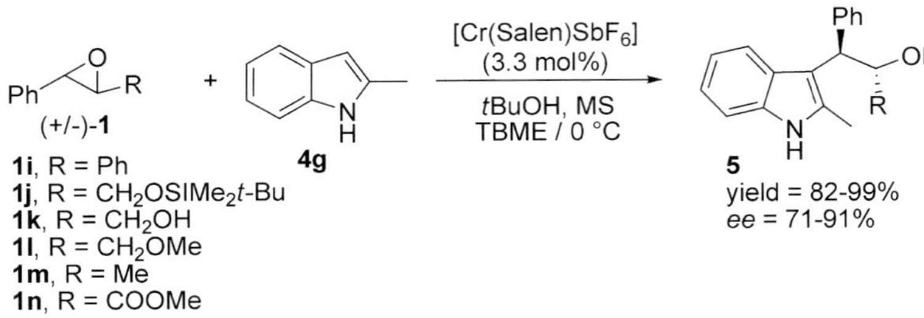

5 Nucleophilic Substitution on Csp³ Carbon Atoms

1i, R = Ph
1j, R = CH$_2$OSIMe$_2$t-Bu
1k, R = CH$_2$OH
1l, R = CH$_2$OMe
1m, R = Me
1n, R = COOMe

yield = 82–99%
ee = 71–91%

Scheme 5.15 Friedel–Crafts reaction of *trans*-racemic epoxides with 2-methylindole **4g** catalyzed by cationic [Cr(Salen)].

concentrated conditions. Interestingly, different activities were observed between *cis*-aromatic epoxides and *trans*-derivatives. Here, the combination of *cis*-epoxides, 2-methylindole **4g** and [Cr(Salen)] led to the opened product in high conversions in 16–24 h at room temperature (yield = 95%, ee = 80% with *cis*-1-phenylpropene oxide; yield = 97%, ee = 83% with 3,4-dihydronaphthalene oxide). In contrast, *trans*-aromatic epoxides were in general found to be less reactive, and a more active cationic [Cr(Salen)SbF$_6$] complex was therefore employed. Starting from racemic epoxides and employing an excess of epoxide (3 equiv.) the opened products **5** were isolated in good yield and high enantiomeric excess (Scheme 5.15).

Studying this reaction with many different substituted epoxides in the presence of just 1 equiv. of indole, it was possible to evaluate the selectivity factor [39], by the use of the Kagan equation [40]. The quantity of indole was therefore adjusted (0.6–0.7 equiv.) to consume one enantiomer of the epoxide, recognized by the [Cr(Salen)] catalyst, leaving the other enantiomer unreacted (Scheme 5.16) [41].

A quite interesting feature of this reaction concerned the absolute configuration of the isolated epoxides, that was opposite with respect to other kinetic resolutions carried out with [Cr(Salen)] in the presence of different nucleophiles (Me$_3$SiN$_3$, thiols). The absolute configuration obtained clearly indicates an S$_N$2-type attack by the indole on the epoxides, activated by the [Cr(Salen)].

It is worth mentioning that the *meso*-stilbene oxide reacted smoothly with a range of indoles in the presence of [Cr(Salen)] giving the corresponding opened products in high enantiomeric excesses (up to 99%).

5.1.3.3 Experiments: Selected Procedures

Preparation of (1R,2R)-trans-2-[(tert-butyldimethylsiloxy)]-3-phenyloxirane (Scheme 5.16) [41]

[Cr(Salen)SbF$_6$], was prepared according to literature procedure [35]. Molecular sieves 4 Å were activated by microwave irradiation (4 × 1 min., 500 W) before use.

A flamed two-necked flask equipped with a magnetic stirring bar was charged with Cr(Salen)SbF$_6$ (30 mg, 0.036 mmol) and activated molecular sieve 4 Å (200 mg). Then TBME (0.4 mL) was added and the resulting solution was stirred under nitrogen at room temperature for 5 min. The solution was cooled to 0 °C and then the epoxide **1j**

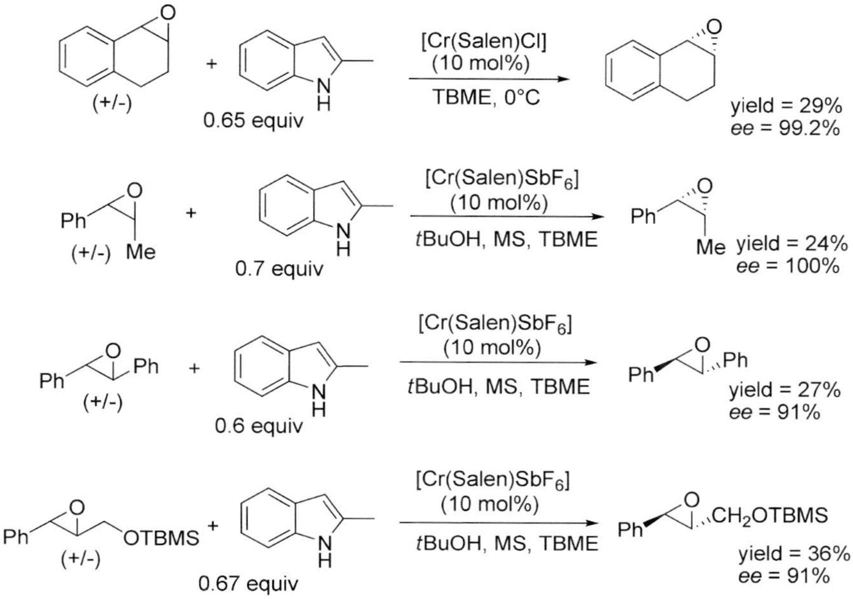

Scheme 5.16 Friedel–Crafts-based kinetic resolution of racemic cis- and trans-epoxides with 2-methylindole catalyzed by [Cr(Salen)] complexes.

(312 mg, 1.18 mmol) and 2-methylindole (93 mg, 0.708 mmol) were added to the reaction mixture. Finally t-BuOH (0.067 mL, 0.708 mmol) was added and the reaction mixture stirred at 0 °C until GC analysis indicated complete conversion of 2-methylindole. After 24 h the reaction goes to completion. The reaction mixture was diluted with Et$_2$O and filtered through celite. The solvent was evaporated and the epoxide was purified by chromatography on silica deactivated with Et$_3$N (eluent, c-Hex:AcOEt:Et$_3$N, 97 : 2 : 1). (112 mg, 36% yield = 96%, on maximum theoretical yield). $[\alpha]_D^{25}$ = 31.5 (c 1.0, CHCl$_3$). ee = 91% by chiral CG analysis. Isotherm 130 °C, (R,R): 46.2 min, (S,S): 47.5 min. ^{1}H NMR (200 MHz, CDCl$_3$) δ = 0.12 (3 H, s); 0.13 (3 H, s); 0.92 (9 H, s); 3.14 (1 H, ddd, J = 1.8, 3.0, 3.9 Hz); 3.21 (1 H, d, J = 1.8 Hz); 3.82 (1 H, dd, J = 3.9, 12.0 Hz); 3.97 (1 H, dd, J = 3.0, 12 Hz); 3.19–3.40 (5 H, m).

5.2
Direct Activation of Alcohols

5.2.1
Introduction

In 2005, the ACS Green Chemistry Institute and the global pharmaceutical corporations developed the ACS GCI Pharmaceutical Roundtable to encourage the integration of green chemistry into the Pharmaceutical industry [42]. The Roundtable has developed a list of key research areas. The substitution of activated alcohols is a frequently used approach for the production of active pharmaceutical ingredients. The OH activation for nucleophilic substitution is considered to be a central issue. The activation is wasteful as it requires additional processing, and produces waste that need to be disposed of. A direct nucleophilic substitution of an alcohol is attractive as it should produce water as the by-product. Unfortunately, hydroxide ion is a poor leaving group, usually requiring activation. Direct activation of benzylic and allylic alcohols may be achieved via an S_N1 reaction.

Recently, a number of interesting methods for the activation of allylic and benzylic alcohols have been demonstrated by the use of a Brønsted acid (4-toluensolfonic acid) [43] or Lewis acids such as $InBr_3$ [44], $In(OTf)_3$ [45], $Bi(OTf)_3$ [46], $FeCl_3$ [47], as well as many others [48]. In recent times there have also been extensive efforts to find suitable conditions for FC alkylations of alcohols or acetate by the use of Lewis acids. Although an extensive mechanistic investigation was not carried out, these reactions can be explained by an S_N1-type mechanism involving carbocationic species as intermediates.

5.2.2
Diastereoselective Reactions

Cationic carbon atoms located in carbenium ion are prostereogenic if the substituents are different. Notoriously, the nucleophilic attack on free carbenium ions occurs stereorandomly, giving the well know racemization of S_N1 reactions. On the other hand, the enantiotopic faces of the plane containing the three substituents of the hybridized Sp2 carbon atom become diastereotopic when the carbocation is chiral (Figure 5.1).

When stereogenic centers are located near to the carbenium ion, a nucleophilic addition leads to the formation of a pair of diastereoisomers that may be formed in nonequal amounts. Chiral benzylic alcohols **26** were readily obtained by Bach *et al.* in optically active form and subjected to conventional chromatographic resolution by reaction with (-)-camphanyl chloride **27** in pyridine (Scheme 5.17) [49].

Saponification of the esters gave the entioenriched *syn* alcohols **26** which were further subjected to FC reactions. Generation of the carbenium ion was carried out with CF_3SO_3H, $BF_3 \cdot Et_2O$ and $HBF_4 \cdot Et_2O$ at rt or $-78\,°C$. As the electron-rich furan

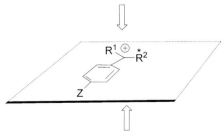

Figure 5.1 The two diastereotopic faces of a chiral benzylic carbocation.

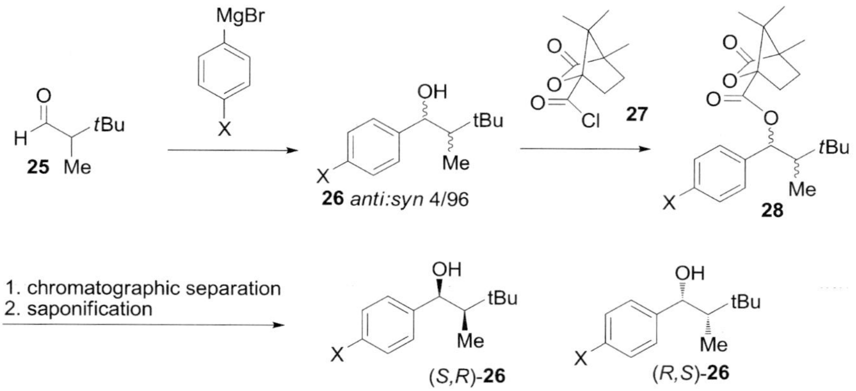

Scheme 5.17 Preparation and separation of enantioenriched precursors of chiral carbocations.

was employed in the optimization of the reaction, the reaction is quite fast and is not sensitive to the reaction temperature. The yields are from moderate to good and, more importantly, a good level of stereoselection (d.r. = 96 : 4 to 97 : 3) was observed. The product distribution was not influenced by the epimeric composition of the starting alcohols, ruling out possible S_N2-type reaction pathways. With other π arene nucleophiles FC diastereoselective reaction takes place giving a good level of stereocontrol.

This diastereoselective FC reaction could be easily interpreted taking into account the Mayr's electrophilicity scale [50]. The tabulated N values for the aromatic rings, used in the FC reactions, range from $N = 2.48$ for the most nucleophilic resorcindimethyl ether, to the less nucleophilic 2-methylthiophene $N = 1.26$. Mayr's scale can be used in a predictive manner, as arenes with $N < 0$ (for example, thiophene) failed to undergo the reaction. The high facial diastereoselective reaction can be explained by a Felkin–Anh transition state, and is correlated with the preferred conformation of the carbocation (Figure 5.2).

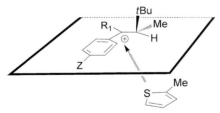

Figure 5.2 Diastereoselective attack of an arene nucleophile on the preferred conformation of a carbocation.

5.2.2.1 BF$_3$-Mediated Reactions

Chung and coworkers have applied the diastereoselective approach described by Bach to solve a synthetic problem connected to the development of a drug containing

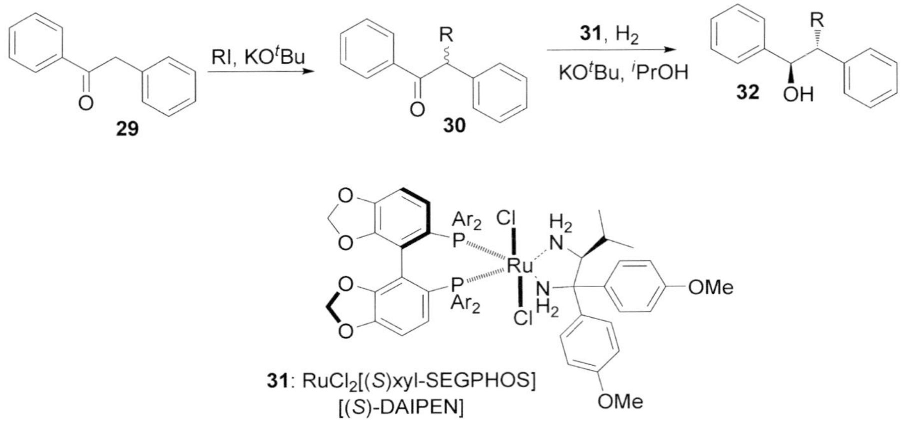

Scheme 5.18 Preparation of enantioenriched phenyl benzyl alcohols 32.

a 1,1,2-triarylalkane fragment [51]. The chiral starting material **32** was prepared by alkylation of desoxybenzoin followed by reduction, or by arylation of ketones, followed by a dynamic kinetic resolution carried out with ruthenium-SEGPHOS **31** (Scheme 5.18) [52]. The successive intermolecular FC reactions were carried out with N-benzensulfonylindole (Table 5.2).

Quite interestingly, in this approach, an *anti* selective nucleophilic addition was observed, in stringent contrast to those reported by Bach in a similar reaction. In these reactions, the phenyl ring is the small group (Figure 5.2). However, as the nucleophile is indole arene–arene interaction between the incoming indole and the phenyl ring could be responsible for the observed selectivity. This hypothesis was also correlated to the scope and limitation of this chemistry. In the case of other arenes examined in the reaction (pyrrole, furan benzofuran, thiophene, benzothiophene, and 1,3-dimethoxybenzene) the electronic properties of the partners significantly alter the observed simple stereoselection, as listed in Table 5.3.

5.2.2.2 Gold(III) Catalysis

In both the above reported methodologies, Brønsted or Lewis acids are used in stoichiometric amounts. The growing demand for a green, sustainable FC methodology requires the use of promoters in catalytic amounts. In this perspective, the recent study reported by Bach is moving in the right direction. The use of gold(III)chloride for catalytic FC alkylation was first reported by Beller [53], and Dyker [54] In these studies, the possibility to use gold salts in FC reactions, involving the generation of carbenium ions, was underlined. Moreover, Bach has shown that gold salts are also efficient catalysts for FC diastereoselective reactions [55]. Compared to other effective catalysts (i.e., $FeCl_3$, $Bi(OTf)_3$) the use of $AuCl_3$ in $MeNO_2$ was superior. In addition, increased yields were found when benzylic acetates **34** were subjected to FC reaction. Various chiral acetates were subjected to an FC Au-catalyzed reaction to produce, cleanly and in excellent yields, the desired products **35** "with high *anti* selectivity" (Scheme 5.19).

Table 5.2 Reaction of N-besyl indole and enantioenriched benzyl alcohols.[a]

A = BF$_3$OEt$_2$ (3 equiv), CH$_2$Cl$_2$, 22 °C
B = TFA (0.2 M), 22 °C

Entry	Benzyl alcohol	Method	anti:syn	Yield (%)[b]
1	32a (Me)	A	1.7 : 1	82
2	32a (Me)	B	2.8 : 1	80
3	32b (Et)	A	4 : 1	85
4	32b (Et)	B	5.3 : 1	80
5	32c (^{n}Pr)	A	5.6 : 1	80
6	32c (^{n}Pr)	B	10 : 1	84

[a] anti:syn ratio was determined by NMR or HPLC.
[b] Isolated yields or determined by HPLC.

Table 5.3 FC alkylation of arene with chiral benzyl alcohol.

Entry	Ar-H	R	Method	anti:syn[a]	Yield (%)[b]
1	benzothiophene-3-H	^{n}Pr	A	89:11	36
2	benzothiophene-3-H	^{n}Pr	B	92:8	56
3	2-methylthiophene-5-H	Me	A	75:25	73
4	2-methylthiophene-5-H	Me	B	—	traces
5	N-Nosyl pyrrole-2-H	^{n}Pr	A	86:14	71
6	N-Nosyl pyrrole-2-H	^{n}Pr	B	90:10	66
7	1,3-dimethoxybenzene	^{n}Pr	A	46:54	51
8	1,3-dimethoxybenzene	^{n}Pr	B	47:53	15

[a] anti:syn ratio was determined by NMR or HPLC.
[b] Isolated yields.

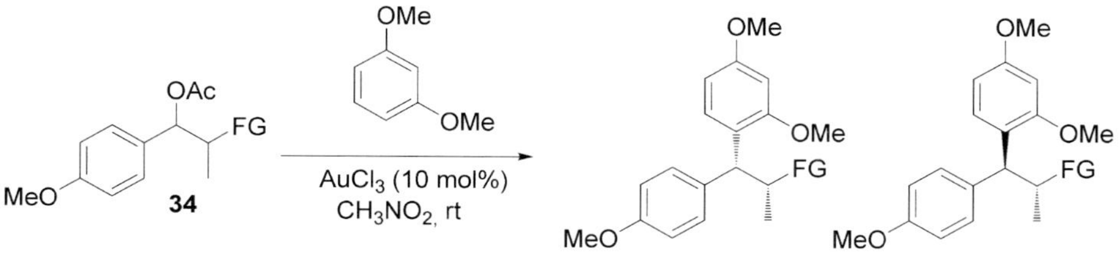

FG = PO(OEt)$_2$, COOMe, CN, Cl, SO$_2$Et, NO$_2$

Scheme 5.19 Diastereoselective gold-catalyzed Friedel–Crafts reaction.

As β-acetoxycarboxylate is a useful building block, accessible in enantioenriched with Evans's chemistry [56], the β-arylation of carboxylate was investigated in the presence of catalytic amounts of AuCl$_3$ (10 mol%). As reported in Table 5.4, compounds **35** were obtained in good yield (63–93%) and in a diastereoisomeric ratio ranging from 85/15 to 97/3.

The observed diastereoselection can be interpreted by a model in which the free carbenium ion adopts a preferred conformation prescribed by the 1,3-allylic strain (Figure 5.2) [57].

5.2.3
Enantioselective Reactions

5.2.3.1 Ruthenium Catalysis

Other types of carbenium ions were investigated in the development of what can be considered a unique example of enantioselective FC reaction with sp3 carbon atoms. Propargyl cation, extensively studied by Olah [58], can be described in the limit form of propargylium and allenylium structures (Figure 5.3a and b). The introduction of a metal in the γ position of a propargyl ion (Figure 5.3c) stabilizes the propargyl ion, enhancing the reactivity of the cation towards FC reactions.

Nishibayashi and coworkers, in a series of interesting papers, have exploited the reactivity of the chalcogeno-bridged diruthenium complexes [59] such as [{Cp*RuCl (η$_2$YR)$_2$} (Cp = η^5-C$_5$Me$_5$; Y = S, Se, Te; R = Me, ^{n}Pr, ^{i}Pr) in the propargylic alkylation of arenes. In all the described catalytic transformations of propargyl alcohols a common intermediate, namely a ruthenium-allenylidene played a key role [60]. The active ruthenium-allenylidene complex is obtained *in situ* by the reaction of propargylic alcohols (**36a**) and the complex **37** in the presence of NH$_4$BF$_4$ (Scheme 5.20a) [61]. Mechanistic information was obtained by performing the reaction with [D]5pyrrole, that gave more than 80% incorporation of deuterium at the C-3 position. The reaction with 2-methylfuran, 2-methylthiophene, N,N-dimethylaniline and indole (CH$_2$Cl$_2$, 60 °C, **37** 5 mol%) gave the corresponding propargylic compound **38a** in 52–94% yields. Aromatics such as benzene, toluene, mesitylene, and methoxy-substituted arenes are not reactive in this reaction.

A valuable intramolecular FC alkylation was also described (Scheme 5.20b). While furan derivative **36b** reacted smoothly providing **38b** in good yield (82%),

Table 5.4 AuCl$_3$-catalyzed diastereoselective alkylation of chiral acetate.

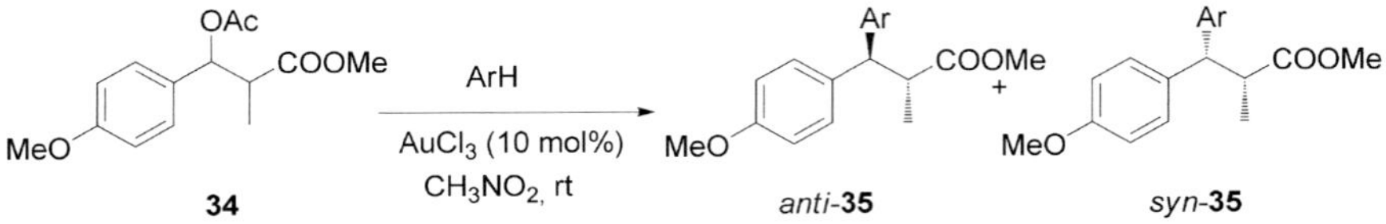

Entry	Ar-H	anti:syn[a]	Yield (%)[b]
1	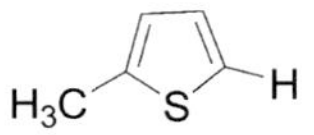 2-methyl-5H-thiophene (H$_3$C–S–H)	90:10	99
2		97:3	90
3	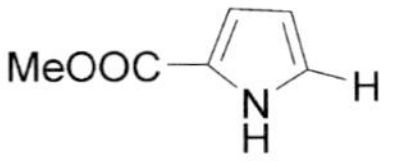 MeOOC-pyrrole-NH	92:8	78
4	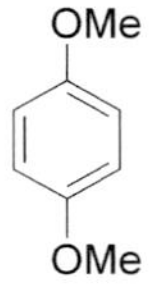 1,4-dimethoxybenzene	91:9	63
5	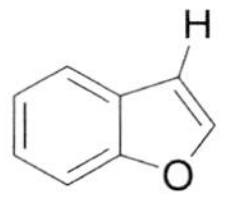 benzofuran-3-H	90:10	99
6	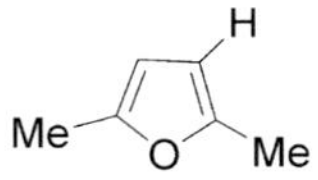 2,5-dimethylfuran	94:6	89
7	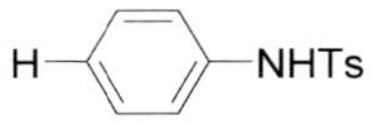 H–C$_6$H$_4$–NHTs	85:15	92

[a] The diastereoisomeric ratio was determined by ^{1}H NMR on the crude product.
[b] Yield of isolated product.

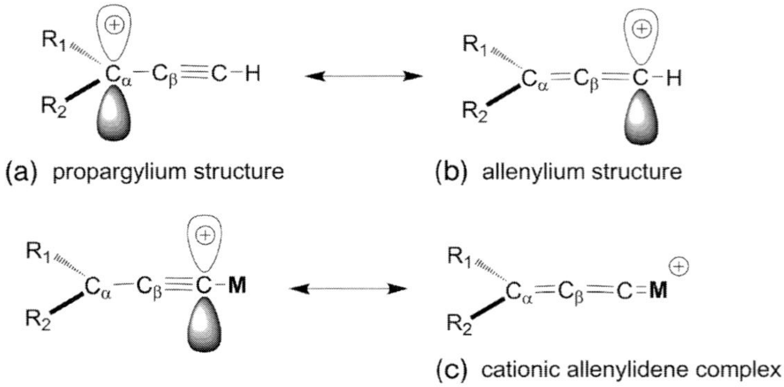

(a) propargylium structure

(b) allenylium structure

(c) cationic allenylidene complex

Figure 5.3 Stabilization of the positive charge of propargyl cations.

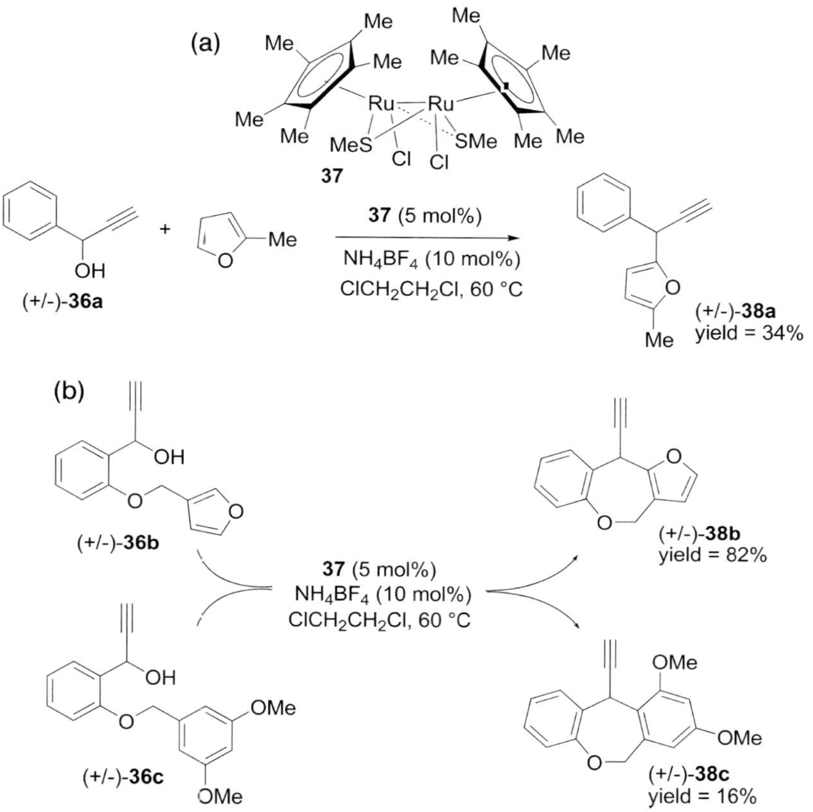

Scheme 5.20 (a) Propargylation of 2-methylfuran;
(b) intramolecular cyclization of propargylic alcohols.

the corresponding derivative **36c**, under better reaction conditions, afforded negligible yield of adduct. Again, explanation for the differences in yields can be ascribed on the difference of the parameter N in the Mayr's scale for the two aromatic compounds.

Enantioselective FC transformations of the ruthenium-allenylidene have been described [62]. In this chemistry the formation of a ruthenium-stabilized propargylic carbocation occurs in a chiral cage, as previously documented in the propargylic enantioselective substitution reaction of acetone (Figure 5.4) [63].

The chiral ruthenium complexes were obtained *in situ* by treating the ruthenium(II) complex [Cp*RuCl]$_4$ with suitable chiral disulfides. At this stage, several disulfides were examined and it turned out that the use of a chiral disulfide, bearing three aromatic phenyl groups, gave the highest enantioselectivity. The formation of a chiral cage and the π–π interactions were crucial in achieving the desired enantioselectivity.

The enantioselective ruthenium-catalyzed propargylation of 2-methylfuran was carried out with several propargylic alcohols, affording the desired compounds in good yield and enantioselectivity (Table 5.5).

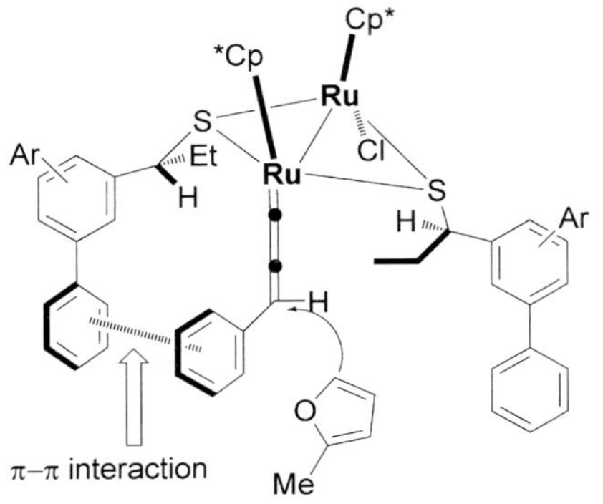

Figure 5.4 Attack of 2-methyl furan on a chiral allenylidene intermediate.

Table 5.5 Ruthenium catalyzed enantioselective propargylation of 2-methylfuran.

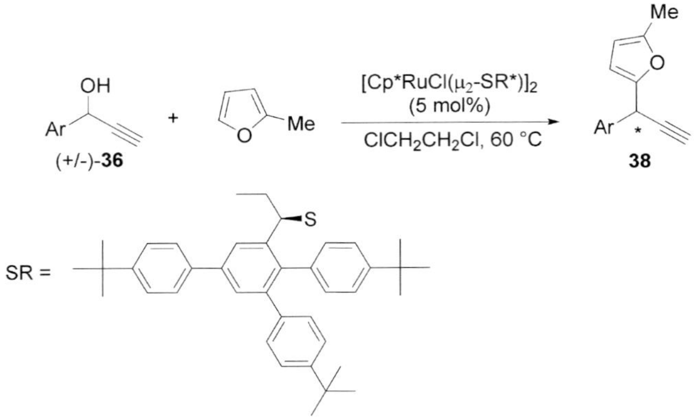

Entry	Ar	Yield (%)[a]	ee (%)[b]
1	Ph	75	77
2	p-MeC$_6$H$_4$	67	82
3	p-ClC$_6$H$_4$	63	68
4	o-PhC$_6$H$_4$	52	94
5	1-naphthyl	59	86
6	2-naphthyl	67	83

[a] Isolated yields after chromatographic purification.
[b] The enantiomeric excesses were determined by HPLC analysis with chiral column.

N,N-Dimethylaniline, less reactive than furan in the propargylation of aromatic compounds, gave the desired adduct in moderate to low yield but in high enantiomeric excess (Figure 5.5).

The enantioselective FC transformation is based on a catalytic cycle in which a starting vinylidene complex, formed by the reaction of the propargylic alcohol with

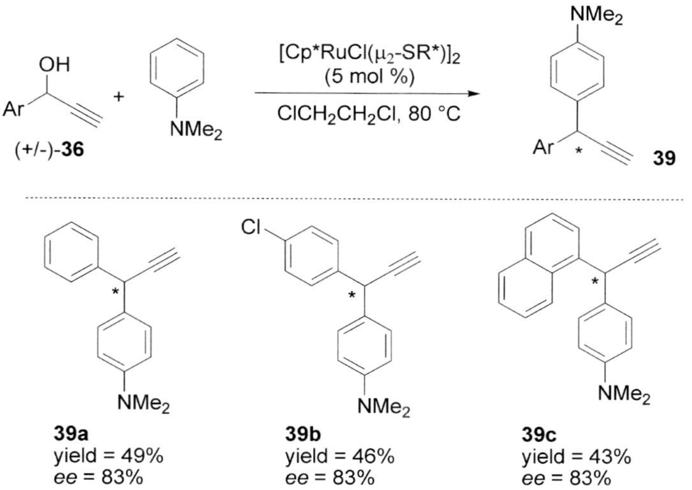

Figure 5.5 Ruthenium catalyzed enantioselective propargylation of N,N-dimethylaniline.

the diruthenium complex, undergoes dehydration in the presence of NH_4BF_4, forming the allenylidene complex.

The same catalyst was also applied to the enantioselective propargylation of indole [64]. Although indole gave the product with low enantiomeric excess, a singular effect of the nature of N-substituents of indole was found. The introduction of a bulky group such as *tert*butyldimethylsilyl or triisopropylsilyl at the nitrogen atom of indole enhanced the enantioselectivity of the process (from 35% *ee* to 82% *ee*). Interestingly, remarkable tolerance toward the presence of substituents at position 5 was recorded. The reaction resulted in high enantiomeric excesses when propargylic alcohols derived from aromatic aldehydes were used. As previously discussed, the results are related to the strong π–π interaction between the phenyl rings of the chiral ligand and the allenylidene ruthenium complex.

5.2.3.2 Brønsted Acid Catalysis

In the scenario of enantioselective FC reactions catalyzed by chiral Brønsted acids, Rueping disclosed an intriguing reaction [65]. The addition of N-methylindole to the α-ketoester **39** catalyzed by the strong Brønsted acid N-triflylphosphoramide **40a** resulted in an interesting adduct **41** that exhibits atropoisomerism (Scheme 5.21).

Application of 5 mol% of a chiral hindered Brønsted acid **40b**, prepared from BINOL, gave the bisindole in an atropoisomeric ratio of 81 : 19. This is a remarkable example of enantioselective catalytic nucleophilic substitution by direct reaction of alcohols. Addition of the second indole molecule is dictated by the formation of a diastereoisomeric ion pair [66].

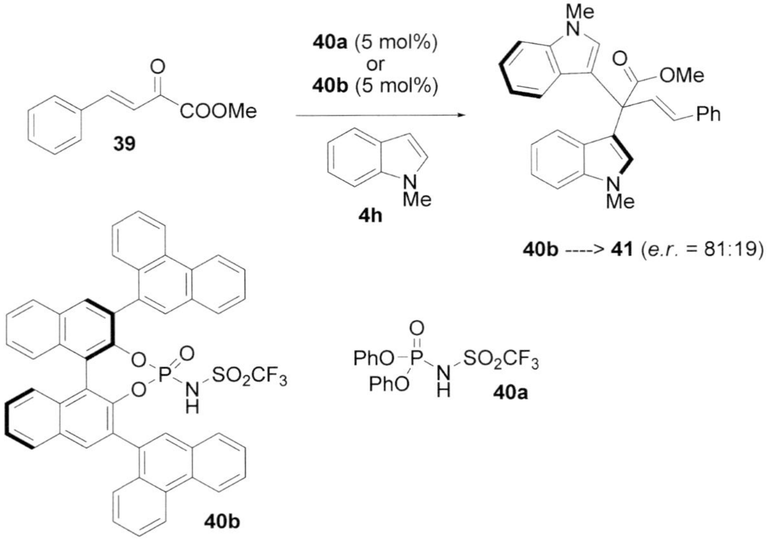

Scheme 5.21 Brønsted acid catalyzed reaction of N-methyl indole with ketoester **39**. Synthesis of atropoisomeric bis-indole **41**.

5.2.3.3 Experiments: Selected Procedures

Ru-catalyzed reaction of furan with propargyl alcohols (Table 5.5) [62]

In a 20 mL round-bottomed flask were placed [Cp*RuCl]$_4$ (5.4 mg, 0.005 mmol) and chiral disulfide (7.6 mg, 0.010 mmol) under N$_2$. Anhydrous THF (1.0 mL) was added, and the mixture was magnetically stirred at room temperature for 12 h. The solvent was evaporated *in vacuo*. Then, NH$_4$BF$_4$ (2.1 mg, 0.020 mmol) and anhydrous ClCH$_2$CH$_2$Cl (5.0 mL) were added under N$_2$, and the mixture was magnetically stirred at room temperature. After the addition of (+/−)-**36** (0.20 mmol, Ar = 1-naphthyl) and 2-methylfuran (0.18 mL, 2.0 mmol), the reaction flask was kept at 60 °C for 3 h. The solvent was concentrated under reduced pressure by an aspirator, and the residue was purified by column chromatography (SiO$_2$) with hexane and ethyl acetate (100 : 1) to give (59 % yield) of **38** isolated as a pale yellow oil. ^{1}H NMR (CDCl$_3$) δ = 2.24 (s, 3H), 2.36 (d, *J* = 2.7 Hz, 1H), 2.37 (s, 3H), 5.15 (d, *J* = 2.7 Hz, 1H), 5.87 (dd, *J* = 1.1, 3.1 Hz, 1H), 6.00 (d, *J* = 3.1 Hz, 1H), 7.11–7.24 (m, 3H), 7.42–7.46 (m, 1H). ^{13}C NMR (CDCl$_3$) δ 13.6, 19.2, 34.1, 71.2, 82.4, 106.2, 107.7, 126.4, 127.4, 128.0, 130.5, 135.7, 136.6, 150.8, 151.9. IR (neat, cm^{-1}) 2120 (C≡C), 3291 (≡C−H). HRMS (EI) Calc. for C$_{15}$H$_{14}$O [M]: 210.1045. Found: 210.1051. The optical purity of **38** (Ar = 1-naphthyl) was determined by HPLC analysis, DAICEL Chiralpak IA, hexane only, flow rate = 0.5 mL min^{-1}, λ = 254 nm, retention time; 11.2 min (minor) and 12.9 min (major), *ee* = 86%.

5.2.3.4 FC Reactions with Chiral Ferrocenyl Compounds

Modern organometallic chemistry began with the discovery of ferrocene and elucidation of its beautiful structure. Ferrocene is widely used as a backbone in asymmetric catalysis, by the incorporation of a stereogenic plane with stereogenic centers, that can act cooperatively in many enantioselective transformations. Materials science uses ferrocene in many applications. Finally, the bio-organometallic chemistry of ferrocene is an expanding field of research. Normally, optically active ferrocene compounds are accessed from Ugi amine [67], able to direct the metallation of ferrocene, followed by nucleophilic displacement of a living group by the amine that occurs with complete retention of configuration [68].

The direct substitution of alcohols would expand the use of ferrocenyl derivatives in organometallic, bioorganic chemistry, and catalysis. However, direct displacement of enantiomerically enriched ferrocenyl alcohols by nucleophiles suffers from extensive racemization, even in the presence of Lewis or Brønsted acids. On the other hand, ferrocenyl carbenium ion represents a special case of carbocation, in which nucleophilic displacements can occur with retention of configuration. Other organometallic chiral alcohols share this property [69], that is determined by the hindrance of one face of the carbocation due to the presence of a metal fragment. Therefore the possibility of using enantioenriched ferrocenyl alcohols in direct FC reactions has been explored.

5.2.3.4.1 Indium-Promoted Nucleophilic Substitution with Ferrocene
Optically active ferrocenyl alcohols are readily prepared by CBS reduction methods or by catalytic hydrogenation with ruthenium complexes [70]. As in the direct substitution of alcohol, water is produced as the by-product; the choice of Lewis acid to generate the carbocation is important. Water-tolerant Lewis acids were selected for the reaction with indium salts, in particular $InBr_3$, showing high efficiency for the reaction. Ferrocenyl alcohol **42a** was prepared and reacted with aromatic substrates such as indole and 1,3-methoxyresorcinol in the presence of indium salts (Scheme 5.22) [71].

In the reaction with indole the 3-arylated product was obtained in high yield, complete selectivity, and complete retention of stereochemical information. Regioisomers arising from attack at the 2-position were not observed. The reaction with

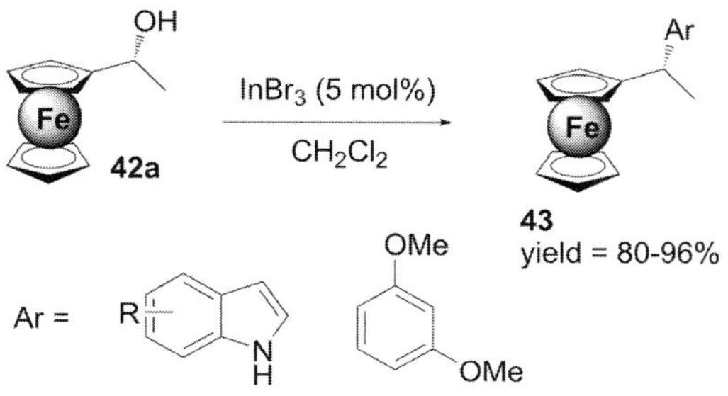

Scheme 5.22 FC reaction promoted by $InBr_3$ with entioenriched ferrocenemethanol **42a**.

pyrrole was somehow complicated by the high reactivity of pyrrole and by the moderate yield obtained (40%). However, in this case also, complete retention of the absolute configuration was observed.

5.2.3.4.2 **"On Water" FC Reactions** The use of water as a reaction medium in organic synthesis has started to attract wide interest [72]. Water is considered cheap, safe and environmentally benign [73]. Nucleophilic reactions between indole and benzylic and allyl halides have been found to be possible in water, despite the fact that carbocations may be generated in the process [74]. These efficient protocols for carbon–carbon bond-forming reactions can be further improved, using alcohols as substrates. As stated before, nucleophilic substitution reactions of ferrocene alcohols are of limited utility due to the poor leaving group ability of the hydroxy group [75]. A breakthrough in the area was reported by Kobayashi [76] in the dehydrative esterification of carboxylic acids [77] and nucleophilic substitution of benzylic alcohols [78] catalyzed by the surfactant Brønsted acid dodecylbenzenesulfonic acid (DBSA). As Kobayashi pointed out, the direct FC reaction between benzydrol and N-methylindole is not promoted in water, even in the presence of strong Brønsted acids such as TFA, TfOH and PTS. In fact, the carbocation derived from benzhydrol is extremely reactive (top of the Mayr's list) with its consequent reaction with the more abundant nucleophile (i.e., water, the solvent) taking place. However, ferrocenyl alcohols can form more stabilized carbocations, lower (at -2.57) [50] in the Mayr list. The lifetime of the carbocation can be enhanced by the introduction of a metal fragment able to stabilize the formation of the positive charge [79]. As the stability of the ferrocenyl cation in water was also proven by the nucleophilic displacements used in the preparation of chiral ferrocene ligands [64], Cozzi explored the reaction of ferrocene alcohols "on water" [80]. The concept of reaction "on water" or "in the presence of water" although sparingly used in Organic Chemistry has recently been firmly established [81]. According to Marcus, the acceleration of organic reaction "on water" is determined by the structure of the water/oil interface in an emulsion, in which protruding OH bonds are available for hydrogen-bonding leading to enhanced chemical rates [82]. We set up "on water" conditions for FC chemistry and surprisingly, the reaction occurred. The reaction conditions developed by Cozzi are quite different from the classical conditions suggested by Sharpless [73]. In fact, the reaction was performed at $80\,°C$ in pure water. In such conditions, the presence of Lewis or Brønsted acids was not necessary for the reaction with indoles and pyrrole (Table 5.6).

Electron-rich and electron-poor indoles could also be employed in the reaction although the conversions observed with indoles substituted by electron-withdrawing groups were inferior (entry 5). 5-Nitroindole was not reactive under the reactions conditions, even with a prolonged reaction time of 36 h. Interestingly, whereas in other studies [68] neat pyrrole was used to effect the reaction (at $-40\,°C$!) in the presence of $InBr_3$, in this case we can efficiently carry out the reaction with 10 equiv. of pyrrole at $80\,°C$. The absence of Lewis acids avoids polymerization of pyrrole or pyrrole derivatives.

Also in this case, the results can be easily explained by Mayr's observations. Many electron-rich π-systems are more nucleophilic than the solvent, such as aqueous

Table 5.6 Reaction of ferrocenyl alcohol **42** with arenes "on water".

42a: R = Me
42b: R = Ph

Entry[a]	Alcohol	Arene	Product	Yield (%)[b]	ee (%)[c]
1	42a	4a	43aa	85	99
2	42a	4h	43ah	68	99
3	42a	4d	43ad	45	99
4	42a	4e	43ae	54	99

Table 5.6 (Continued)

Entry[a]	Alcohol	Arene	Product	Yield (%)[b]	ee (%)[c]
5	42a	4c	43ac	24	99
6	42a	4k	43ak	68	99
7[d]	42a	pyrrole	43a	68	99
8[d]	42b	pyrrole	43b	60	94

[a] All the reactions were carried out in 1.0 mL of water at 80 °C for 24–36 h. 2 equiv. of indole were used in all reactions.
[b] Isolated yield after purification.
[c] The enantiomeric excesses were evaluated by chiral HPLC.
[d] The reaction was performed with 10 equiv. of pyrrole.

acetone or acetonitrile [83]. Therefore, in slightly basic or neutral conditions the intermediate of the S_N1 reaction could be trapped by electron-rich π-systems.

5.2.3.4.3 Experiments: Selected Procedures

General procedure for the FC reaction catalyzed by indium tribromide (Scheme 5.22) [71]

To a reaction flask containing CH_2Cl_2 (2 mL), (1-hydroxyethyl)ferrocene **42c** (0.6 mmol, 138 mg) and the 5-bromoindole (2 equiv., 1.2 mmol) were added, immediately followed by $InBr_3$ (0.1 equiv., 0.06 mmol). The reaction flask was stirred at room temperature until disappearance of the ferrocenyl alcohol (checked by TLC). The reaction was quenched with water and the organic phase separated. The aqueous phase was extracted with diethyl ether (2 × 3 mL). The organic phases were collected, dried over Na_2SO_4 then evaporated under reduced pressure to give a mixture purified by chromatography. (R)-1-[3-(5-bromoindole)ethyl]ferrocene (**43ae**). $R_f = 0.31$ (c-Hex: Et_2O 8:2) $[\alpha]_D = +52.7°$ (c 0.11, $CHCl_3$). Yield = 86%. Dark yellow solid, mp = 163 °C. ^{1}H-NMR ($CDCl_3$, 300 MHz) δ: 1.69 (d, $J = 5.7$ Hz, 3H), 4.06 (bs, 1H), 4.37 (m, 9H), 6.74 (s, 1H), 7.15 (m, 2H), 7.31–7.78 (s, 1H), 7.87 (bs, 1H). ^{13}C-NMR ($CDCl_3$, 75 MHz) δ: 21.9, 30.8, 67.2, 67.7, 68.4, 68.9, 69.7 (5C), 95.7, 102.6, 112.7, 112.9, 122.2, 123.3, 124.9, 128.5, 135.1. ESI MS: 409, 407 (M). IR: 3414, 3093, 1458 cm^{-1}. HPLC analysis AD: isocratic, flow 0.7 mL min^{-1} (hexane: i-PrOH) 85:15. Rt = 10.97 min; Rt = 11.71 min. ee = 86%.

General procedure for the FC reaction "on water" (Table 5.6) [80a]

Ferrocene alcohol **42a** (0.2 mmol, 46 mg) and indole (0.4 mmol, 46 mg) were introduced into a reaction flask under air, and deionized water (pH = 6.52, 3.0 mL) was added. The flask was sealed and stirred under air at 80 °C for 24 h, then allowed to cool at room temperature. Diethyl ether (10 mL) was added and the organic phase was separated, dried over anhydrous $MgSO_4$ and evaporated at reduced pressure to give an oil purified by flash chromatography (c-Hex:Et_2O 9:1). (S)-1-(3-indolethyl)ferrocene **43a**. Yellow solid, mp = 133 °C. Yield = 85%. $R_f = 0.3$ (c-Hex/Et_2O 8/2) $[\alpha]_D = -53°$ (c 0.6, $CHCl_3$). ^{1}H-NMR ($CDCl_3$, 200 MHz) δ: 1.73 (d, $J = 7.0$ Hz, 3H). 4.35–4.05 (m, 10 H), 6.79 (s, 1H), 7.16 (dq, $J = 1.4, 7.2$ Hz, 2H), 7.35 (dd, $J = 1.4, 7.2$ Hz, 1H), 7.69 (d, $J = 7.2$ Hz, 1H), 7.84 (bs, 1H). ^{13}C-NMR ($CDCl_3$, 50 MHz) δ: 21.8. 30.8, 67.0, 67.3, 67.8, 68.6, 69.3 (5C), 95.6, 136.4, 111.3, 119.2, 119.4, 120.8, 121.9, 123.4, 126.6, IR: 3440, 3089, 1618 cm^{-1}. ESI MS: 330 (M + 1), 329 (M), 213. HPLC analysis AD: isocratic, flux 0.8 mL min^{-1} (hexane:^{i}PrOH) 85:15. Rt (major) = 10.41 min; Rt (major) = 11.47 min. ee = 99%.

Acknowledgments

Financial support was provided by the MIUR (Rome), Consorzio C.I.N.M.P.I.S. and University of Bologna.

Abbreviations

Amb-15	Amberlyst-15
Amb-Na	Amberlyst-Na
Bs	besyl
DCE	dichloroethane
EDTA	ethylenediaminetetraacetic acid
FC	Friedel–Crafts
LAs	Lewis acids
OTf	triflate

References

1 Olah, G.A., Krishnamurti, R. and Surya, G.K. (1991) *Friedel–Crafts Alkylations in Comprehensive Organic Synthesis*, Pergamon Press, Oxford, UK, pp. 293–335.

2 Milstein, N. (1968) *Journal of Heterocyclic Chemistry*, **5**, 337–338.

3 Nakajima, T., Nakamoto, Y. and Suga, S. (1975) *Bulletin of the Chemical Society of Japan*, **48**, 960–965.

4 Brauman, J.I. and Solladié-Cavallo, A. (1968) *Journal of the Chemical Society – Chemical Communications*, 1124–1125.

5 (a) Dyker, G. (1997) Polycyclic Ring Systems via Palladacycles as Reactive Intermediates, in *Organic Synthesis via Organometallics* (OSM 5, Heidelberg), ed. G. Helmchen, Vieweg Verlag, Braunschweig, Wiesbaden, pp. 129–137; (b) Pastor, I.M. and Yus, M. (2005) *Current Organic Chemistry*, **9**, 1–29; (c) Pineschi, M. (2006) *Journal of Organic Chemistry*, **71**, 4979–4988 and references therein; (d) Schneider, C. (2006) *Synthesis*, 3919–3944; (e) Nielsen, L.P.C. and Jacobsen, E.N. (2006) *Aziridines and Epoxides in Organic Synthesis* (ed. A.K. Yudin), Wiley-VCH, Weinheim; (f) Crotti, P. and Pineschi, M. (2006) Epoxides in complex molecule synthesis, in *Aziridines and Epoxides in Organic Synthesis* (ed. A.K. Yudin), Wiley-VCH, Weinheim, 271–313.

6 Kotsuki, H., Nishiuchi, M., Kobayashi, S. and Nishizawa, H. (1990) *The Journal of Organic Chemistry*, **55**, 2969–2972.

7 Kotsuki, H., Hayashida, K., Shimanouchi, T. and Nishizawa, H. (1996) *The Journal of Organic Chemistry*, **61**, 984–990.

8 Rickborn, B. (1991) *Acid-catalyzed Rearrangements of Epoxides in Comprehensive Organic Synthesis*, vol. 3 (eds B.M. Trost and I. Fleming), Pergamon, Oxford, pp. 733–771.

9 Fráter, G., Müller, U. and Kraft, P. (1999) *Helvetica Chimica Acta*, **82**, 1656–1665.

10 Linares-Palomino, P.J., Prakash, G.K.S. and Olah, G.A. (2005) *Helvetica Chimica Acta*, **88**, 1221–1225.

11 Prakash, G.K.S., Linares-Palomino, P.J., Glinton, K., Chacko, S., Rasul, G., Mathew, T. and Olah, G.A. (2007) *Synlett*, 1158–1162.

12 Bertolini, F., Crotti, P., Macchia, F. and Pineschi, M. (2006) *Tetrahedron Letters*, **47**, 61–64.

13 (a) Pineschi, M., Bertolini, F., Crotti, P. and Macchia, F. (2006) *Organic Letters*, **8**, 2627–2630; (b) Bertolini, F., Crotti, P., Di Bussolo, V., Macchia, F. and Pineschi, M. (2007) *The Journal of Organic Chemistry*, **72**, 7761–7764.

14 Hu, X.E. (2004) *Tetrahedron*, **60**, 2701–2743.

15 (a) Yadav, J.S., Reddy, B.V.S., Rao, R.S., Veerrndhar, G. and Nagaiah, K. (2001) *Tetrahedron Letters*, **42**, 8067–8070; (b) Sun, X., Sun, W., Fan, R. and Wu, J. (2007) *Advanced Synthesis and Catalysis*, **349**, 2151–2155; (c) Wang, Z., Sun, X. and Wu, J. (2008) *Tetrahedron*, **64**, 5013–5018.

16 Farr, R.N., Alabaster, R.J., Chung, J.Y.L., Craig, B., Edwards, J.S., Gibson, A.W., Ho, G.-J., Humphrey, G.R., Johnson, S.A. and Grabowski, E.J.J. (2003) *Tetrahedron: Asymmetry*, **14**, 3503–3515.

17 Westermaier, M. and Mayr, H. (2008) *Chemistry – A European Journal*, **14**, 1638–1647.

18 Smith, S.G., Fainberg, A.H. and Winstein, S. (1961) *Journal of the American Chemical Society*, **83**, 618–625.

19 Minegishi, S., Kobayashi, S. and Mayr, H. (2004) *Journal of the American Chemical Society*, **126**, 5174–5181.

20 Lakhdar, S., Westermaier, M., Terrier, F., Goumont, R., Boubaker, T., Ofial, A.R. and Mayr, H. (2006) *The Journal of Organic Chemistry*, **71**, 9088–9085.

21 Bandini, M., Cozzi, P.G., Melchiorre, P. and Umani-Ronchi, A. (2002) *The Journal of Organic Chemistry*, **67**, 5386–5389.

22 Heitbaum, M., Glorius, F. and Escher, I. (2006) *Angewandte Chemie – International Edition*, **45**, 4732–7362.

23 (a) Blossey, E.C., Turner, L.M. and Neckers, D.C. (1973) *Tetrahedron Letters*, **14**, 1823–1826; (b) Blossey, E.C., Turner, L.M.

and Neckers, D.C. (1975) *The Journal of Organic Chemistry*, **40**, 959–960.
24. (a) Clark, J.H. (2002) *Accounts of Chemical Research*, **79**, 1–7; (b) Corma, A. and García, H. (2003) *Chemical Reviews*, **103**, 4307–4366.
25. Bandini, M., Fagioli, M., Melloni, A. and Umani-Ronchi, A. (2004) *Advanced Synthesis and Catalysis*, **346**, 573–578.
26. (a) Ranu, B.C. (2000) *Journal of Organic Chemistry*, **65**, 2347–2356; (b) Chauhan, K.K. and Frost, C.G. (2000) *Journal of the Chemical Society – Perkin Transactions 1*, 3015–3019.
27. Shi, Z. and He, C. (2004) *Journal of the American Chemical Society*, **126**, 5964–5965.
28. Liu, Y., Li, X., Lin, G., Xiang, Z., Xiang, J., Zhao, M., Chen, J. and Yang, Z. (2008) *The Journal of Organic Chemistry*, **73**, 4625–4629.
29. (a) Kharasch, M.S. and Isbell, H.S. (1931) *Journal of the American Chemical Society*, **53**, 3053–3059; (b) Fuchita, Y., Utsunomiya, Y. and Yasutake, M. (2001) *Journal of The Chemical Society – Dalton Transactions*, 2330–2334.
30. (a) Shi, Z. and He, C. (2004) *Journal of the American Chemical Society*, **126**, 13596–13597; (b) Reetz, M. T. and Sommer, K. (2003) *European Journal of Chemistry*, 3485–3496.
31. Gumpta, K.C. and Sutar, A.K. (2008) *Coordination Chemistry Reviews*, **252**, 1420–1450.
32. Darensbourg, D.J. (2007) *Chemical Reviews*, **107**, 2388–2410.
33. Bandini, M., Cozzi, P.G. and Umani-Ronchi, A. (2002) *Chemical Communications*, 919–927.
34. Hansen, K.B., Leighton, J.L. and Jacobsen, E.N. (1996) *Journal of the American Chemical Society*, **118**, 10924–10925.
35. (a) Jacobsen, E.N. and Schaus, S.E. (1999) PCT Int Appl, WO 9936375; (b) Ruck, R. and Jacobsen, E.N. (2002) *Journal of the American Chemical Society*, **124**, 2882–2883.
36. (a) Schaus, S.E., Larrow, J.F. and Jacobsen, E.N. (1997) *The Journal of Organic Chemistry*, **62**, 4197–41999; (b) Larrow, J.F., Schaus, S.E. and Jacobsen, E.N. (1996) *Journal of the American Chemical Society*, **118**, 7420–7421; (c) Lebel, H. and Jacobsen, E.N. (1999) *Tetrahedron Letters*, **40**, 7303–7306.
37. Nielsen, L.P., Stevenson, C.P., Blackmond, D.G. and Jacobsen, E.N. (2004) *Journal of the American Chemical Society*, **126**, 1360–1362.
38. Brandes, B.D. and Jacobsen, E.N. (2001) *Synlett*, 1013–1015.
39. Keith, J.M., Larrow, J.F. and Jacobsen, E.N. (2001) *Advanced Synthesis and Catalysis*, **343**, 5–26.
40. The lower limit of the selectivity factor s for the epoxides was estimated using the equation: $s = \ln[1 - c(1 + ee)]/\ln[1 - c(1 - ee)]$ in the reaction with 2-methylindole (ee is the percent enantiomeric excess of the ring-opened products; yields of opened products were used for conversion c.
41. Bandini, M., Cozzi, P.G., Melchiorre, P. and Umani-Ronchi, A. (2003) *Angewandte Chemie – International Edition*, **43**, 84–87.
42. Constable, D.J.C., Dunn, P.J., Hayler, J.D., Humphrey, G.R., Leazer, J.L. Jr, Linderman, R.J., Lorenz, K., Manley, J., Pearlman, B.A., Wells, A., Zaks, A. and Zhang, T.Y. (2007) *Green Chemistry*, **9**, 411–420.
43. (a) Sanz, R., Miguel, D., Martinez, A., Alvarez-Gutiérrez, J.M. and Rodrìguez, F. (2007) *Organic Letters*, **9**, 2027–2030; (b) Sanz, R., Miguel, D., Martinez, A., Alvarez-Gutiérrez, J.M. and Rodrìguez, F. (2007) *Organic Letters*, **9**, 727–730.
44. (a) Saito, T., Nishimoto, Y., Yasuda, M. and Baba, A. (2006) *The Journal of Organic Chemistry*, **71**, 8516–8522; (b) Yasuda, M., Somyo, T. and Baba, A. (2006) *Angewandte Chemie – International Edition*, **45**, 793–796.
45. Huang, W., Wang, J., Shen, Q. and Zhou, X. (2007) *Tetrahedron Letters*, **48**, 3969–3973.
46. Rueping, M., Nachtsheim, B.J. and Kuenkel, A. (2007) *Organic Letters*, **9**, 825–828.

47 Kischel, J., Mertins, K., Michalik, D., Zapf, A. and Beller, M. (2007) *Advanced Synthesis and Catalysis*, **349**, 865–870.

48 For recent reports of catalyzed C–C, C–N and C–O bond formation through direct substitution of allylic or propargylic alcohols with nucleophiles, see: (a) Terrasson, V., Marque, S., Campagne, J.-M. and Prim, D. (2006) *Advanced Synthesis and Catalysis*, **348**, 2063–2067; (b) Zhan, Z., Wang, W., Yang, R., Yu, J., Li, J. and Liu, H. (2006) *Chemical Communications*, 3352–3354; (c) Nishibayashi, Y., Shinoda, A., Miyake, Y., Matsuzawa, H. and Sato, M. (2006) *Angewandte Chemie – International Edition*, **45**, 4835–4839; (d) Motokura, K., Fujita, N., Mori, K., Mizugaki, T., Ebitani, K. and Kaneda, K. (2006) *Angewandte Chemie – International Edition*, **45**, 2605–2609; (e) Qin, H., Yamagiwa, N., Matsunaga, S. and Shibasaki, M. (2007) *Angewandte Chemie – International Edition*, **46**, 409–413; (f) Sanz, R., Martínez, A., Miguel, D., Álvarez-Guitiérrez, J.M. and Rodríguez, F. (2006) *Advanced Synthesis and Catalysis*, **348**, 1841–1845; (g) Noji, M., Konno, Y. and Ishii, K. (2007) *The Journal of Organic Chemistry*, **72**, 5161–5167; (m) Jana, U., Biswas, S. and Maiti, S. (2007) *Tetrahedron Letters*, **48**, 4065–4069; (n) Le Bras, J. and Muzart, J. (2007) *Tetrahedron*, **63**, 7942–7948.

49 Muhlthau, F., Stadler, D., Goeppert, A., Olah, G.A., Surya Prakash, G.K. and Bach, T. (2006) *Journal of the American Chemical Society*, **128**, 9668–9675.

50 Mayr, H., and Ofial, Ar.R. (2003) *Accounts of Chemical Research*, **36**, 66–77.

51 Chung, J.Y.L., Manchero, D., Dormer, P.G., Variankaval, N., Ball, R.G. and Tsou, N.N. (2008) *Organic Letters*, **10**, 3037–3040.

52 Wan, X., Sun, Yanhui, Luo, Y., Li, D. and Zhang, Z. (2005) *The Journal of Organic Chemistry*, **70**, 1070–1072 and references therein.

53 Mertins, K., Iovel, I., Kischel, J., Zapf, A. and Beller, M. (2006) *Advanced Synthesis and Catalysis*, **348**, 691–695.

54 Liu, J., Muth, E., Flörke, U., Henkel, G., Merz, K., Sauvageau, J., Schwake, E. and Dyker, G. (2006) *Advanced Synthesis and Catalysis*, **348**, 456–462.

55 Rubenbauer, P. and Bach, T. (2008) *Advanced Synthesis and Catalysis*, **350**, 1125–1130.

56 Evans, D.A., Takacs, J.M., McGee, L.R., Ennis, M.D., Mathre, D.J. and Bartroli, J. (1981) *Pure and Applied Chemistry*, **53**, 1109–1127.

57 Hoffmann, R.W. (1989) *Chemical Reviews*, **89**, 1841–1860.

58 (a) Krishnamurthy, V.V., Prakash, G.K.S., Iyer, P.S. and Olah, G.A. (1986) *Journal of the American Chemical Society*, **108**, 1575–1579 and references therein; (b) Olah, G.A., Krishnamurthy, V.V. and Prakash, G.K.S. (1990) *The Journal of Organic Chemistry*, **55**, 6061–6062 and references therein.

59 The thiolate-bridged diruthenium complexes represent a formidable synthetic tool for activation and successive reactions of various terminal alkynes: (a) Nishibayashi, Y., Yamanashi, M., Wakiji, I. and Hidai, M. (2000) *Angewandte Chemie – International Edition*, **39**, 2909–2911 and references therein; (b) Nishibayashi, Y., Imajima, H., Onodera, G., Hidai, M. and Uemura, S. (2004) *Organometallics*, **23**, 26–30; (c) Nishibayashi, Y., Imajima, H., Onodera, G., Inada, Y., Hidai, M. and Uemura, S. (2004) *Organometallics*, **23**, 5100–5103.

60 Nishibayashi, Y., Milton, M.D., Inada, Y., Yoshikawa, M., Wakiji, I., Hidai, M. and Uemura, S. (2005) *Chemistry – A European Journal*, **11**, 1433–1451.

61 Inada, Y., Yoshikawa, M., Milton, M.D., Nishibayashi, Y. and Uemura, S. (2006) *Journal of Organic Chemistry*, **71**, 881–890.

62 Matsuzawa, H., Miyake, Y. and Nishibayashi, Y. (2007) *Angewandte Chemie – International Edition*, **46**, 6488–6491.

63 Inada, H.Y., Nishibayashi, Y. and Uemura, S. (2005) *Angewandte Chemie – International Edition*, **44**, 7715–7717.

64 Matsuzawa, H., Kanao, K., Miyake, Y. and Nishibayashi, Y. (2007) *Organic Letters*, **9**, 5561–5564.

65 Rueping, M., Nachtsheim, B.J., Moreth, S.A. and Bolte, M. (2008) *Angewandte Chemie – International Edition*, **47**, 593–593.

66 Mayer, S. and List, B. (2006) *Angewandte Chemie – International Edition*, **45**, 4193–4195.

67 (a) Marquarding, D., Klusacek, H., Gokel, G.W., Hoffmann, P. and Ugi, I. (1970) *Journal of the American Chemical Society*, **92**, 5389–5393; (b) Gokel, G., Marquarding, D. and Ugi, I. (1972) *The Journal of Organic Chemistry*, **37**, 3052–3058.

68 Togni, A., Breutel, C., Schnyder, A., Spindler, F., Landert, H. and Taijani, A. (1994) *Journal of the American Chemical Society*, **116**, 4062–4066.

69 Gleiter, R., Bleiholder, C. and Rominger, F. (2007) *Organometallics*, **26**, 4850–4859.

70 (a) CBS reduction: Corey, E.J., Bakshi, R.K. and Shibata, S. (1987) *Journal of the American Chemical Society*, **109**, 5551–5552; (b) Kloetzin, R.J., Lotz, M. and Knochel, P. (2003) *Tetrahedron:Asymmetry*, **14**, 255–264; Ruthenium catalyzed hydrogenation: Lam, W.-S., Kok, S.H.L., Au-Yeung, T.T.-L., Wu, J., Cheung, H.-Y., Lam, F.-L., Yeung, C.-H. and Chan, A.S.C. (2006) *Advanced Synthesis and Catalysis*, **348**, 370–374.

71 Vicennati, P. and Cozzi, P.G. (2007) *Journal of Organic Chemistry*, **72**, 2248–2253.

72 (a) Li, C.-J. and Chan, T.-H. (1997) *Organic Reaction in Aqueous Media*, Wiley, New York; (b) (1998) *Organic Synthesis in Water*, (ed P.A. Grieco), Blackie Academic and Professional, London; (c) Li, C.-J. (2005) *Chemical Reviews*, **105**, 3095–3166; (d) Li, C.-J. and Chen, L. (2006) *Chemical Society Reviews*, **35**, 68–82; (e) (2007) *Organic Reaction in Water* (ed. U.M. Lindström), Blackwell Publishing, Oxford.

73 Narayan, S., Muldoon, J., Finn, M.G., Fokin, V.V., Kolb, H.C. and Sharpless, K.B. (2005) *Angewandte Chemie – International Edition*, **44**, 3275–3279.

74 Hofmann, M., Hampel, N., Kanzian, T. and Mayr, H. (2004) *Angewandte Chemie – International Edition*, **43**, 5402–5405.

75 (a) Gullickson, G.C. and Lewis, D.E. (2003) *Australian Journal of Chemistry*, **56**, 385–388; (b) Bisaro, F., Prestat, G., Vitale, M. and Poli, G. (2002) *Synlett*, 1823–1826; (c) Coote, S.J., Davies, S.G., Middlemiss, D. and Naylor, A. (1989) *Tetrahedron Letters*, **30**, 3581–3584.

76 (a) Kobayashi, S. and Ogawa, C. (2006) *Chemistry - A European Journal*, **12**, 5954–5960; (b) for reaction "on water" see: Pirrung, M.C. and Das Sarma, K. (2004) *Journal of the American Chemical Society*, **126**, 444–446; (c) Price, B.K. and Tour, J. (2006) *Journal of the American Chemical Society*, **128**, 12899–12904; (d) Gonzàles-Cruz, D., Tejedor, D., de Armas, P. and Garcìa-Telaldo, F. (2007) *Chemistry – A European Journal*, **13**, 4823–4832.

77 Manabe, K., Iimura, S., Sun, X.-M. and Kobayashi, S. (2002) *Journal of the American Chemical Society*, **124**, 11971–11978.

78 Shirakawa, S. and Kobayashi, S. (2007) *Organic Letters*, **9**, 311–314.

79 (a) Richard, P., Amyes, T.L. and Toteva, M.M. (2001) *Accounts of Chemical Research*, **34**, 981–988; The 4,4′-bis(dimethylamino) diphenylmethane carbocation has a half-life of 10–20 s; see: (b) McCLelland, R.A., Kanagasabapathy, V.M., Banait, N.S. and Steenken, S.J. (1989) *Journal of the American Chemical Society*, **111**, 3966–3972.

80 (a) Cozzi, P.G. and Zoli, L. (2007) *Green Chemistry*, **9**, 1292–1295; (b) Cozzi, P.G. and Zoli, L. (2008) *Angewandte Chemie – International Edition*, **47**, 4162–4166.

81 (a) Brogan, A.P., Dickerson, T.J. and Janda, K.D. (2006) *Angewandte Chemie – International Edition*, **45**, 8100–8102; (b) Hayashi, Y. (2006) *Angewandte Chemie – International Edition*, **45**, 8103–8104; for other interesting reactions "on water", see: (c) Pirrung, M.C. and Das Sarma, K. (2004) *Journal of the American Chemical Society*, **126**, 444–445;

(d) Price, B.K. and Tour, J. (2006) *Journal of the American Chemical Society*, **128**, 12899–12904; (e) Gonzàles-Cruz, D., Tejedor, D., de Armas, P. and Garcìa-Telaldo, F. (2007) *Chemistry – A European Journal*, **13**, 4823–4832; (f) Zhang, H.-B., Liu, L., Wang, Y.-J.C.D. and Li, C.-J. (2006) *Journal of Organic Chemistry*, **71**, 869–873.

82 (a) Jung, Y. and Marcus, R.A. (2007) *Journal of the American Chemical Society*, **129**, 5492–5502; (b) Vilotijevic, I. and Jamison, T. (2007) *Science*, **317**, 1189–1192.

83 Hofmann, M., Hampel, N., Kanzian, T. and Mayr, H. (2004) *Angewandte Chemie – International Edition*, **43**, 5402–5405.

6
Unactivated Alkenes
Ross A. Widenhoefer

Summary

Since 2000, a number of transition metal-catalyzed methods for the enantioselective alkylation of arenes with unactivated (non-Michael acceptor) C=C bonds have been reported. Early examples include the enantioselective ortho-alkylation of benzamide with norbornene catalyzed by iridium(I) bis(phosphine) complexes and the atropselective alkylation of heterobiaryl compounds with ethylene catalyzed by rhodium(I) mono(phosphine) complexes. More recently, a number of effective protocols have been developed for the intramolecular alkylation of arenes with electronically unactivated C=C bonds. For example, Rh(I) mono (phosphoramidite) complexes catalyze the imine-directed 5-endo hydroarylation of 2-propenylarenes. Although high catalyst loadings are required, Rh(I)-catalyzed enantioselective hydroarylation has been employed in complex molecular environments, providing functionalized arenes with up to 96% *ee*, and has been applied to the synthesis of (+)-lithospermic acid and biologically-active dihydropyrroloindoles. Mechanistically distinct from these Ir(I)- and Rh(I)-catalyzed methods is the 6- and 7-exo hydroarylation of 2-alkenylindoles catalyzed by cationic platinum(II) bis(phosphine) complexes and the 6- and 7-exo hydroarylation of 2-allenylindoles catalyzed by cationic bis(gold) phosphine complexes. These latter transformations likely occur via an outer-sphere mechanism involving attack of indole on a metal-complexed C=C bond and provide tricyclic indole derivatives with up to 92% *ee*.

6.1
Introduction

The transition metal-catalyzed addition of the C–H bond of an arene across a C=C double bond of an unactivated alkene (hydroarylation) has attracted considerable attention as an atom-economical approach to the functionalization of arenes with potential application to both the large-scale alkylation of simple arenes and the

Catalytic Asymmetric Friedel–Crafts Alkylations. Edited by M. Bandini and A. Umani-Ronchi
Copyright © 2009 WILEY-VCH Verlag GmbH & Co. KGaA, Weinheim
ISBN: 978-3-527-32380-7

synthesis of complex molecules [1]. A number of catalysts have been identified for the hydroarylation of arenes including Ir(III) acetonate [2], Ru(II) phosphine [3], and Ru(II) hydridotris(pyrazolyl)borate [4], cationic Ru(III) [5], rhodium(I) phosphine [6–9], neutral [10, 11] and cationic platinum(II) phosphine [12], platinum (IV) [13], Bi (III) [14], Fe(III) [15], and cationic zirconocene complexes [16]. Although a number of these complexes catalyze the hydroarylation of unfunctionalized arenes [2, 4, 5, 13], these transformations typically require forcing conditions and a large excess of the arene. Rather, selective alkene hydroarylation is restricted to arenes that are predisposed to formation of a metal–carbene intermediate, arenes that possess a suitable directing group, or electron-rich heteroaromatic compounds.

In comparison to the numerous advances made in the catalytic hydroarylation of alkenes with achiral catalysts, effective processes for the enantioselective hydroarylation of alkenes remain scarce. Although a number of highly enantioselective Lewis or Brønsted acid-catalyzed procedures for the hydroarylation of electron-deficient alkenes with electron-rich arenes have been developed [17], these procedures are not effective for electronically unactivated alkenes. Rather, the catalytic enantioselective hydroarylation of electronically unactivated (non-Michael acceptor) alkenes was unknown prior to 2000 and early examples were of modest efficiency. However, there has been steady development in the area of enantioselective alkene hydroarylation over the past five years and a number of effective processes have been disclosed. This account summarizes the progress made in the enantioselective hydroarylation of electronically-unactivated (non-Michael acceptor) alkenes including simple alkenes, strained alkenes, and allenes. Chiral ligands employed in the enantioselective hydroarylation of alkenes are collected in Figure 6.1.

6.2
Early Studies

6.2.1
Enantioselective Hydroarylation of Norbornene

In the course of developing more effective catalysts for the enantioselective intermolecular hydroamination of norbornene, Togni identified several electron-rich iridium(I) bis(phosphine) cyclopentadienyl complexes that catalyzed the enantioselective ortho-hydroarylation of benzamide with norbornene. Although rather inefficient and low yielding, these represent some of the first examples of the enantioselective hydroarylation of non-Michael acceptor alkenes [18]. For example, heating a toluene solution of norbornene and benzamide with a catalytic amount of CpIr(R)-biphep (Cp = η^5-C$_5$H$_5$; 1 mol%) at 100 °C for 72 h led to the isolation of 2-(exo-norbornyl)benzamide (**1**) in 16% yield with 87% *ee* (Table 6.1, entry 1). The yield of **1** increased to 35% when 10 mol% of CpIr(R)-biphep was employed without reduction in enantioselectivity (Table 6.1, entry 2). Reaction of norbornene and

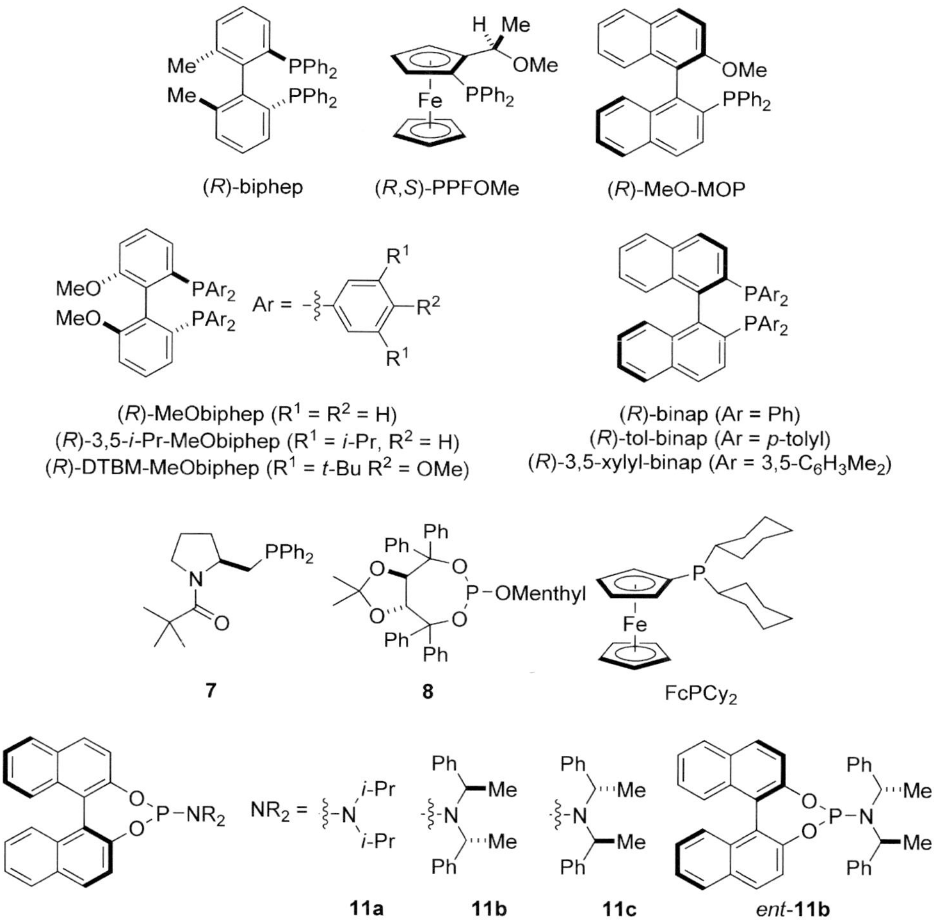

Figure 6.1 Phosphine ligands employed in the enantioselective hydroarylation of alkenes.

benzamide catalyzed by CpIr(R)-MeObiphep led to isolation of **1** in diminished yield, but with 94% *ee* (Table 6.1, entry 3). In comparison, reaction of norbornene and benzamide catalyzed by the iridium bis(phosphine) chloride dimer {Ir[(R)-biphep]Cl}$_2$ led to formation of a mixture of **1** and the hydroamination product N-phenyl-*exo*-aminobornane (**2**) in low yield with 85% *ee* for the hydroarylation product (Table 6.1, entry 4). Employment of related iridium chloride dimers as catalysts for the reaction of benzamide with norbornene led predominantly to hydroamination (Table 6.1, entries 5 and 6). The mechanism of iridium-catalyzed hydroarylation remains unclear as the 18-electron iridium cyclopentadienyl complexes presumably must undergo either ligand dissociation or, more likely, η^3–η^3 ring slippage to generate the requisite vacant site for C−H bond activation.

Table 6.1 Iridium(I)-catalyzed enantioselective hydroarylation and hydroamination of norbornene with benzamide.

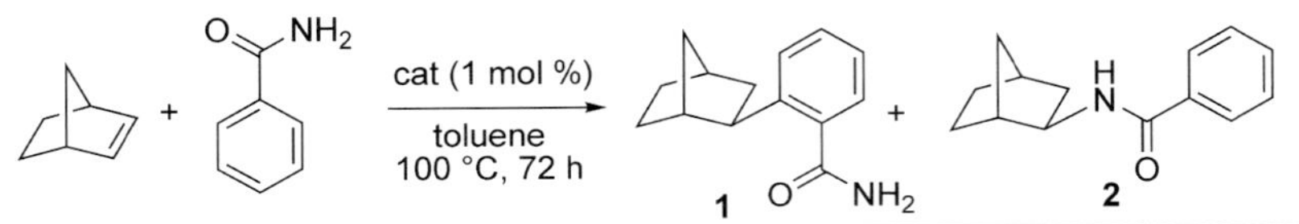

Entry	Cat	Yield 1 (%)	Yield 2 (%)	ee1 (%)	ee2 (%)
1	CpIr(R)-biphep	16	—	87	—
2[a]	CpIr(R)-biphep	35	—	87	—
3	CpIr(R)-MeObiphep	12	—	94	—
4	{Ir[(R)-biphep]Cl}$_2$	10	5	85	65
5	{Ir(R)-MeO-biphep]Cl}$_2$	—	50	—	79
6	{Ir(S)-binap]Cl}$_2$	4	30	81	70

[a]10 mol% catalyst employed.

6.2.2
Atropselective Alkylation of Biaryls

Murai and coworkers have reported a rhodium(I)-catalyzed protocol for the atropselective alkylation of 2-(1-naphthyl)-3-methylpyridine (**3**) and 1-(1-naphthyl)isoquinoline (**4**) with ethylene [19]. Efficient atropselective alkylation requires facile interconversion of the atropisomers of the parent biaryls but slow interconversion of the corresponding α-alkylated derivatives, conditions that are potentially met via rhodium-catalyzed alkylation (Scheme 6.1). For example, reaction of **3** with ethylene (100 psi) catalyzed by a mixture of [RhCl(coe)$_2$]$_2$ (coe = cyclooctene; 5 mol%) and the chiral ferrocenylphophine ligand (R,S)-PPFOMe (30 mol%) in toluene at 120 °C for 20 h led to isolation of ortho-alkylated product **5** in 37% yield with 49% *ee* (Equation 6.1). Under similar conditions, reaction of **4** with ethylene

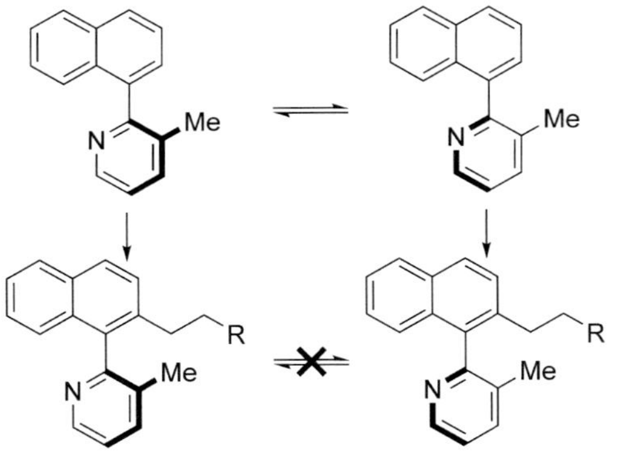

Scheme 6.1 Atropisomerization of heterobiaryl compounds.

$$
\begin{array}{c}
\text{[RhCl(coe)}_2\text{]}_2 \text{ (5 mol \%)} \\
\text{(}R,S\text{)-PPFOMe (30 mol \%)} \\
\hline
\text{toluene, 120 °C, 20 h} \\
\text{37\%, 49\% ee}
\end{array}
\quad (6.1)
$$

3 + H$_2$C=CH$_2$ (100 psi) → 5

formed **6** in 33% yield with 22% *ee* (Equation 6.2). Rhodium-catalyzed hydroarylation of **3** and **4** employing (*R*)-binap or (*R*)-MeO-MOP gave either no conversion or no asymmetric induction, respectively.

$$
\begin{array}{c}
\text{[RhCl(coe)}_2\text{]}_2 \text{ (5 mol \%)} \\
\text{(}R,S\text{)-PPFOMe (30 mol \%)} \\
\hline
\text{toluene, 120 °C, 20 h} \\
\text{33\%, 22\% ee}
\end{array}
\quad (6.2)
$$

4 + H$_2$C=CH$_2$ (100 psi) → 6

6.3
Rh(I)-Catalyzed Enantioselective Hydroarylation of Iminoarenes

6.3.1
Catalyst Control

6.3.1.1 Alkylation of Aromatic Ketimines

Bergman and Ellman have developed an effective protocol for the enantioselective imine-directed 5-endo hydroarylation of 2-propenylarenes catalyzed by a mixture of the rhodium chloride dimer [RhCl(coe)$_2$]$_2$ and a chiral, monodentate phosphine [20]. Aromatic ketimines were targeted as substrates for enantioselective alkylation owing to the effectiveness of the ketimine moiety as a directing group in the corresponding achiral transformations [7]. The directing group presumably facilitates C—H bond activation to form the key rhodium arenyl hydride species **1** (Scheme 6.2) [8]. Hydrometallation of the pendant alkene of **1** followed by C—C reductive elimination releases the alkylated arene. Chiral monodentate phosphines were targeted as supporting ligands for enantioselective alkylation because chelating bis(phosphines) performed poorly in the corresponding achiral transformations, presumably due to the intermediacy of catalytically active rhodium mono(phosphine) complexes (Scheme 6.2).

A number of chiral mono(phosphine) ligands, including (*R*)-MeO-MOP, (*R,S*)-PPFOMe, 2-diphenylphosphino methylpyrrolidine **7**, and taddol-derived phosphite **8** were screened for their effectiveness in the rhodium-catalyzed conversion of aromatic ketimine **9** to heterobicycle **10** (Table 6.2 and Figure 6.1). From this ligand screen, the (*S*)-BINOL-derived phosphoramidites **11a**, **11b**, and **11c** were identified as efficient ligands for the enantioselective conversion of **9** to **10** (Table 6.2, entries 5–7) [20]. For example, reaction of **9** with a catalytic 1:3 mixture of [RhCl(coe)$_2$]$_2$

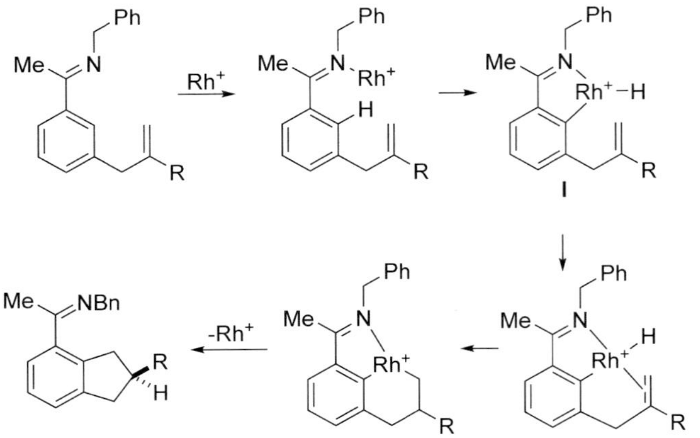

Scheme 6.2 Proposed mechanism of Rh-catalyzed hydroarylation of alkenes with aromatic ketimines.

and **11b** (M : L = 1 : 1.5) at 125 °C for 2 h formed **10** in quantitative yield with 88% *ee* (Table 6.2, entry 6). The configuration of the dialkyl amino group of ligand **11b** had no significant effect on the enantioselectivity of hydroarylation and reaction of **9** catalyzed by a 1 : 3 mixture of [RhCl(coe)$_2$]$_2$ and **11c** gave **10** in 87% *ee* (Table 6.2, entry 7). Ligand to metal ratios of 1 : 1–1.5 : 1 were optimal; employment of higher L/M ratios led to loss of activity without loss of enantioselectivity, consistent with the intermediacy of rhodium mono(phosphine) complexes.

The rhodium complexes generated from [RhCl(coe)$_2$]$_2$ and phosphoramidite ligands **11** were considerably more active hydroarylation catalysts than were rhodium PPh$_3$ complexes. The enhanced reactivity of the rhodium phosphoramidite complexes is presumably due to the reduced σ-donation and enhanced π-acceptor ability of the phosphoramidite ligands compared to a triaryl phosphine,

Table 6.2 Effect of phosphine ligand on the enantioselective hydroarylation of ketimine **9**.

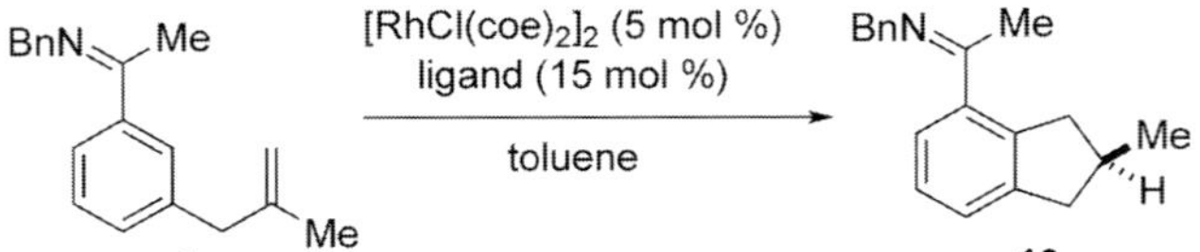

Entry	Ligand	Temp. (°C)	Time (h)	Yield (%)	*ee* (%)
1	(*R*)-MeO-MOP	125	20	48	8
2	(*R*,*S*)-PPFOMe	75	6	99	35
3	7	100	6	56	23
4	8	75	20	91	38
5	11a	125	<2	100	83
6	11b	125	<2	100	88
7	11c	125	<2	99	87

Table 6.3 Enantioselective intramolecular hydroarylation of aromatic ketimines catalyzed by a mixture of [RhCl(coe)$_2$]$_2$ (5 mol %) and ligand (10–15 mol%) in toluene-d_8.

Entry	Alkenyl arene	Ligand	Temp. (°C)	Time (h)	Product	Yield (%)	ee (%)
1	**9** (R = Me)	**11b**	50	9		94	95
2	R = Ph	**11b**	75	3		96	90
3	R = SiMe$_2$Ph	**11a**	125	0.3		91	70
4	**12**	**11b**	125	1		90	70
5	**13**	**11b**	rt	23		95	96

which facilitates turnover-limiting C–C reductive elimination [20]. The high activity of the [RhCl(coe)$_2$]$_2$/**11b** catalyst system allowed enantioselective hydroarylation to be conducted at significantly lower temperatures than with the corresponding achiral catalyst system, leading to enhanced asymmetric induction. For example, cyclization of **9** catalyzed by a 1 : 3 mixture of [RhCl(coe)$_2$]$_2$ and **11b** at 50 °C led to formation of **10** in 94% yield with 95% ee (Table 6.3, entry 1). Rhodium-catalyzed enantioselective hydroarylation was of limited scope, but was effective for 1-imino-3-(2-propenyl) arenes that possessed a phenyl or silyl group at the internal position of the alkene (Table 6.3, entries 2 and 3). Rhodium-catalyzed enantioselective hydroarylation was also effective for cyclization of the indole derivative **12** and the vinyl ether **13**. The latter substrate was particularly effective, cyclizing at 23 °C with 96% ee (Table 6.3, entries 4 and 5).

6.3.1.2 Alkylation of Aromatic Aldimines

Bergman and Ellman have applied rhodium-catalyzed enantioselective hydroarylation to the synthesis of dihydropyrroloindole **14**, a protein kinase C inhibitor selective for isozyme β [21]. The key hurdle in the proposed synthesis (Scheme 6.3) was the

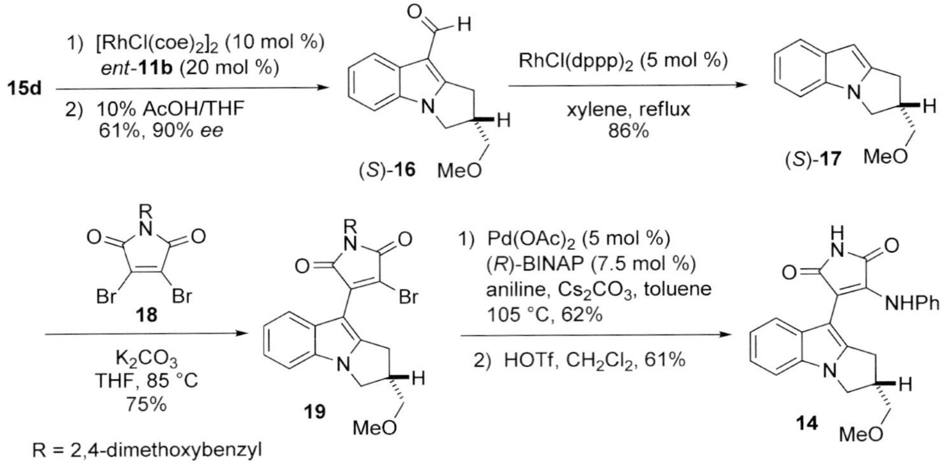

Scheme 6.3 Synthesis of PKC inhibitor 14 via Rh-catalyzed enantioselective hydroarylation.

R = 2,4-dimethoxybenzyl

necessity to perform rhodium-catalyzed enantioselective cyclization on an aromatic aldimine, substrates that cyclized poorly employing the conditions used to good effect for the cyclization of aromatic ketimines. For example, reaction of iminoindole **15a** with a catalytic 1:2 mixture of [RhCl(coe)$_2$]$_2$ and **11b** followed by hydrolysis formed the desired tricyclic aldehyde **16** in only 31% yield, albeit with good enantioselectivity (Table 6.4, entry 1).

Brookhart has demonstrated that rhodium(I)-catalyzed hydroarylation of aromatic ketones is facilitated by electron-withdrawing groups on the metal-bound ketone, presumably owing to the lower barrier for rate-limiting C—C reductive elimina-

Table 6.4 Effect of N-directing group on the enantioselective hydroarylation of aromatic aldimines **15**.

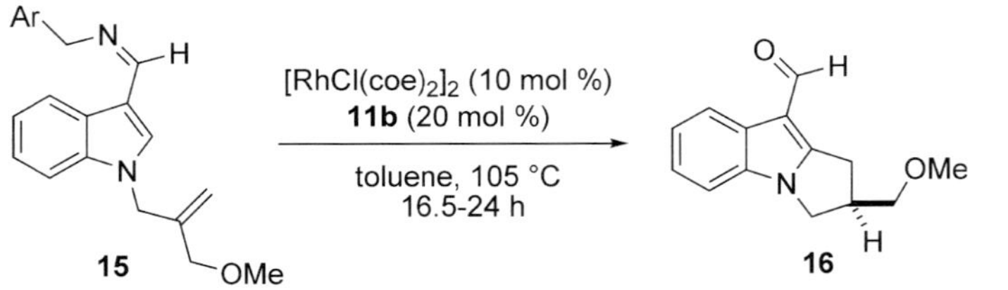

Entry	Substrate	Ar	Yield (%)	ee (%)
1	15a	Ph	31	84
2	15b	4-C$_6$H$_4$OMe	28	80
3	15c	4-C$_6$H$_4$CF$_3$	44	86
4	15d	3,5-C$_6$H$_3$(CF$_3$)$_2$	57	86
5[a]	15d	3,5-C$_6$H$_3$(CF$_3$)$_2$	65	90
6	15e	4-C$_6$H$_4$NO$_2$	22	56

[a] Reaction run at 90 °C.

tion [6]. Guided by this observation, Bergman and Ellman targeted aryl aldimines that possessed electron-deficient N-arylmethyl groups for enantioselective hydroarylation [21]. A screen of substituted aryl aldimines identified the N–CH$_2$C$_6$H$_3$(CF$_3$)$_2$ substituted iminoindole **15d** as a more effective substrate for rhodium-catalyzed hydroarylation (Table 6.4). For example, treatment of **15d** with a catalytic 1 : 2 mixture of [RhCl(coe)$_2$]$_2$ and **11b** at 90 °C followed by hydrolysis formed tricyclic aldehyde **16** in 65% yield with 90% ee (Table 6.4, entry 5).

The optimized directing group strategy was then applied to the synthesis of PKC inhibitor **14** [21]. Towards this objective, reaction of **15d** with a catalytic mixture of [RhCl(coe)$_2$]$_2$ and the enantiomer of ligand **11b** (*ent*-**11b**) followed by hydrolysis led to isolation of aldehyde (*S*)-**16** in 61% yield with 90% ee (Scheme 6.3). Rhodium-catalyzed decarbonylation of (*S*)-**16** gave indole **17** in 86% yield. Direct alkylation of **17** with dibromomaleimide **18** required 5.5 days but provided **19** in 75% yield. Palladium-catalyzed cross-coupling of **19** with aniline followed by deprotection with triflic acid gave **14** in 38% yield. From **15d**, the synthesis required eight linear steps and provided **14** in 15% overall yield (Scheme 6.3).

6.3.2
Directing Group Control

6.3.2.1 Synthesis of (+)-Lithospermic Acid

Effective rhodium-catalyzed enantioselective hydroarylation of aromatic imines was restricted to alkenes that lacked terminal substitution. For substrates not amenable to enantioselective hydroarylation employing the [RhCl(coe)$_2$]$_2$/**11** catalyst system, Bergman and Ellman developed a complementary protocol for enantioselective hydroarylation that employed achiral rhodium catalysts in conjunction with chiral imine directing groups [22]. Towards the goal of substrate-based chirality transfer, the chiral, non-racemic aldimine derivatives **20a–20g** were treated with a catalytic mixture of [RhCl(coe)$_2$]$_2$ and the achiral ferrocenyl phosphine FcPCy$_2$ in toluene-d_8 at 75 °C (Table 6.5). From this substrate screen, aminoindane derivative **20f** was identified as a suitable substrate for enantioselective intramolecular hydroarylation, providing **21f** in 63% yield and 76% ee (Table 6.5, entry 6).

The optimized directing group protocol was then applied to the synthesis of (+)-lithospermic acid [(+)-**22**], a potential HIV medication that inhibits HIV-1 integrase [22]. Treatment of aldimine **20f** with a catalytic mixture of [RhCl(coe)$_2$]$_2$ (10 mol %) and FcPCy$_2$ (30 mol %) followed by *in situ* hydrolysis gave *cis*-dihydrobenzofuran **23** in 56% yield and 99% ee after a single recrystallization (Scheme 6.4). Knoevenagel condensation of **23** with malonic acid occurred with concomitant epimerization of carbon C20 and formed cinnamic acid **24** as a 10 : 1 mixture of diastereomers favoring the desired anti-diastereomer. Esterification of **24** with alcohol **25** formed triester **26** in 80% yield. Global deprotection of **26** proved troublesome but was achieved via hydrolysis of the C10 ester group with Me$_3$SnOH followed by treatment with TMSI-quinoline adduct to form (+)-**22** in 33% yield. The overall conversion of **20f** to (+)-**22** was achieved in nine linear steps with 5.9% overall yield.

Table 6.5 Effect of imine directing group on the enantioselective cyclization of aromatic aldimines **20**.

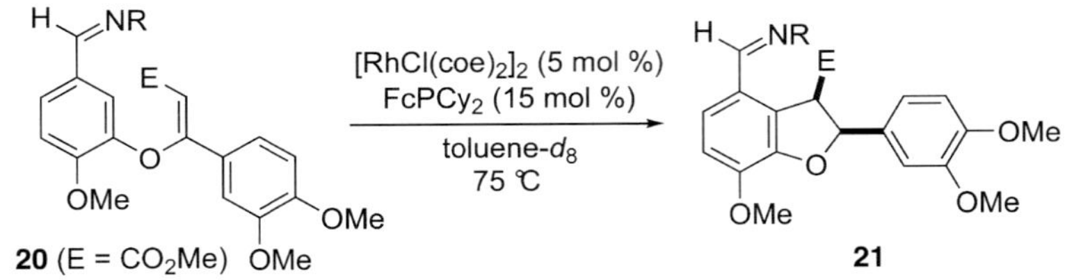

Entry	NR	Derivative	Time (h)	Yield (%)	ee (%)
1	NCH(Ph)Me	a	6	53	48
2	NCH(Ph)Et	b	4	68	50
3	NCH(1-naphthyl)Me	c	22	<10	18
4	NCH(t-Bu)Me	d	24	0	—
5	NCH(Ph)CO$_2$t-Bu	e	4	37	50
6	(1-indanyl-N)	f	6	63	76
7	(1,2,3,4-tetrahydronaphthyl-N)	g	20	3	—

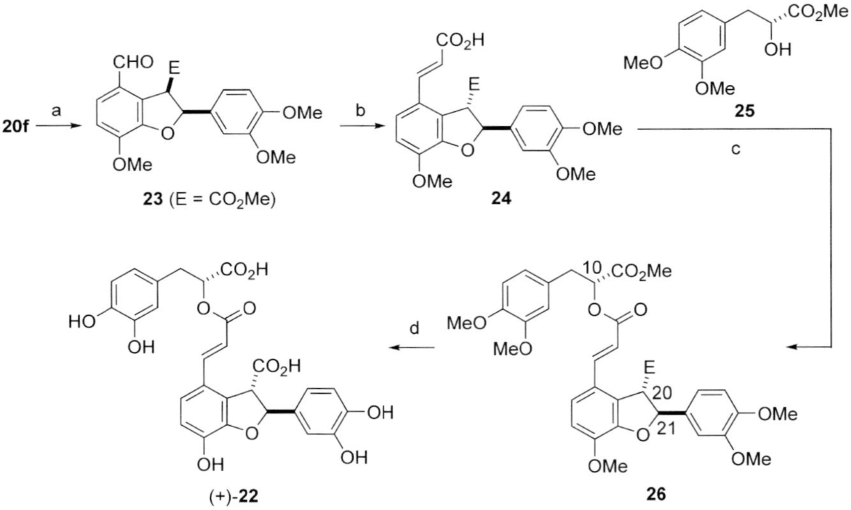

a) i) [RhCl(coe)$_2$]$_2$ (10 mol %), FcPCy$_2$ (30 mol %), toluene, 75 °C; ii) HCl/H$_2$O; 56%, 99% ee (after recrystallization). b) malonic acid, pyridine, 100 °C, 85%, 10:1 dr. c) **25**, EDC, DMAP, CHCl$_3$, 80%. d) i) MeSnOH, DCE, ii) TMSI-quinoline, 100 °C, 33%.

Scheme 6.4 Synthesis of (+)-lithospermic acid [(+)-**22**] via Rh-catalyzed enantioselective hydroarylation.

6.3.2.2 Optimization and Scope

Given the superiority of the aminoindane group of **20f** relative to other chiral N-directing groups in the rhodium-catalyzed enantioselective hydroarylation of aromatic aldimines **20** (Table 6.5), further effort was directed towards the application of 7-substituted aminoindanes as chiral N-directing groups for catalytic hydroarylation [23]. Indeed, rhodium-catalyzed hydroarylation of the 7-methyl, 7-phenyl, and 7-fluoroaminoindane derivatives **20h–20j** in each case provided higher enantioselectivies than were obtained with **20f** (Table 6.6). In a notable example, treatment of the 7-fluoroaminoindane derivative **20j** with a mixture of [RhCl(coe)$_2$]$_2$ (10 mol %) and FcPCy$_2$ (30 mol %) at 60 °C for 36 h followed by hydrolysis gave aldehyde **23** in 70% yield with 90% ee (Table 6.6, entry 4).

The scope of the rhodium-catalyzed cyclization of chiral aldimine derivates with respect to substitution about the alkenyl C=C bond was also investigated. Aryl aldimines **27** that possessed an unadorned phenyl group at the internal alkene carbon atom cyclized to form bicycles **28** in good yield but with diminished enantioselectivity (Table 6.6, entries 5 and 6). Similarly, substrates that possessed a single methyl group at the internal alkene carbon atom underwent rhodium-catalyzed cyclization in quantitative yield but with modest enantioselectivity (Table 6.7, entries 1 and 2). Aromatic aldimines that possessed a single methyl group at the terminal alkene carbon atom cyclized in poor yield with modest enantioselectivity (Table 6.7, entries 3 and 4).

Table 6.6 Effect of aminoindane and arene substitution on the enantioselective hydroarylation of aromatic aldimines.

Entry	R	Ar	Substrate	Product	Yield (%)	ee (%)
1	H	3,4-C$_6$H$_3$(OMe)$_2$	20f	23	76	76
2	Me	3,4-C$_6$H$_3$(OMe)$_2$	20h	23	62	80
3	Ph	3,4-C$_6$H$_3$(OMe)$_2$	20i	23	90	83
4[a]	F	3,4-C$_6$H$_3$(OMe)$_2$	20j	23	70	90
5	H	Ph	27a	28	80	65
6	F	Ph	27b	28	81	70

[a]Reaction run at 60 °C for 36 h.

Table 6.7 Effect of alkene substitution on the rhodium-catalyzed enantioselective alkylation of aromatic aldimines.

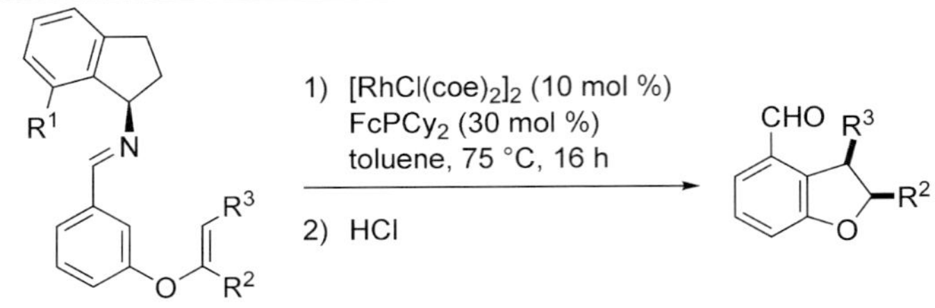

Entry	R¹	R²	R³	Yield (%)	ee (%)
1	H	Me	H	>99	58
2	F	Me	H	>99	68
3	H	H	Me	34	55
4	F	H	Me	50	58

6.4
Pt(II)-Catalyzed Enantioselective Hydroarylation of Alkenylindoles

Han and Widenhoefer have reported an effective platinum(II)-catalyzed protocol for the enantioselective intramolecular hydroarylation of unactivated C=C bonds with indoles [24]. In contrast to the rhodium-catalyzed protocols described in Section 6.3, platinum-catalyzed hydroarylation likely occurs via an outer-sphere mechanism involving initial coordination of the alkene to the electrophilic Pt(II) center, nucleophilic attack of the pendant indole on the platinum-complexed alkene, and protonolysis of the resulting Pt–C σ-bond (Scheme 6.5) [10]. For this reason, cationic platinum bis(phosphine) complexes were targeted as catalysts for enantioselective hydroarylation. The cationic nature of these complexes was anticipated to both enhance the electrophilicity of the platinum catalyst and provide an open coordination site required for alkene complexation.

In an initial experiment, reaction of 2-alkenylindole **29** with a 1:1 mixture of [(S)-binap]PtCl₂ and AgOTf at 60 °C in dioxane for 22 h led to isolation of tetrahydrocarbazole **30** in good yield, but with no significant asymmetric induction (Table 6.8, entry 1). A screen of chiral bis(phosphine) ligands revealed that the enantioselectivity of platinum-catalyzed hydroarylation increased with the increasing

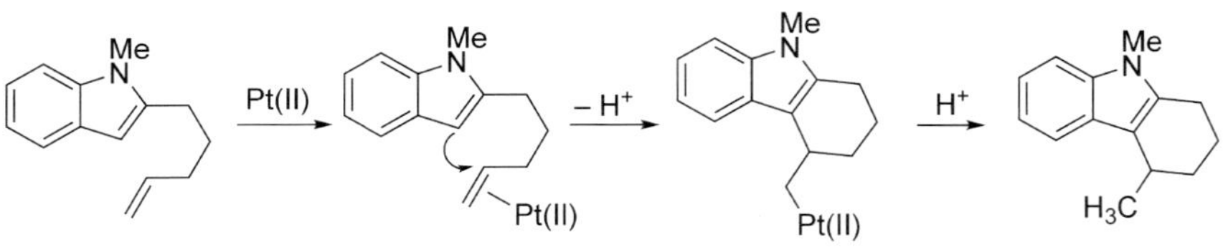

Scheme 6.5 Proposed mechanism for Pt(II)-catalyzed hydroarylation of alkenylindoles.

Table 6.8 Effect of ligand and solvent on the platinum-catalyzed intramolecular hydroarylation of 29.

Entry	P-P	Solvent	Yield (%)	ee (%)
1	(S)-binap	dioxane	82	5
2	(S)-tol-binap	dioxane	83	21
3	(R)-3,5-xylyl-binap	dioxane	81	41
4	(R)-3,5-i-Pr-MeObiphep	dioxane	86	63
5	(S)-DTBM-MeObiphep	dioxane	88	87
6	(S)-DTBM-MeObiphep	THF	92	89
7	(S)-DTBM-MeObiphep	acetone	83	90
8	(S)-DTBM-MeObiphep	MeOH	93	90

steric bulk of the P-bound aryl groups. In particular, employment of sterically-hindered (S)-DTBM-MeObiphep as a supporting ligand for the intramolecular hydroarylation of 29 formed 30 in 88% yield with 87% ee (Table 6.8, entry 5). Employment of polar, weakly-coordinating solvents led to improvement in both the yield and enantioselectivity of hydroarylation (Table 6.8, entries 6–8) and, in an optimized protocol, reaction of 29 with a catalytic 1:1 mixture of [(S)-DTBM-MeObiphep]PtCl$_2$ and AgOTf (10 mol %) at 60 °C in methanol for 20 h led to isolation of tetrahydrocarbazole 30 in 93% yield with 90% ee (Table 6.8, entry 8).

As was the case with the rhodium-catalyzed enantioselective hydroarylation of aromatic ketimines, the scope of platinum-catalyzed enantioselective hydroarylation was modest. N-allyl and N-benzyl-2-(4-pentenyl)indoles and 5-fluoro- and 5-methoxy-2-(4-pentenyl)indoles underwent enantioselective hydroarylation in good yield with 87–88% ee (Table 6.9, entries 1–4) [10–24]. While the protocol tolerated substitution at the C1 or C2 position of the 4-pentyl chain (Table 6.9, entries 5–8), 2-(4-pentenyl)indoles that lacked substitution along the alkyl chain or that possessed substitution at the internal or terminal alkene carbon atom failed to undergo efficient enantioselective cyclization. 2-(5-Hexenyl)indoles underwent platinum-catalyzed enantioselective hydroarylation to form the corresponding 7-membered heterocyclic product in modest yield and enantioselectivity (Table 6.9, entry 9).

Han and Widenhoefer also evaluated the effect of a homoallylic sp^3 stereocenter on the efficiency and stereoselectivity of platinum-catalyzed hydroarylation. In particular, platinum-catalyzed hydroarylation of alkenylindole 31, which possessed a single carbomethoxy group at the C2 position of the 4-pentenyl chain, led to isolation of a 9:1 mixture of cis-32 with ≤5% ee and trans-32 with ≥90% ee in 94% combined yield

Table 6.9 Enantioselective hydroarylation of 2-alkenylindoles catalyzed by a mixture of [(S)-DTBM-MeObiphep]PtCl$_2$ and AgOTf in methanol at 60 °C (E = CO$_2$Me).

Entry	Alkenyl indole	Tricyclic product	Yield (%)	ee (%)
1	R^1 = H, R^2 = allyl		91	87
2	R^1 = H, R^2 = Bn		93	88
3	R^1 = F, R^2 = Me		93	88
4	R^1 = OMe, R^2 = Me		96	87
5	R = CO$_2$t-Bu		56	86
6	R = CH$_2$OH		82	73
7	R = Me		90	68
8			93	71
9			69	74

(Equation 6.3). The high diastereoselectivity of the transformation coupled with the negligible enantioselectivity of the major diastereomer (cis-**32**) point to predominant substrate control of stereochemical induction.

$$
\mathbf{31}\ (E = CO_2Me) \xrightarrow[\substack{\text{MeOH, 60 °C, 64 h} \\ 94\%\ (cis:trans = 9:1)}]{\substack{[(S)\text{-DTBM-MeObiphep}]PtCl_2\ (10\ \text{mol \%}) \\ AgOTf\ (10\ \text{mol \%})}} cis\text{-}\mathbf{32}\ (5\%\ ee) + trans\text{-}\mathbf{32}\ (90\%\ ee)
$$

(6.3)

6.5
Au(I)-Catalyzed Enantioselective Hydroarylation of Allenylindoles

One of the key limitations of platinum-catalyzed enantioselective hydroarylation was the failure of the protocol to tolerate substitution at the alkenyl carbon atoms and, as a result, the procedure was restricted to the synthesis of polycyclic indoles that possessed an exocyclic methyl group on the newly formed saturated ring. As a means of circumventing this limitation and allowing the enantioselective synthesis of more highly functionalized polycyclic indole derivatives, Liu and Widenhoefer developed an effective protocol for the enantioselective intramolecular hydroarylation of 2-allenylindoles catalyzed by mixtures of chiral, non-racemic bis(gold) phosphine complexes of the form (P–P)(AuCl)$_2$ (P–P = bidentate phosphine) and silver salts [25]. Development of enantioselective gold(I)-catalyzed allene hydroarylation was guided both by the identification of the achiral mono(gold) phosphine complex Au[P(t-Bu)$_2$(o-biphenyl)]Cl as a highly active precatalyst for the intramolecular hydroarylation of 2-allenylindoles [26], and by the effective application of bis(gold) phosphine complexes as catalysts for the enantioselective functionalization of allenes [27].

In an initial experiment, reaction of the 2-(4,5-hexadienyl)indole **33** with a 1:2 mixture of [(S)-binap](AuCl)$_2$ and AgOTf in dioxane for 3.5 h led to complete conversion to form **34** as the exclusive product with 32% ee (Table 6.10, entry 1). Subsequent optimization highlighted the importance of the ligand, counterion, and solvent in the enantioselective hydroarylation of allenylindoles. Substitution of

Table 6.10 Effect of ligand, silver salt, and solvent on the gold-catalyzed enantioselective hydroarylation of **33**.

Entry	P-P	X	Solvent	Time (h)	ee (%)
1	(S)-binap	OTf	dioxane	3.5	32
2	(S)-3,5-xylyl-binap	OTf	dioxane	2	42
3	(S)-MeObiphep	OTf	dioxane	3.5	48
4	(S)-MeObiphep	ClO$_4$	dioxane	24	55
5	(S)-MeObiphep	SbF$_6$	dioxane	24	44
6	(S)-MeObiphep	BF$_4$	dioxane	24	55
7	(S)-DTBM-MeObiphep	BF$_4$	dioxane	24	76
8	(S)-DTBM-MeObiphep	BF$_4$	THF	47	70
9	(S)-DTBM-MeObiphep	BF$_4$	MeCN	6	63
10	(S)-DTBM-MeObiphep	BF$_4$	MeOH	3	67
11	(S)-DTBM-MeObiphep	BF$_4$	toluene	2	81

[(S)-binap](AuCl)$_2$ with [(S)-MeObiphep](AuCl)$_2$ increased the enantioselectivity of hydroarylation from 32 to 48% ee and subsequent substitution of AgOTf with AgBF$_4$ further increased the enantioselectivity to 55% ee (Table 6.10, entries 3 and 6). Employing AgBF$_4$ as a co-catalyst, substitution of [(S)-MeObiphep](AuCl)$_2$ with the sterically hindered [(S)-DTBM-MeObiphep](AuCl)$_2$ and substitution of dioxane with toluene increased the enantioselectivity of the conversion of **33** to **34** to 81% ee (Table 6.10, entries 7 and 11).

The enantioselectivity of the conversion of **33** to **34** catalyzed by [(S)-DTBM-MeObiphep](AuCl)$_2$ and AgBF$_4$ increased with decreasing temperature and, in a preparative-scale reaction, treatment of **33** with a catalytic 1 : 2 mixture of [(S)-DTBM-MeObiphep](AuCl)$_2$ and AgBF$_4$ in toluene at −10 °C for 22 h led to isolation of **34** in 88% yield with 92% ee (Table 6.11, entry 1) [25]. In addition to **33**, both the 5-methoxy- and 5-fluoro-2-(4,5-hexadienyl)indoles underwent enantioselective hydroarylation, albeit with diminished enantioselectivity, and the protocol tolerated unprotected hydroxymethyl groups at the C2 position of the 4,5-hexadienyl chain (Table 6.11, entries 2–4). Gold(I)-catalyzed enantioselective hydroarylation was also effective for the hydroarylation of a 2-(γ-allenyl)indole that possessed a terminally disubstituted allenyl group and for the 7-exo cyclization of a 2-(5,6-heptadienyl)indole, in both cases providing product with >90% ee (Table 6.11, entries 5 and 6).

Enantioselective hydroarylation of 2-γ-allenylindoles that possessed either an sp^3 stereocenter along the alkyl tether or an axially chiral allenyl group occurred with high diastereoselectivity and low enantioselectivity in a substrate-controlled process [25]. For example, gold(I)-catalyzed cyclization of the 2-(4,5-undecadienyl) indole **35** led to isolation of a 28:1 mixture of (E)-**36** and (Z)-**36** in 80% combined yield with 9% and ∼60% ee, respectively (Equation 6.4a). Similarly, gold-catalyzed cyclization of the 2-(2-methoxycarbonyl-4,5-hexadienyl)indole **37** led to isolation of a 9 : 1 mixture of cis-**38** and trans-**38** in 80% combined yield with ∼15% and ∼60% ee, respectively (Equation 6.4b).

Table 6.11 Enantioselective hydroarylation of 2-allenylindoles catalyzed by a mixture of [(S)-DTBM-MeObiphep](AuCl)$_2$ (2.5 mol %) and AgBF$_4$ (5 mol %) in toluene at $-10\,^\circ$C for 18–24 h (E = CO$_2$Me).

Entry	Allenyl indole	Product	Yield (%)	ee (%)
1	R = H		88	92
2	R = F		90	75
3	R = OMe		85	78
4	R = CH$_2$OH		50	72
5			82	91
6[a]			80	91

[a]Reaction run at room temperature.

6.6
Conclusions and Outlook

A number of methods for the enantioselective hydroarylation of unactivated alkenes, strained alkenes, and allenes have been developed in the past decade. Notable among these are the rhodium-catalyzed alkylation of aryl imines and the platinum-catalyzed hydroarylation of 2-allenylindoles. The rhodium-catalyzed protocols have been employed in complex molecular environments and have been applied to the synthesis of (+)-lithospermic acid [(+)-**22**] and PKC inhibitor **14**. These applications strongly suggest that development of more general and more effective enantioselective hydroarylation protocols will lead to widespread application to the synthesis of complex organic molecules. With that said, enantioselective hydroarylation of unactivated alkenes remains problematic. Existing protocols require high catalyst loadings (up to 20 mol% in metal) and suffer from limited scope, particularly with respect to alkenyl substitution. Also lacking are effective methods for the enantioselective hydroarylation of simple electron-rich arenes and for the enantioselective intermolecular hydroarylation of unactivated alkenes. The restrictions with respect to alkene substitution can be circumvented to a degree through employment of allenes as substrates for enantioselective hydroarylation, as was demonstrated for the Au(I)-catalyzed enantioselective hydroarylation of allenylindoles, but at the cost of greater complexity and enhanced sensitivity of the substrate.

6.7
Experimental: Selected Procedures

Rhodium-catalyzed enantioselective cyclization of 9 (Table 6.2) [20]

In a nitrogen-filled glove box, analytically pure **9** (150.0 mg, 0.597 mmol) was added to a premixed solution of [RhCl(coe)$_2$]$_2$ (5 mol%) and **11b** (15 mol%) in toluene (0.1 M). The reaction mixture was stirred at 50 °C for 2.5 h, concentrated, treated with 1 N HCl (aq), and stirred vigorously for 3 h. The resulting mixture was extracted three times with ethyl acetate and the combined extracts were dried, filtered, concentrated, and chromatographed (SiO$_2$; hexanes-ether = 20 : 1) to give (S)-(-)-1-(2-methyl-indan-4-yl)ethanone (68 mg, 65% yield, 88% *ee*) as a colorless oil.

Rhodium-catalyzed conversion of 20j to 23 (Table 6.6) [23]

Under a nitrogen atmosphere, a solution of [RhCl(coe)$_2$]$_2$ (3.59 mg, 5 mmol, 0.1 equiv) and FcPCy$_2$ (5.73 mg, 15.0 mmol, 0.3 equiv) in toluene-d_8 (0.25 mL) was added to a solution of **20j** (50.0 mmol, 1.0 equiv) in toluene (0.25 mL). The reaction mixture was transferred to an NMR tube, which was sealed and heated at 70 °C for 36 h. The resulting solution was cooled to room temperature, transferred to a round-bottom flask, and concentrated under vacuum. The residue was treated with 1 N HCl (aq), stirred for 20 min, and then extracted with EtOAc. The combined extracts were dried, filtered, concentrated, and chromatographed (SiO$_2$; hexanes-EtOAc = 50 : 1 → 1 : 1) to give **23** (70%, 90% *ee*) as a white solid.

Platinum-catalyzed enantioselective conversion of 29 to 30 (Table 6.8) [10]

Under a nitrogen atmosphere, alkenylindole **29** (67 mg, 0.21 mmol) was added to a suspension of [(S)-DTBM-MeObiphep]PtCl$_2$ (30 mg, 2.1 × 10^{-2} mmol) and AgOTf (5 mg, 2.1 × 10^{-2} mmol) in MeOH (0.82 mL) and the reaction mixture was stirred at 60 °C for 20 h. The resulting suspension was cooled to room temperature, filtered through a plug of silica gel, and eluted with ether. The combined eluants were concentrated under vacuum and the residue was chromatographed (SiO$_2$; hexanes–EtOAc = 10 : 1) to give **30** (62 mg, 93%, 90% *ee*) as a white solid.

Gold-catalyzed enantioselective conversion of 33 to 34 (Table 6.10) [27]

Under a nitrogen atmosphere, a solution of allenylindole 33 (41 mg, 0.13 mmol) in toluene (0.3 mL) was added via syringe to a pre-mixed suspension of [(S)-DTBM-MeObiphep](AuCl)$_2$ (5.1 mg, 3.1×10^{-3} mmol) and AgBF$_4$ (1.2 mg, 6.3×10^{-3} mmol) in toluene (0.2 mL) at $-20\,°C$. The resulting mixture was stored in a $-10\,°C$ freezer for 17 h, warmed to room temperature, and chromatographed (SiO$_2$; hexanes–EtOAc = 10 : 1 $\rightarrow$ 5 : 1) to give 34 (71 mg, 87%, 92% ee) as a pale yellow oil.

References

1 Goj, L.A. and Gunnoe, T.B. (2005) *Current Organic Chemistry*, **9**, 671–685.
2 (a) Matsumoto, T., Taube, D.J., Periana, R.A., Taube, H. and Yoshida, H. (2000) *Journal of the American Chemical Society*, **122**, 7414–7415; (b) Matsumoto, T., Periana, R.A., Taube, D.J. and Yoshida, H. (2002) *Journal of Molecular Catalysis A – Chemical*, **180**, 1–18; (c) Bhalla, G., Oxgaard, J., Goddard, W.A. and Periana, R.A. (2005) *Organometallics*, **24**, 3229–3232; (d) Oxgaard, J., Periana, R.A. and Goddard, W.A. (2004) *Journal of the American Chemical Society*, **126**, 11658–11665.
3 (a) Murai, S., Kakiuchi, F., Sekine, S., Tanaka, Y., Kamatani, A., Sonoda, M. and Chatani, N. (1993) *Nature*, **366**, 529–531; (b) Kakiuchi, F. and Murai, S. (2002) *Accounts of Chemical Research*, **35**, 826–834.
4 (a) Lail, M., Arrowood, B.N. and Gunnoe, T.B. (2003) *Journal of the American Chemical Society*, **125**, 7506–7507; (b) Foley, N.A., Lail, M., Gunnoe, T.B., Cundari, T.R., Boyle, P.D. and Petersen, J.L. (2007) *Organometallics*, **26**, 5507–5516; (c) Foley, N.A., Lail, M., Lee, J.P., Gunnoe, T.B., Cundari, T.R. and Petersen, J.L. (2007) *Journal of the American Chemical Society*, **129**, 6765–6781.
5 Youn, S.W., Pastine, S.J. and Sames, D. (2004) *Organic Letters*, **6**, 581–584.
6 Lenges, C.P. and Brookhart, M. (1999) *Journal of the American Chemical Society*, **121**, 6616–6623.
7 Thalji, R.K., Ahrendt, K.A., Bergman, R.G. and Ellman, J.A. (2001) *Journal of the American Chemical Society*, **123**, 9692–9693.
8 Jun, C.-H., Hong, J.B., Kim, Y.H. and Chung, K.Y. (2000) *Angewandte Chemie – International Edition*, **39**, 3440–3442.
9 (a) Tan, K.L., Bergman, R.G. and Ellman, J.A. (2001) *Journal of the American Chemical Society*, **123**, 2685–2686; (b) Tan, K.L., Park, S., Ellman, J.A. and Bergman, R.G. (2004) *Journal of Organic Chemistry*, **69**, 7329–7335; (c) Ahrendt, K.A., Bergman, R.G. and Ellman, J.A. (2003) *Organic Letters*, **5**, 1301–1303; (d) Tan, K.L., Bergman, R.G. and Ellman, J.A. (2002) *Journal of the American Chemical Society*, **124**, 13964–13965; (e) Tan, K.L., Bergman, R.G. and Ellman, J.A. (2002) *Journal of the American Chemical Society*, **124**, 3202–3203; (f) Tan, K.L., Vasudevan, A., Bergman, R.G., Ellman, J.A. and Souers, A.J. (2003) *Organic Letters*, **5**, 2131–2134; (g) Wiedemann, S.H., Bergman, R.G. and Ellman, J.A. (2004) *Organic Letters*, **6**, 1685–1687; (h) Wiedemann, S.H., Ellman, J.A. and Bergman, R.G. (2006) *Journal of Organic Chemistry*, **71**, 1969–1976; (i) Wiedemann, S.H., Lewis, J.C., Ellman, J.A. and Bergman, R.G. (2006) *Journal of the American Chemical Society*, **128**, 2452–2462; (j) Lewis, J.C., Wiedemann, S.H., Bergman, R.G. and Ellman, J.A. (2004) *Organic Letters*, **6**, 35–38.
10 Liu, C., Han, X., Wang, X. and Widenhoefer, R.A. (2004) *Journal of*

the American Chemical Society, **126**, 3700–3701.

11 (a) Liu, C., Bender, C.F., Han, X. and Widenhoefer, R.A. (2007) *Chemical Communications*, 3607–3618; (b) Zhang, Z., Wang, X. and Widenhoefer, R.A. (2006) *Chemical Communications*, 3717–3719.

12 (a) Karshtedt, D., Bell, A.T. and Tilley, T.D. (2004) *Organometallics*, **23**, 4169–4171; (b) Cucciolito, M.E., D'Amora, A., Tuzi, A. and Vitagliano, A. (2007) *Organometallics*, **26**, 5216–5223; (c) Karshtedt, D., McBee, J.L., Bell, A.T. and Tilley, T.D. (2006) *Organometallics*, **25**, 1801–1811.

13 Pastine, S.J., Youn, S.W. and Sames, D. (2003) *Organic Letters*, **5**, 1055–1058.

14 Rueping, M., Nachtsheim, B.J. and Scheidt, T. (2006) *Organic Letters*, **8**, 3717–3719.

15 Kischel, J., Jovel, I., Mertins, K., Zapf, A. and Beller, M. (2006) *Organic Letters*, **8**, 19–22.

16 Jordan, R.F. and Taylor, D.F. (1989) *Journal of the American Chemical Society*, **111**, 778–779.

17 (a) Bandini, M., Melloni, A. and Umani-Ronchi, A. (2004) *Angewandte Chemie – International Edition*, **43**, 550–556; (b) Jørgensen, K.A. (2003) *Synthesis*, 1117–1125; (c) Palomo, C., Oiarbide, M., Kardak, B.G., Garcia, J.M. and Linden, A. (2005) *Journal of the American Chemical Society*, **127**, 4154–4155; (d) Evans, D.A., Fandrick, K.R. and Song, H.J. (2005) *Journal of the American Chemical Society*, **127**, 8942–8943; (e) Jia, Y.-X., Zhu, S.-F., Yang, Y. and Zhou, Q.-L. (2006) *Journal of Organic Chemistry*, **71**, 75–80; (f) Wang, Y.-Q., Song, J., Hong, R., Li, H. and Deng, L. (2006) *Journal of the American Chemical Society*, **128**, 8156–8157; (g) Terada, M. and Sorimachi, K. (2007) *Journal of the American Chemical Society*, **129**, 292–293.

18 Aufdenblatten, R., Diezi, S. and Togni, A. (2000) *Monatshefte für Chemie*, **131**, 1345–1350.

19 Kakiuchi, F., Le Gendre, P., Yamada, A., Ohtaki, H. and Murai, S. (2000) *Tetrahedron: Asymmetry*, **11**, 2647–2651.

20 Thalji, R.K., Ellman, J.A. and Bergman, R.G. (2004) *Journal of the American Chemical Society*, **126**, 7192–7193.

21 Wilson, R.M., Thalji, R.K., Bergman, R.G. and Ellman, J.A. (2006) *Organic Letters*, **8**, 1745–1747.

22 O'Malley, S.J., Tan, K.L., Watzke, A., Bergman, R.G. and Ellman, J.A. (2005) *Journal of the American Chemical Society*, **127**, 13496–13497.

23 Watzke, A., Wilson, R.M., O'Malley, S.J., Bergman, R.G. and Ellman, J.A. (2007) *Synlett*, 2383–2389.

24 Han, X. and Widenhoefer, R.A. (2006) *Organic Letters*, **8**, 3801–3804.

25 Liu, C. and Widenhoefer, R.A. (2007) *Organic Letters*, **9**, 1935–1938.

26 (a) Zhang, Z., Liu, C., Kinder, R.E., Han, X., Qian, H. and Widenhoefer, R.A. (2006) *Journal of the American Chemical Society*, **128**, 9066–9073; (b) Watanabe, T., Oishi, S., Fujii, N. and Ohno, H. (2007) *Organic Letters*, **9**, 4821–4824; (c) Tarselli, M.A. and Gagné, M.R. (2008) *Journal of Organic Chemistry*, **73**, 2439–2441.

27 (a) Zhang, Z. and Widenhoefer, R.A. (2007) *Angewandte Chemie – International Edition*, **46**, 283–285; (b) Munoz, M.P., Adrio, J., Carretero, J.C. and Echavarren, A.M. (2005) *Organometallics*, **24**, 1293–1300; (c) Tarselli, M.A., Chianese, A.R., Lee, S.J. and Gagné, M.R. (2007) *Angewandte Chemie – International Edition*, **46**, 6670–6673; (d) Luzung, M.R., Mauleon, P. and Toste, F.D. (2007) *Journal of the American Chemical Society*, **129**, 12402–12403; (e) LaLonde, R.L., Sherry, B.D., Kang, E.J. and Toste, F.D. (2007) *Journal of the American Chemical Society*, **129**, 2452–2453; (f) Hamilton, G.L., Kang, E.J., Mba, M. and Toste, F.D. (2007) *Science*, **317**, 496–499; (g) Johansson, M.J., Gorin, D.J., Staben, S.T. and Toste, F.D. (2005) *Journal of the American Chemical Society*, **127**, 18002–18003.

7
Catalytic Asymmetric Friedel–Crafts Alkylations in Total Synthesis

Gonzalo Blay, José R. Pedro, and Carlos Vila

Summary

This chapter surveys the application of the catalytic asymmetric Friedel–Crafts reaction in the total synthesis of known natural products or targeted bioactive compounds. It is organized in four parts according to the aromatic core of the nucleophilic reaction partner. The first part surveys the synthesis of indole-containing products such as β-indolyl-propanoic acids, tryptamine and tryptophan derivatives, polycyclic indoles and 2-aminomethyl indoles. The second presents the synthesis of pyrrolizine and indolizidine alkaloids as well as the synthesis of pyrrolo[1,2-a]pyrazines, using the FC reaction with pyrroles. The FC alkylation of furans and its application to the synthesis of butenolides is dealt with in the third part. The last part presents the FC alkylation of arenes to give mandelic acid derivatives, amino acids and chromanes.

7.1
Introduction

Since the publication in 1990 by Erker and van der Zeijden [1] of the first catalytic asymmetric Friedel–Crafts (FC) reaction, a large number of papers dealing with this reaction, either with aromatic or heteroaromatic substrates, have appeared in the literature. Most of the work has addressed the identification of catalysts, reaction conditions and appropriate electrophile/nucleophile partners that can provide the reaction products with high enantiomeric excess (*ee*). This work, which is still under continuous development, is reviewed in the chapters of this book and will not be considered here except in special cases. Simultaneously with this development, the first examples of application in synthesis have appeared, showing the potential of a reaction that will become an important synthetic tool in the near future. Here, we will survey the application of the catalytic asymmetric FC reaction in the total synthesis of known natural products or particular bioactive compounds. We will also consider the FC reactions carried out with less common substrates that provide access to structural motifs characteristic of a determined group of natural or bioactive compounds.

7.2
Total Synthesis of Indole-Containing Compounds

The indole nucleus is present in more than 3000 isolated natural products. It is also a privileged structure from the biological point of view, being present in nearly 1500 bioactive products of natural and synthetic origin, from which more than 40 are medicinal agents of diverse therapeutic action. Because of this, the synthesis of these kinds of products, especially in an enantiomerically enriched form, has attracted much attention.

7.2.1
Synthesis of β-Indolyl-Propanoic Acids

The Michael addition of indoles to unsaturated carbonyl compounds provides access to β-(3-indolyl)-propanoic acids. MacMillan *et al.* [2] have reported a two-step synthesis of the COX-2 inhibitor **4** developed by the Merck laboratories [3]. The key step in this synthesis is an organocatalytic FC reaction between the 5-methoxy-2-methylindole **1** and (*E*)-crotonaldehyde **2** promoted by imidazolidinone **3** and DCA as the co-catalyst. Subsequent oxidation of the FC product with $AgNO_3$ afforded compound **4** in 87% *ee* and in 82% yield for the two steps (Scheme 7.1).

A different approach to these compounds, developed by Evans *et al.* [4], uses α,β-unsaturated 2-acyl imidazoles and PyBOX-Sc(III) complexes. The reaction between *N*-methylindole **5** and compound **6** in the presence of 1 mol % of complex **7** afforded compound **8** in 97% yield and with 98% *ee*. Interestingly, the reaction afforded better selectivities with lower catalyst loadings, which is tentatively explained in terms of the formation of a 1 : 1 : 1 substrate:product:catalyst complex that is favored at lower catalyst loading. Such an aggregate is considered more enantioselective than the corresponding 1 : 1 substrate:catalyst complex which would be favored at higher catalyst loading. The protecting group can be easily removed and transformed into a carboxylic acid by methylation of the imidazole group followed by

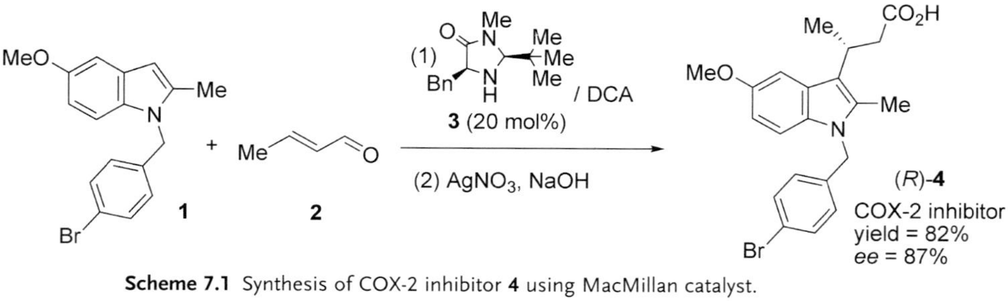

Scheme 7.1 Synthesis of COX-2 inhibitor **4** using MacMillan catalyst.

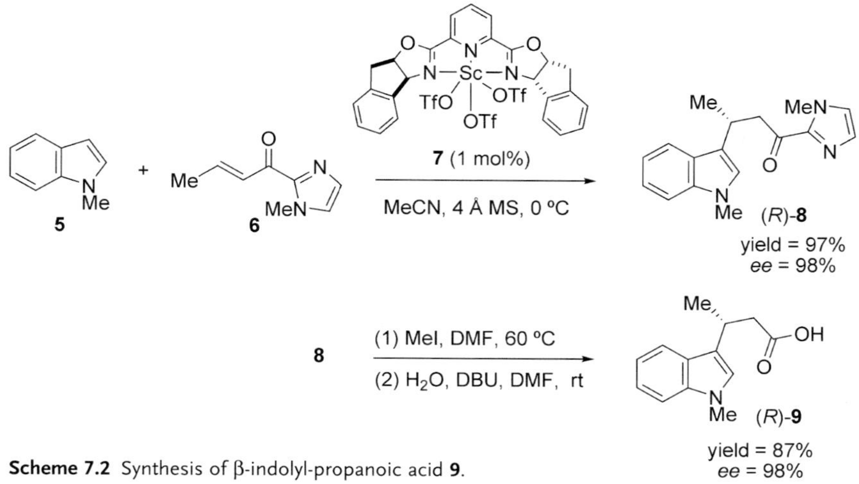

Scheme 7.2 Synthesis of β-indolyl-propanoic acid **9**.

treatment with H₂O/DBU in DMF in a one-pot operation. The antibacterial and algicide acid **9** [5] was obtained with 91% ee following this protocol (Scheme 7.2).

7.2.2
Synthesis of Tryptamine and Tryptophan Analogs

β-Indolyl nitroalkanes represent ideal synthetic precursors for numerous natural indole-based compounds. Reduction of the nitro group to amine allows the preparation of tryptamine and melatonin analogs. In particular, the enantioselective alkylation of indoles with β-substituted-α,β-unsaturated nitroalkanes is an appealing strategy for the synthesis in a stereocontrolled way of tryptamine derivatives having a benzylic stereogenic center. In fact, tryptamine derivatives having this structural feature have already been successfully tested as gonadotropin-releasing hormone (GnRH) antagonists [6]. In the outlined example [7], the reaction between indole **10** and nitrostyrene **11** in the presence of a chiral phosphoric acid **12** provides the expected FC adduct **13** in quantitative yield and with 91% *ee*. Reduction of this compound with NaBH₄ and NiCl₂·6H₂O gave tryptamine **14**, which was transformed into the melatonin analog **15** after acetylation with acetyl chloride (Scheme 7.3). These products were obtained in good yields and without racemization.

Bandini et al. [8] have also described a [SalenAl]-catalyzed FC reaction of 2-phenylindole with nitroalkenes that gives enantiomerically enriched tryptamine precursors containing the 2-arylindole core. This class of molecules has shown high affinity and selectivity for the h5-HT2A and h5-HT2C receptors [9].

Tryptophan is an essential amino acid with an indole moiety that has received considerable attention on account of its varied biological activities. Tryptophan analogs are also important building blocks for the synthesis of bioactive and natural products. Liu and Chen have developed a straightforward procedure for the synthesis of tryptophan analogs based on a catalytic asymmetric FC alkylation of indoles **10** with

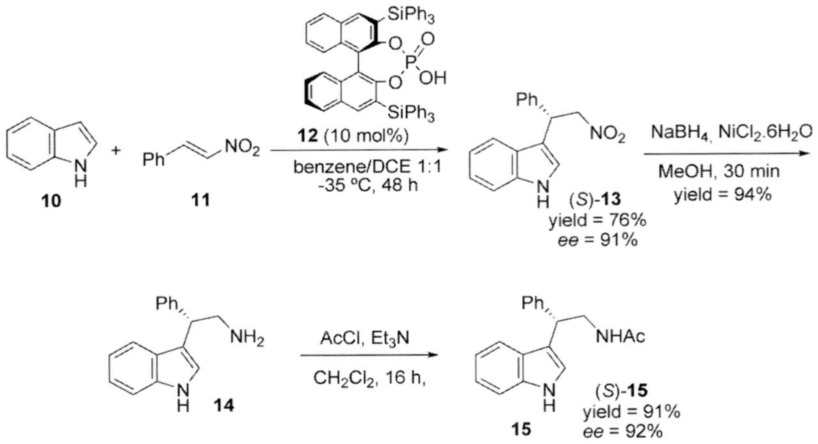

Scheme 7.3 Synthesis of tryptamine and melatonin analogs.

nitroacrylates **16** catalyzed by the (4R,5S)-diPh-BOX-Cu(OTf)$_2$ complex **17**-Cu(OTf)$_2$ [10]. The reaction provides tryptophan nitro-precursors **18** in moderate diastereoselectivities and good enantioselectivities. The alkylation products **18** are transformed into tryptophan analogs by reduction of the nitro group with Zn/H$^+$. This sequence has been applied to the synthesis of **19** (Scheme 7.4), a methylated analog of a phosphodiesterase inhibitor compound [11]. Tryptophan analogs can also be elaborated from indolyl diesters such as **22**, which are obtained in up to 98% *ee* from the FC reaction of indoles and alkylidene malonates **20** catalyzed by the pseudo C$_3$-symmetric TOX **21**-copper (II) complexes [12]. Interestingly, the absolute stereochemistry of the reaction is solvent dependent, providing the opposite enantiomers in isobutyl alcohol or acetone–ether, and in 1,1,2,2-tetrachloroethane (TTCE). Treating the alkylated product **22** with KOH in THF:EtOH afforded the hemiacid ester **23** in high yield, which was transformed into the somastatin agonist **24** by using documented procedures (Scheme 7.4) [13].

King and Meng [14] have carried out the enantioselective synthesis of the highly potent and selective serotonin reuptake inhibitor (BMS-594726, **30**) [15]. This kind of drug has found widespread utility in the treatment of depression and other mental illness, and is being studied for the treatment of premature ejaculation in men. Compound **30** is a homotryptamine derivative with conformational restriction in the side chain, a structural feature that optimizes potency and selectivity in a number of selective serotonin reuptake inhibitors. The key step in the synthesis of **30** is the enantioselective alkylation of 5-iodoindole **25** with 1-formyl-cyclopentene **26** which was accomplished utilizing MacMillan's imidazolidinone catalyst **27**. Under optimal conditions, **28** was isolated in good yield (75%), good diastereoselectivity (trans/cis ratio 24 : 1) and with 84% *ee* (Scheme 7.5). The choice of both the catalyst and the indole substrate was crucial. As a matter of fact, catalyst **3** (Scheme 7.1) was completely ineffective with the β-branched aldehyde **26**, while 5-cyanoindole was poorly reactive at −25 °C and gave decomposition mixtures at higher temperatures. Reductive amination of aldehyde **28** provided **29** in quantitative yield. Finally, substitution of the iodide by the cyano group using NaCN and CuI worked smoothly

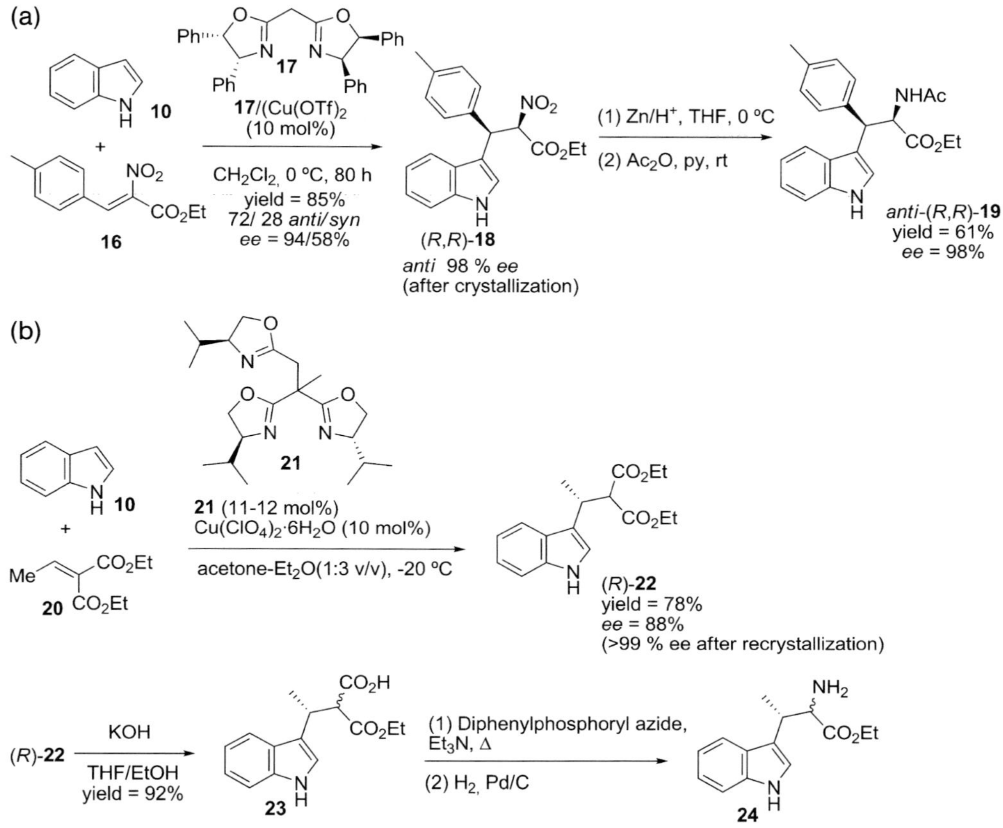

Scheme 7.4 Synthesis of tryptophan analogs.

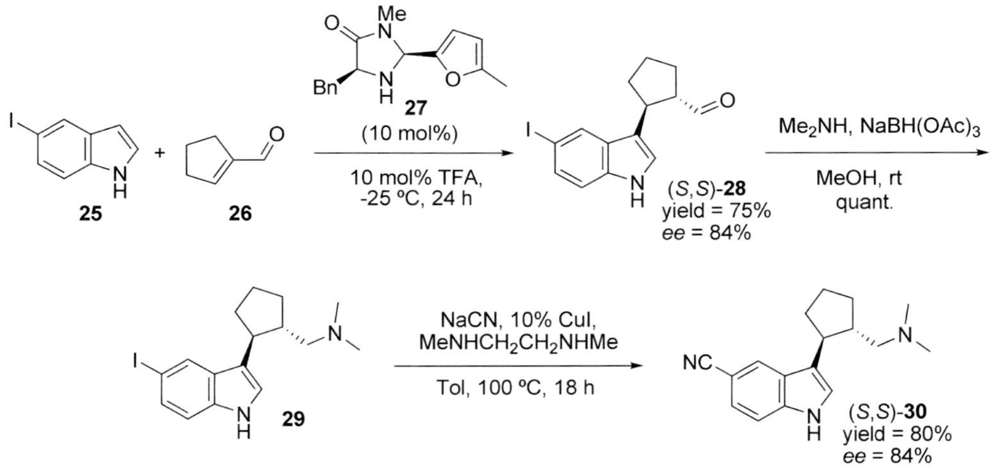

Scheme 7.5 Synthesis of the serotonin reuptake inhibitor BMS-594726 (30).

(yield = 80%). Stereochemical integrity was maintained throughout these two steps. Enantiomerically pure **30** ($ee = 100\%$) was obtained after crystallization of the final product as the HCl salt. On the other hand, very recently, Rueping et al. have described a procedure for the synthesis of homotryptophanes that involves a catalytic enantioselective FC alkylation of indoles with β,γ-unsaturated α-ketoesters in the presence of a chiral N-triflylphosphoramide (yield = 43–88%, ee = 80–90%) followed by reductive amination of the ketone carbonyl with Hantzsch dihydripyridine and p-anisidine to give N-PMP protected homotryptophane (yield = 43%) as a 1.3 : 1 diastereomeric mixture [16].

7.2.3
Synthesis of Polycyclic Indoles

Over the past years, considerable effort has been directed toward the development of new routes to polycyclic indoles through catalytic enantioselective intramolecular alkylation reactions, including the Pictet–Spengler (PS) reaction. 1,2,3,4-Tetrahydro-β-carbolines (THBCs) represent an important class of indole-containing alkaloids with important biological and medicinal activities. The intramolecular condensation of tryptamine or tryptophan derivatives with carbonyl compounds is a primary route to this kind of compound. The PS reaction of chiral tryptamine derivatives prepared as described in Section 7.2.2 allows THBCs, bearing stereogenic centers at the C-1 and C-4 positions, to be obtained [7]. On the other hand, several chiral catalysts which provide access to chiral THBCs, bearing a stereogenic center at the C-1 position of the carboline ring, have been developed [17]. However several challenging tasks still need to be solved: Catalyst loading, substrate scope, removal of the N-protecting group or improved enantioselectivity.

A PS reaction with nitrones has been developed by Nagakawa et al. [17a] which, however, requires stoichiometric amounts of chiral Lewis acid. List et al. have described a chiral Brønsted acid-catalyzed reaction which requires the presence of a geminal diester functionality in the vicinity of the imine electrophile [17b]. Hiemstra et al. have recently reported a catalytic asymmetric synthesis of THBCs based on the PS condensation of N-sulfenyltryptamines with a wide range of aldehydes. The reaction is catalyzed by phosphoric acid **12** providing the PS products with up to 87% ee [17c]. Finally, Jacobsen et al. have developed an enantioselective catalytic acyl-PS reaction based on the use of urea-derived catalysts which takes place with high ee, although it is limited to imines derived from aliphatic aldehydes [17d, e]. The reaction involves the participation of reactive N-acyliminium ions that are generated in situ. This strategy has been applied to the synthesis of (+)-harmicine **35** (Scheme 7.6) [17e]. Treatment of tryptamine **31** with succinic anhydride, followed by reduction with $NaBH_4$, gave a hydroxylactame intermediate **32** which was cyclized via an acyliminium ion (i) upon treatment with TMSCl in the presence of catalyst **33** (**34**, overall yield = 65%, ee = 97%). Finally, reduction of the amide group with $LiAlH_4$ gave (+)-harmicine [18].

Xiao et al. have used an enantioselective organocatalyzed ring-closing FC reaction of amino- or ether-linked enals **36** to give THBCs **37** (X = NTs) and tetrahydropyrano [3,4-b]indoles (THPIs) **37** (X = O) bearing a stereogenic center on the C-4 position

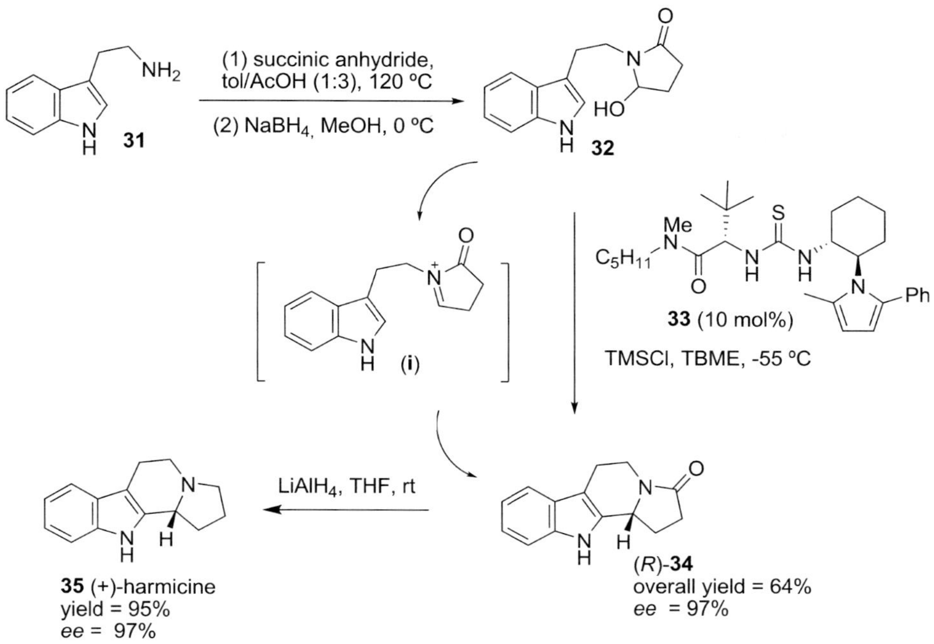

Scheme 7.6 Synthesis of (+)-harmicine (**35**) via asymmetric organocatalyzed PS reaction.

(Scheme 7.7a) [19]. A related intramolecular strategy has been used by Bandini *et al.* [20] to access THBCs as well as tetrahydro-γ-carbolines (THGCs) starting from 2-formyl- **38** or 3-formylindole **42**, respectively. The key step in the synthetic sequences is the Pd-catalyzed intramolecular allylic alkylation of the indole nucleus (Scheme 7.7b).

Intramolecular FC alkylations of tethered indoles bearing electrophilic double bonds have also been applied to the synthesis of chiral tetrahydrocarbazole derivatives [4, 19] with a stereogenic center at the C-1 position of the cyclohexane ring. Application of Evans catalyst **7** to compound **47**, prepared from **46** in two steps, allows the synthesis of the bioactive tetrahydrocarbazolyl acetic acid **48** in a straightforward manner (Scheme 7.8).

Formation of the cyclohexane ring fused to indole can also be accomplished with unactivated alkenes by using a cationic chiral phosphine-Pt(II) complex [21a] or from 2-allenyl indoles via gold(I)-catalyzed intramolecular enantioselective hydroarylation of the allene [21b]. However, the products obtained from these reactions have little possibility of synthetic manipulation.

7.2.4
Synthesis of 2-Aminomethyl Indoles

The direct alkylation of indole at the C-2 position cannot be carried out directly on account of the higher reactivity of the C-3 carbon in this system. Based on the initial work of Saraçoglu [22], C-2 alkylated indoles can be attained after alkylation

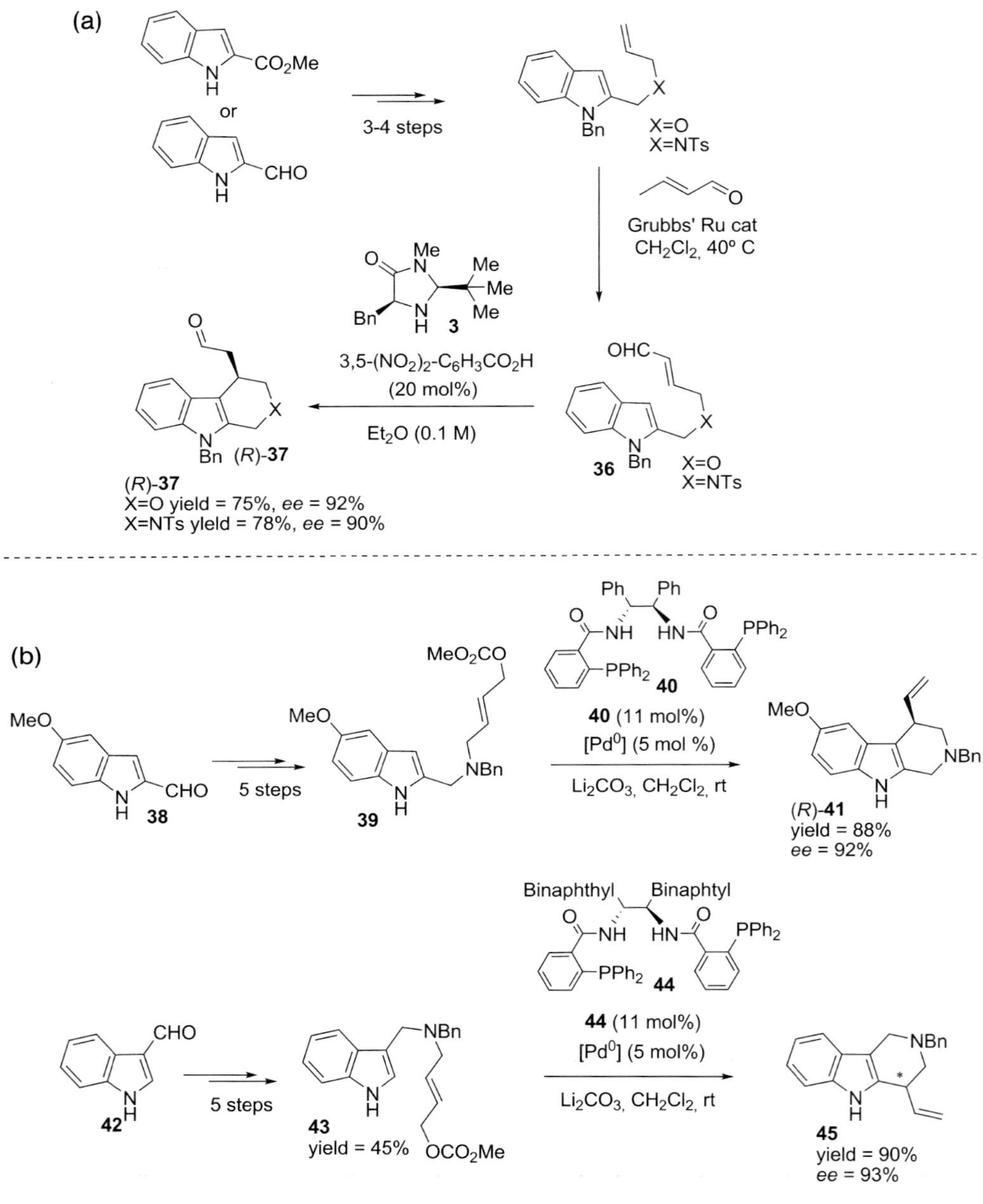

Scheme 7.7 Intramolecular FC alkylation of indoles to give THBCs, THPIs and THGCs.

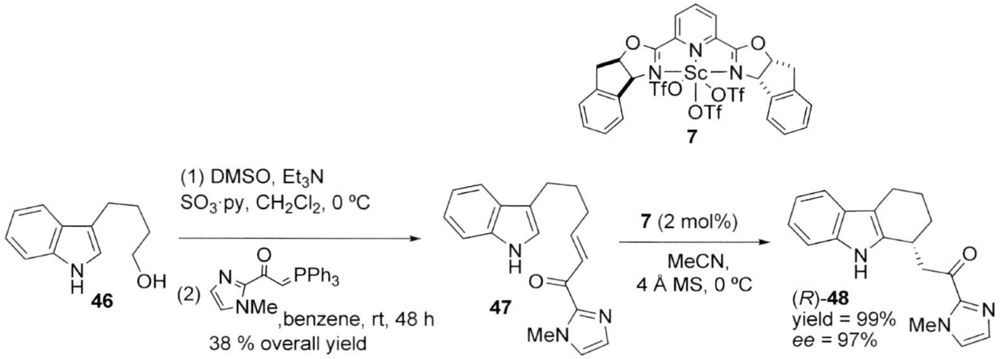

Scheme 7.8 Synthesis of tetrahydrocarbazole derivatives.

of 4,7-dihydroindoles with carbon electrophiles followed by oxidation of the corresponding products. The first catalytic enantioselective versions of this reaction have been realized by Evans et al. [23] and Pedro et al. [24]. You et al. have extended this methodology to the enantioselective FC reaction of 4,7-dihydroindoles with imines using a chiral phosphoric acid as catalyst [25]. The reaction yields 2-indolyl methanamine derivatives which are a popular structure core in many biologically active natural and unnatural products. As a representative example, Scheme 7.9 shows the preparation of compound **52**, a synthetic intermediate for the preparation of HIV protease inhibitors [26]. Alkylation of 1,7-dihydroindole **49** with benzaldehyde N-tosylimine **50** and re-aromatization by oxidation are carried out in a one-pot method to give compound **51** in high yield (88%) and with high ee (99%). The tosyl group can be readily removed and converted to the carbobenzoxy **52** group in excellent yield without loss of optical purity.

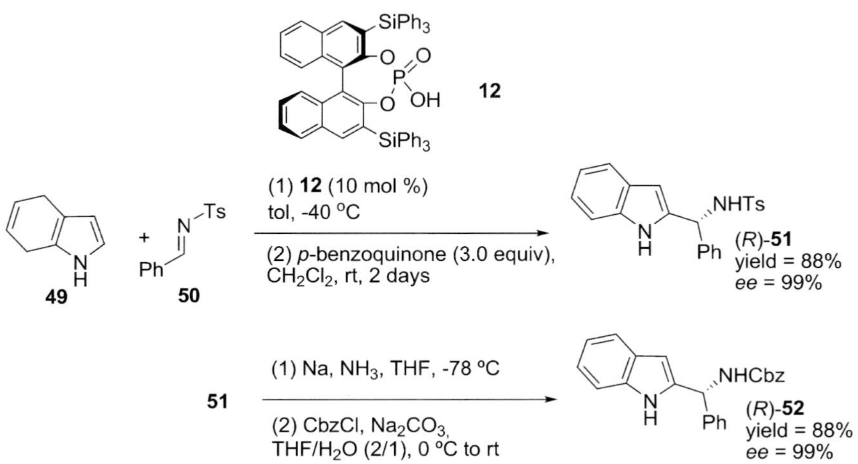

Scheme 7.9 Synthesis of 2-aminomethyl indole **52**.

7.2.5
Experiments: Selected Procedures

Synthesis of the COX-2 inhibitor 4 (Scheme 7.1)

To 1-(4-bromo-benzyl)-5-methoxy-2-methyl-1H-indole (**1**, 110 mg, 0.333 mmol) in a 2-dram amber vial was added CH_2Cl_2 (0.60 mL), IPA (0.066 mL), dichloroacetic acid (5.5 µL, 0.066 mmol) and (2S,5S)-**3** (16.4 mg, 0.066 mmol). This solution was stirred for 10 min at rt, then placed in a $-70\,°C$ bath for an additional 10 min. Crotonaldehyde (**2**, 82 µL, 1.0 mmol) was then added and the reaction mixture was stirred at $-70\,°C$ for 9 h. The reaction mixture was then transferred cold through a silica plug into a flask and concentrated to provide, after silica gel chromatography (hexanes: EtOAc = 80 : 20), the intermediate aldehyde as a colorless oil (111 mg, yield = 84%). $[\alpha]_D = -20.8$ (c 1.0, $CHCl_3$). $ee = 87\%$. HPLC (analysis of the alcohol, obtained after reduction with $NaBH_4$): Chiralcel OD–H, 96/4 hexanes/EtOH, 1 mL min^{-1}, t_S 45.1 min (minor), $t_R = 35.9$ min (major). ^{1}H NMR (300 MHz, $CDCl_3$) δ 9.69 (dd, $J = 1.8, 1.8$ Hz, 1H), 7.38 (dt, $J = 2.4, 9.0$ Hz, 2H), 7.12 (d, $J = 2.1$ Hz, 1H), 7.05 (d, $J = 9.0$ Hz, 1H), 6.79–6.75 (m, 3H), 5.19 (s, 2H), 3.88 (s, 3H), 3.66 (dt, $J = 7.2, 22.2$ Hz, 1H), 3.02 (ddd, $J = 1.8, 8.1, 16.5$ Hz, 1H); 2.85 (ddd, $J = 2.1, 6.6, 16.5$ Hz, 1H), 2.30 (s, 3H) 1.48 (d, $J = 7.2$ Hz, 3H). ^{13}C NMR (75 MHz, $CDCl_3$) δ 202.5, 153.6, 137.0, 132.7, 132.0, 131.9, 127.6, 126.6, 121.1, 114.5, 110.0, 109.8, 102.3, 56.2, 50.6, 46.2, 26.4, 21.4, 10.9. HRMS (CI): m/z $C_{21}H_{22}BrNO_2$ requires 399.0834, found 399.0833. IR (film) 2930, 2823, 2730, 1722, 1618, 1581, 1530, 1483, 1452, 1405, 1229, 1156, 1073, 1037, 1011, 902, 798, 476 cm^{-1}. A solution of the above aldehyde (110 mg, 0.250 mmol) and silver nitrate (59.7 mg, 0.275 mmol) in 1.3 mL absolute ethanol was treated with a solution of 5 N NaOH in ethanol (1 : 5, 0.9 mL, 0.75 mmol NaOH). After 45 min this was treated with 10 mL water, acidified to pH 3 and extracted with $CHCl_3$ (5 × 20 mL) rinsing each extract with brine. The combined organics were dried over Na_2SO_4 and concentrated *in vacuo* to provide, after silica gel chromatography compound (R)-**4** (101 mg, yield = 97%). Pale yellow solid. $[\alpha]_D = -30.9$ (c 1.0, $CHCl_3$). ^{1}H NMR (300 MHz, $CDCl_3$) δ 7.36 (d, $J = 9.0$ Hz, 2H), 7.11 (d, $J = 2.4$ Hz, 1H), 7.04 (d, $J = 8.7$ Hz, 1H), 6.77–6.73 (m, 3H), 5.18 (s, 2H), 3.86 (s, 3H), 3.56 (dt, $J = 7.2, 21.9$ Hz, 1H), 2.86 (d, $J = 3.6$ Hz, 1H), 2.83 (d, $J = 3.3$ Hz, 1H), 2.27 (s, 3H), 1.49 (d, $J = 7.2$ Hz, 3H). ^{13}C NMR (75 MHz, $CDCl_3$) δ 178.0, 153.7, 137.3, 133.8, 133.0, 132.0, 127.7, 126.7, 121.2, 118.8, 114.6, 110.1, 109.9, 102.4, 56.3, 46.3, 41.5, 28.6, 21.1, 10.9. HRMS (CI): m/z $C_{21}H_{23}BrNO_3$ (M + 1) requires 416.0861, found 416.0867. IR (film) 3425, 2961, 2934, 2833, 1706, 1483, 1451, 1405, 1228, 1156, 1010, 796, 755 cm^{-1}.

Preparation of the Sc(III) triflate complex 7 (Scheme 7.2)

To an oven dried 2-dram vial in a dry-box was added an appropriate amount $Sc(OTf)_3$ and 1.2 equiv. of (S,R)-IndapyBOX ligand. The vial was capped with a septum and purged with 1 mL of CH_2Cl_2. The catalyst was allowed to age for 2 h at rt with moderate magnetic stirring. The CH_2Cl_2 was removed with a steady stream of N_2 to yield a white solid.

Synthesis of compound 9

To complex **7** (0.0026 mmol) in a 2-dram vial was added acetonitrile (1 mL) and cooled to −40 °C before **6** (36 µL, 40 mg, 0.26 mmol) and N-methylindole (**5**, 40 µL, 41 mg, 0.31 mmol) were added to the vial. After 16 h of moderate stirring, the mixture was purified by flash chromatography to afford 71 mg (97%) of compound (R)-**8**. $R_f = 0.39$ (benzene:EtOAc 1 : 1). $[\alpha]^{25}{}_D = -4.4$ (c 1.0, CH_2Cl_2). ee = 98%, HPLC: Chiralcel OD-H, 90/10 hexane/IPA, 0.8 mL min^{-1}, $t_R = 17.6$ min (major), $t_S = 19.3$ min (minor). ^{1}H NMR (500 MHz, $CDCl_3$) δ 7.67 (d, J = 8.1 Hz, 1H), 7.27 (m, 1H), 7.21 (t, J = 7.0 Hz, 1H), 7.15 (s, 1H), 7.09 (d, J = 7.0 Hz, 1H), 7.00 (s, 1H), 6.94 (s, 1H), 3.94 (s, 3H), 3.89–3.83 (m, 1H), 3.73 (s, 3H), 3.56 (dd, J = 16.1 and 6.2 Hz, 1H), 3.46 (dd, J = 15.7 and 8.1 Hz, 1H), 1.43 (d, J = 7.0 Hz, 3H). ^{13}C NMR (100 MHz, $CDCl_3$) δ 192.3, 143.4, 137.0, 128.9, 126.9, 126.8, 125.0, 121.4, 119.9, 119.4, 118.5, 109.1, 46.8, 36.1, 32.6, 27.1, 21.9. HRMS (CI): m/z $C_{17}H_{19}N_3O$ [M + H]$^+$ requires 282.1606, found 282.1599. IR (film): 3107.6, 3055.2, 2959.9, 1670.7, 1467.7, 1406.1, 1372.4, 1328.1, 1270.8, 1239.5, 1155.0, 1132.1, 1106.0, 1003.6, 980.0, 915.0, 740.6, 695.8, 642.8 cm^{-1}.

Hydrolysis of the imidazole moiety: 200 µL of methyl iodide (3.2 mmol, 9.7 equiv.) was added to a solution of compound (R)-**8** (90 mg, 0.33 mmol) in DMF (0.2 mL) at rt. The excess methyl iodide was removed *in vacuo*. To the yellow reaction mixture in DMF was added distilled water (0.2 mL) and DBU (150 µL, 1 mmol) at rt. After 2 h stirring at rt, the reaction was acidified with 1 N HCl (20 mL) and extracted with EtOAc (3 × 20 mL). The combined organic layers were dried over anhydrous Na_2SO_4, filtered and concentrated *in vacuo*. The residue was purified by column chromatography to produce 62 mg (87% yield) of compound (R)-**9**. $R_f = 0.30$ (hexanes:EtOAc = 1 : 1). ^{1}H NMR (500 MHz, $CDCl_3$) δ 7.69 (d, J = 7.8 Hz, 1H), 7.35–7.32 (m, 1H), 7.30–7.26 (m, 1H), 7.18–7.14 (m, 1H), 6.90 (s, 1H), 3.77 (s, 3H), 3.70–3.62 (m, 1H), 2.92 (dd, J = 15.1, 5.9 Hz, 1H), 2.65 (dd, J = 15.1, 8.8 Hz, 1H), 1.49 (d, J = 6.8 Hz, 3H). ^{13}C NMR (125 MHz, $CDCl_3$) δ 179.2, 137.1, 126.6, 124.8, 121.6, 119.1, 119.0, 118.7, 109.3, 42.3, 32.6, 27.6, 21.2. HRMS (CI) $C_{13}H_{15}NO_2$ [M + H]$^+$: 218.1181, found: 218.1180. IR (film): 3051.6, 2963.5, 2927.4, 2679.0, 1705.7, 1616.2, 1477.2, 1419.7, 1375.3, 1328.4, 1292.1, 1238.8, 1205.5, 1156.5, 1133.2, 1101.6, 929.4, 805.3, 739.6 cm^{-1}.

Synthesis of compound 15 (Scheme 7.3)

To a suspension of activated powder MS 3 Å (20 mg), (R)-3,3′-bis(triphenylsilyl)-1,1′-binaphthylphosphate (R)-**12** (17.3 mg, 0.0200 mmol) and nitrostyrene (**11**, 149.3 mg, 1.000 mmol) in benzene (0.5 mL) and DCE (0.5 mL) was added indole (**10**, 23.4 mg, 0.200 mmol) at −35 °C. After being stirred at this temperature for 48 h, the reaction mixture was purified by column chromatography (hexane:EtOAc) to give 44.7 mg (76%) of product (S)-**13**. $R_f = 0.4$ (hexane/EtOAc = 3 : 1). $[\alpha]^{23}{}_D$ + 22.5 (c 0.9, CH_2Cl_2). ee = 91%, HPLC: Chiralpak AD-H, 90/10 Hexane/IPA, 0.75 mL min^{-1}, $t_R = 32.9$ min (minor), $t_S = 36.4$ min (major). ^{1}H NMR (400 MHz, $CDCl_3$) δ 8.09 (brs, 1H), 7.44 (d, J = 7.9 Hz, 1H), 7.37–7.18 (m, 7H), 7.09–7.03 (m, 2H), 5.19 (t, J = 8.0 Hz, 1H), 5.07 (dd, J = 12.5, 7.5 Hz, 1H), 4.94 (dd, J = 12.5, 8.2 Hz, 1H). ^{13}C

NMR (100 MHz, CDCl$_3$) δ 139.1, 136.4, 128.9, 127.7, 127.5, 126.6, 122.7, 121.6, 119.9, 118.9, 114.4, 111.3, 79.5, 41.5.

Reduction of the nitro group: To a suspension of compound (*S*)-**13** (139.0 mg, 0.52 mmol) and NiCl$_2$·6H$_2$O (126.2 mg, 0.53 mmol) in methanol (3.8 mL) was added NaBH$_4$ (102.2 mg, 2.70 mmol) at 0 °C and the mixture was stirred for 30 min. The reaction mixture was quenched by addition of satd aq NH$_4$Cl at 0 °C and extracted with CH$_2$Cl$_2$. The combined organic layers were washed with brine and dried over Na$_2$SO$_4$. After filtration, the filtrate was concentrated and purified by column chromatography (CH$_2$Cl$_2$:MeOH = 20 : 1 to 10 : 1) to give 116.5 mg (94%) of product (*S*)-**14**. R_f = 0.3 (CH$_2$Cl$_2$:MeOH = 5 : 1). [α]$^{22}_D$ = +4.3 (*c* 1.0, CHCl$_3$). ^{1}H NMR (300 MHz, DMSO-d_6): δ 10.97 (brs, 1H), 7.38–7.23 (m, 7H), 7.09–7.03 (m, 2 H), 7.14 (t, *J* = 7.0 Hz, 1H), 7.02 (t, *J* = 7.6 Hz, 1H), 6.88 (t, *J* = 7.3 Hz, 1 H), 4.20 (brs, 1 H), 3.28 (brs, 1H), 3.12 (brs, 3H). ^{13}C NMR (75 MHz, DMSO-d_6): δ 144.3, 136.5, 128.4, 128.2, 127.0, 126.2, 122.1, 121.2, 118.8, 118.4, 116.3, 111.6, 46.7, 45.8.

Acetylation of the amino group: To a suspension of compound (*S*)-**14** (38.9 mg, 0.16 mmol) in CH$_2$Cl$_2$ (3.8 mL) was added Et$_3$N (36 µL, 0.26 mmol), AcCl (18 µL, 0.25 mmol) at rt and the mixture was stirred for 15.5 h. The reaction was quenched by addition of H$_2$O at 0 °C and extracted with CH$_2$Cl$_2$. The combined organic layers were washed with brine, dried over Na$_2$SO$_4$, filtered and concentrated. The crude product was purified by TLC (EtOAc) to give 41.7 mg (91%) of product (*S*)-**15**. Solid, m.p. 193–194 °C. R_f 0.4 (EtOAc). [α]$^{21}_D$ = +38.8 (*c* 1.0, MeOH). *ee* = 92%, HPLC: chiralpak AD-H, 90/10 hexane/IPA, 1.0 mL min^{-1}, t_S = 20.6 min (major), t_R = 22.3 min (minor). ^{1}H NMR (400 MHz, CDCl$_3$): δ 8.49 (brs, 1H), 7.34 (d, *J* = 7.9 Hz, 1H), 7.24 (d, *J* = 8.1 Hz, 1H), 7.20–7.04 (m, 6H), 6.95–6.91 (m, 2 H), 5.53 (brs, 1H), 4.32 (t, *J* = 7.6 Hz, 1H), 3.95 (ddd, *J* = 13.4, 7.6, 6.0 Hz, 1H), 3.70 (ddd, *J* = 13.4, 7.6, 5.7 Hz, 1H), 1.78 (s, 1H). ^{13}C NMR (100 MHz, CDCl$_3$): δ 170.2, 142.1, 136.5, 128.6, 128.0, 126.7, 126.7, 122.1, 121.6, 119.4, 119.3, 116.4, 111.2, 44.2, 42.6, 23.3. MS(DI): *m/z* 278(M$^+$, 2), 220(10), 219(58), 207(15), 206(100), 204(30), 179(11), 178(19), 128(7), 102(6), 77(13). Anal. C$_{18}$H$_{18}$N$_2$O requires C, 77.67; H, 6.52; N, 10.06, found: C, 77.41; H, 6.50; N, 9.77. IR (CHCl$_3$) 3477, 3445, 3020, 1665, 1518, 1493, 1456, 1417, 1371, 1213, 1097 cm^{-1}.

Synthesis of compound 19 (Scheme 7.4a)

Cu(OTf)$_2$ (5.4 mg, 0.015 mmol) and ligand **17** (7.6 mg, 0.0165 mmol) were charged in a dried tube under argon, followed by addition of CH$_2$Cl$_2$ (0.9 mL). The solution was stirred at rt for 30 min and nitroacrylate **16** (35 mg, 0.15 mmol) was added. The mixture was stirred for 10 min at rt, then for 15 min at 0 °C. Indole (**10**, 21 mg, 0.18 mmol) was added. After stirring for 80 h, water (6 mL) was added, followed by extraction with CH$_2$Cl$_2$. The combined organic phases were dried with Na$_2$SO$_4$ and the solvent was removed under reduced pressure. The residue was purified by flash chromatography (petroleum ether:EtOAc = 4 : 1) to afford 45 mg (85%) of **18**. The *dr* value (*anti*: *syn* = 72 : 28) was determined by ^{1}H NMR signals at δ 1.02 ppm and 0.91 ppm. The diastereomers were separated by flash chromatography on silica gel (petroleum ether:

EtOAc = 4 : 1); *anti-(R,R)-*18: white solid, m.p. = 146–148 °C. [α]$_D^{25}$ = 58.2 (*c* 0.55, CHCl$_3$). *ee* = 99% (after recrystallization), HPLC: Chiralpak AD-H, 90 : 10 hexane/IPA, 0.5 mL min^{-1}, *t* = 58.1 min (major), *t* = 77.2 min (minor). ^{1}H NMR (300 MHz, CDCl$_3$) δ 8.05 (br s, 1H), 7.48 (d, *J* = 7.9 Hz, 1H), 7.32 (d, *J* = 9.2 Hz, 1H), 7.28–7.22 (m, 2H), 7.18–7.13 (m, 2H), 7.09–7.03 (m, 3H), 5.89 (d, *J* = 11.4 Hz, 1H), 5.34 (d, *J* = 11.4 Hz, 1H), 4.07–4.01 (m, 2H), 2.27 (s, 3H), 1.02 (t, *J* = 7.1 Hz, 3H). ^{13}C NMR (75 MHz, CDCl$_3$): δ 163.5, 137.5, 136.3, 134.4, 129.5, 128.4, 126.3, 122.7, 120.5, 119.9, 118.9, 114.0, 111.3, 91.8, 62.9, 44.2, 21.1, 13.6. HRMS (FAB): *m/z* C$_{20}$H$_{20}$N$_2$O$_4$ (M$^+$) requires 352.1417, found: 352.1414. FTIR (KBr): 3420, 2924, 1745, 1561, 1358, 1097, 1013, 742 cm^{-1}. *syn-(2S,3R)-*18: white solid, m.p. = 145–146 °C. *ee* = 58%, HPLC: Chiralpak 90 : 10 AD-H, hexane/IPA, 0.5 mL min^{-1}, *t* = 46.3 min (major), *t* = 77.7 min (minor). ^{1}H NMR (300 MHz, CDCl$_3$): δ 8.05 (br s, 1H), 7.59 (d, *J* = 7.9 Hz, 1H), 7.33–7.26 (m, 4H), 7.18–7.08 (m, 4H), 5.89 (d, *J* = 11.6 Hz, 1H), 5.32 (d, *J* = 11.6 Hz, 1H), 4.03–3.94 (m, 2H), 2.27 (s, 3H), 0.91 (t, *J* = 7.1 Hz, 3H). ^{13}C NMR (75 MHz, CDCl$_3$): δ 163.4, 137.3, 136.0, 135.6, 129.5, 127.5, 126.2, 122.7, 121.7, 120.0, 119.0, 113.2, 111.6, 91.9, 62.9, 43.5, 21.0, 13.4. HRMS (EI): *m/z* C$_{20}$H$_{20}$N$_2$O$_4$ (M$^+$) requires 352.1423, found: 352.1425. FTIR (KBr): 3419, 2984, 1744, 1561, 1457, 1182, 1031, 744 cm^{-1}.

Reduction of the nitro group: To a solution of *anti-(R,R)-*18 (39 mg, 0.11 mmol) in a mixture of THF (9 mL), conc. HCl (0.37 mL), AcOH (2 mL), and water (3.7 mL) was added Zn dust (190 mg, 2.9 mmol) at 0 °C and the mixture stirred for 3 h at this temperature. The Zn dust was filtered off, and the reaction mixture was diluted with CH$_2$Cl$_2$ and washed with water and satd aq NaHCO$_3$. The combined organic layer was dried over Na$_2$SO$_4$. After evaporation, conventional acetylation (by Ac$_2$O in pyridine) of the residue was carried out, followed by concentration and purification by flash chromatography on silica gel (petroleum ether/EtOAc 1 : 1) to afford 24 mg (61%) of the tryptophan analog *anti-(R,R)-*19. White solid, m. p. = 70–73 °C. [α]$_D^{25}$ 58.7 (*c* 0.4, CHCl$_3$). *ee* = 99%, HPLC: Chiralpak AD-H, 70 : 30 hexane/IPA, 0.5 mL min^{-1}, *t* = 10.0 min (major), *t* = 21.4 min (minor). ^{1}H NMR (300 MHz, CDCl$_3$): δ 8.22 (br s, 1H), 7.34 (d, *J* = 8.6 Hz, 2H), 7.20–7.12 (m, 3H), 7.06–7.01 (m, 4H), 5.98 (d, *J* = 8.6 Hz, 1H), 5.31 (t, *J* = 8.6 Hz, 1H), 4.65 (d, *J* = 8.6 Hz, 1H), 3.96–3.87 (m, 2H), 2.27 (s, 3H), 1.89 (s, 3H), 0.95 (t, *J* = 7.1 Hz, 3H). ^{13}C NMR (75 MHz, CDCl$_3$): δ 172.0, 169.9, 136.9, 136.6, 136.4, 129.1, 128.2, 127.1, 122.3, 121.9, 119.7, 119.1, 114.7, 111.3, 61.2, 56.3, 45.6, 23.2, 21.0, 13.7. HRMS (EI) C$_{22}$H$_{24}$N$_2$O$_3$ (M$^+$) requires 364.1787, found 364.1790. FTIR (KBr): 3305, 2926, 1733, 1659, 1516, 1455, 1373, 1233, 1106, 1025, 742 cm^{-1}.

Synthesis of BMS-594726 (30, Scheme 7.5)

Trifluoroacetic acid (0.63 mL, 8.2 mmol) was added to a −35 °C solution of 1-cyclopentene-1-carboxaldehyde (**26**, 12 g, 125 mmol) and catalyst **27** (2.2 g, 8.2 mmol) in 85 : 15 CH$_2$Cl$_2$-IPA (80 mL). After 15 min, 5-iodoindole (**25**, 20 g, 82 mmol) in 85 : 15 CH$_2$Cl$_2$/IPA (80 mL) was added. The reaction was stirred at −30 °C for 18 h. The reaction was diluted with CH$_2$Cl$_2$ (400 mL) and washed with aqueous NaHCO$_3$ (400 mL), 1 N HCl (2 × 200 mL), and brine (2 × 200 mL). The aqueous layer was re-extracted each time with 40 mL of CH_2Cl_2 and the extract recombined with the main organic layer. The organic layer was dried over $MgSO_4$ and solvent was removed under vacuum. Silica gel chromatography (hexane:EtOAc = 100:0 to 80:20) gave 21 g (75%) of trans-(1S,2S)-2-(5-iodo-1H-indol-3-yl)-cyclopentanecarbaldehyde (S,S)-**28**. ee = 84% (determined after reduction of the aldehyde with $NaBH_4$ in MeOH), HPLC: Chiralcel OD, 90/10 hexane/EtOH, 0.8 mL min^{-1}, t = 15.64 min (major). ^{1}H NMR (400 MHz, ACN-d_3) δ 1.74–1.93 (m, 3 H), 2.00–2.06 (m, 2 H), 2.20–2.26 (m, 1 H), 2.95 (ddd, J = 16.7, 8.5, 3.2 Hz, 1 H), 3.56 (q, J = 8.6 Hz, 1 H), 7.13 (d, J = 2.5 Hz, 1 H), 7.26 (d, J = 8.6 Hz, 1 H), 7.42 (dd, J = 8.6, 1.7 Hz, 1 H), 7.97 (d, J = 1.7 Hz, 1 H), 9.23 (br s, 1 H), 9.64 (d, J = 2.9 Hz, 1 H). ^{13}C NMR (400 MHz, ACN-d_3), δ 204.2, 136.3, 130.1, 129.8, 128.1, 122.8, 117.4, 114.2, 82.1, 58.5, 38.2, 34.5, 26.5, 25.0. HRMS (EI), $C_{14}H_{14}INO$ (M-H) requires 338.0042, found 338.0048. Anal. $C_{14}H_{14}INO$ requires C, 49.57; H, 4.16; N, 4.13; I, 37.41, found: C, 49.49; H, 74.05; N, 4.03; I, 37.30.

Reduction of the aldehyde: To a stirred solution of (S,S)-**28** (17.5 g, 51.6 mmol) in MeOH (200 mL) at rt was added dimethylamine (64 mL of a 2.0 M solution in THF, 128 mmol) followed by acetic acid (0.5 mL). After 15 min, $NaBH(OAc)_3$ (12 g, 57 mmol) was added slowly. The resulting mixture was stirred for 18 h and then concentrated under vacuum. The residue was partitioned between 400 mL EtOAc and 300 mL aqueous $NaHCO_3$. The organic layer was washed twice with aqueous $NaHCO_3$ (250 mL). The aqueous layer was re-extracted each time with EtOAc (50 mL) and the extract recombined with the main organic layer. The combined organic extracts were dried over $MgSO_4$. Solvent was removed under vacuum and the residue was dried under high vacuum overnight to give 29.4 g (100%) of crude trans-(S,S)-**29** which was used directly in the next reaction without further purification. An analytical sample was prepared by crystallization from EtOAc. ^{1}H NMR (400 MHz, MeOH-d_4): δ 1.50–1.40 (m, 1H), 1.86–1.70 (m, 3H), 2.14–2.05 (m, 2 H) 2.16 (s, 6 H) 2.37–2.21 (m, 3 H) 2.78 (q, J = 8.56 Hz, 1 H) 7.03 (s, 1 H), 7.13 (d, J = 8.56 Hz, 1H), 7.30 (dd, J = 8.56, 1.71 Hz, 1H), 7.86 (d, J = 1.71 Hz, 1H). ^{13}C NMR (400 MHz, MeOH-d_4): δ 137.6, 131.1, 130.5, 128.9, 123.2, 119.0, 114.6, 82.3, 66.2, 46.0, 45.6, 43.4, 35.0, 33.0, 25.2. HRMS (EI): m/z $C_{16}H_{21}IN_2$ (M + H) requires 369.0828, found 369.0836. Anal. $C_{16}H_{21}IN_2$: C, 52.18; H, 5.74; N, 7.60; I, 34.46, found: C, 52.31; H, 5.68; N, 7.49; I, 34.22.

Substitution of iodide by a cyano group: To a mixture of trans-(S,S)-**29** (19.4 g, 53 mmol), sodium cyanide (3.1 g, 63 mmol), and copper(I) iodide (1 g, 5.3 mmol) in deoxygenated anhydrous toluene (50 mL) was added N,N'-dimethylethylenediamine (5.66 mL, 53 mmol). The resulting mixture was heated in an oil bath to 125 °C under N_2 for 18 h. The reaction was cooled, diluted with 400 mL EtOAc, and heated to reflux for 5 min. The resulting suspension was transferred to a separatory funnel and washed with aqueous $NaHCO_3$ (3 × 300 mL). The aqueous layer was re-extracted each time with EtOAc (50 mL) and the extracts recombined with the main organic layer. The organic layer was dried over $MgSO_4$. Solvent removal and column chromatography (CH_2Cl_2 : 2.0 M NH_3 in MeOH = 100:0 to 93:7) gave 11.3 g (80%) of trans-(S,S)-**30** (80%) as free base. ^{1}H NMR (400 MHz, $CDCl_3$) δ 1.53–1.43 (m, 1 H) 1.84–1.63 (m, 3 H)

2.18–2.08 (m, 3H) 2.20 (s, 6 H) 2.38–2.22 (m, 3 H) 2.84 (q, $J = 8.40$ Hz, 1H) 7.15 (s, 1H) 7.36 (t, $J = 1.10$ Hz, 2H) 7.94 (t, $J = 1.1$ Hz, 1H).

Synthesis of **30** *hydrochloride:* Compound (S,S)-**30** (1.76 g, 6.6 mmol) was dissolved in MeOH (100 mL) and treated with 2.0 M HCl/ether (3.3 mL) at rt. After stirring for 30 min, the solvent was removed under vacuum and the residue was sonicated in MeOH (3 mL), cooled, and filtered to obtain a white solid which was recrystallized from ethanol to yield 1.1 g (55%) of *trans*-(S,S)-**30**·HCl. White crystals. $[\alpha]^{20}_D = 55.18$ (c 2.8, MeOH). $ee = 100\%$, HPLC: Chiralpak AD, 95/5 : 0.15 hexane/EtOH/DEA, 1.0 mL min^{-1}, $t = 13.54$ min (major). ^{1}H NMR (400 MHz, MeOH-d_4): δ 1.75–1.53 (m, 2H), 1.93–1.78 (m, 2H), 2.20–2.13 (m, 1H), 2.32–2.24 (m, 1H), 2.56–2.45 (m, 1H), 2.66 (s, 3H) 2.70 (s, 3H), 3.05–2.84 (m, 3H) 7.32 (dd, $J = 8.6$, 1.5 Hz, 1H), 7.36 (s, 1H), 7.40 (dd, $J = 8.3$, 0.7 Hz, 1H) 7.84 (dd, $J = 1.59$, 0.61 Hz, 1H). ^{13}C NMR (400 MHz, MeOH-d_4): δ 140.4, 128.0, 125.6, 125.5, 125.4, 122.0, 118.7, 113.8, 102.5, 63.8, 44.0, 43.6, 43.1, 35.0, 31.7, 24.9. HRMS (EI): m/z C$_{17}$H$_{21}$N$_3$ (M + H) requires 268.1814, found 268.1811. Anal. C$_{17}$H$_{22}$ClN$_3$: C, 67.20; H, 7.29; N, 13.83; Cl, 11.66, found: C, 67.03; H, 7.50; N, 13.78; Cl, 11.53.

Synthesis of (+)-harmicine (35, Scheme 7.6)

A solution of tryptamine (**31**, 1.6 g, 10.0 mmol) and succinic anhydride (1.0 g, 10.0 mmol) in toluene (10 mL) and acetic acid (20 mL) was refluxed for 24 h. The flask was cooled to 0 °C. Hexanes (100–150 mL) were added and the product was removed by filtration, and washed with an additional 200 mL hexanes. Excess acetic acid was removed via a heptane azeotrope affording 1.75 g (72%) of 1-[2-(1*H*-Indol-3-yl)-ethyl]-pyrrolidine-2,5-dione. ^{1}H NMR (500 MHz, CDCl$_3$): δ 8.06 (br s, 1H), 7.67 (d, $J = 7.8$ Hz, 1H), 7.35 (d, $J = 8.3$ Hz, 1H), 7.19 (td, $J = 7.6$, 1.0 Hz, 1H), 7.13 (td, $J = 7.6$, 1.0 Hz, 1H), 7.07 (d, $J = 2.4$ Hz, 1H), 3.83 (dd, $J = 7.6$, 7.6 Hz, 2H), 3.06 (dd, $J = 7.6$, 7.6 Hz, 2H), 2.61 (s, 4H). ^{13}C NMR (100 MHz, 5% DMSO-d_6/CDCl$_3$): δ 175.69, 134.94, 125.76, 121.13, 119.72, 117.09, 116.60, 110.04, 109.38, 37.65, 26.63, 21.85. LRMS (ApCI): 243.051 (100%) [M + H]$^+$. FTIR (CH$_2$Cl$_2$ thin film): 3400, 2944, 1693, 1404, 1340, 1296, 1162, 745 cm^{-1}. NaBH$_4$ (942 mg, 24.8 mmol) was added portion wise over 5–10 min to a solution of the above imide (200 mg, 0.826 mmol) in anhydrous MeOH (41.3 mL), contained in a flask capped with a rubber septum, pierced with a 21G needle (for venting), and cooled to 0 °C. The reaction was stirred vigorously for 2 h at 0 °C. Upon complete conversion of the imide, the reaction mixture was poured onto vigorously stirring 200 mL 1 : 1 CH$_2$Cl$_2$/NaHCO$_3$ (satd aq) which had been pre-cooled to 0 °C. The biphasic slurry was stirred at 0 °C for 5 min, and the layers partitioned in a separatory funnel. The aqueous layer was extracted with CH$_2$Cl$_2$ (1 × 25 mL). and the combined organic layers were dried over Na$_2$SO$_4$, filtered, and concentrated *in vacuo* until about1 mL. The remaining solvent was removed with a steady stream of N$_2$ gas, until a viscous oil remained, which was immediately subjected to high vacuum for 5 min, affording 201 mg (99%) of the hydroxylactam **32**. Off-white foam. ^{1}H NMR (600 MHz, CDCl$_3$): δ 8.07 (br s, 1H), 7.63 (d, $J = 7.9$ Hz, 1H), 7.36 (d, $J = 8.2$ Hz, 1H), 7.20 (td, $J = 7.0$, 0.9 Hz, 1H),

7.12 (td, $J = 7.9$, 0.9 Hz, 1H), 7.03 (d, $J = 2.4$ Hz, 1H), 4.98 (m, 1H), 3.80 (ddd, $J = 14.1, 7.0, 7.0$ Hz, 1H), 3.56 (ddd, $J = 14.1, 7.0, 7.0$ Hz, 1H), 3.05 (m, 2H), 2.54 (ddd, $J = 17.0, 9.4, 7.6$ Hz, 1H), 2.51 (br m, 1H), 2.27 (ddd, $J = 17.0, 10.0, 4.1$ Hz, 1H), 2.18 (m, 1H), 1.76 (dddd, $J = 13.8, 9.4, 4.1, 2.3$ Hz, 1H). ^{13}C NMR (100 MHz, CDCl$_3$) δ 174.97, 136.20, 127.27, 122.07, 121.94, 119.38, 118.61, 112.96, 111.26, 83.66, 40.77, 28.93, 28.26, 23.63.

Cyclization of hydroxylactam 32: A 100 mL round-bottom flask was charged with the crude **32** (230 mg, 0.942 mmol) and a stir bar. Freshly distilled *tert*-butylmethyl ether (94.2 mL) was added, followed by catalyst **33** (48 mg, 94.2 μmol), and the reaction sealed with a rubber septum. The reaction was stirred vigorously at rt for 1 min, and then cooled to −78 °C. Freshly distilled TMSCl (241 μL, 1.88 mmol) was added via a syringe. The reaction was warmed to −55 °C, and stirred rapidly at this temperature for 48 h. The reaction was quenched with pre-cooled TEA (−55 °C, 650 μL, 4.71 mmol), and poured onto 100 mL saturated aq. NaHCO$_3$. The aqueous phase was extracted with CH$_2$Cl$_2$ (3 × 75 mL), dried over Na$_2$SO$_4$, filtered, and concentrated *in vacuo*. Flash chromatography (CH$_2$Cl$_2$:MeOH = 98:2) afforded 191 mg (90%) of (*R*)-**34**. Off-white solid. $[\alpha]^{25}_D = +249.5$ (*c* 1.0, CHCl$_3$). *ee* = 97%, SFC: Chiralpak AS-H, 4.0 mL min^{-1}, 20/80 MeOH/hexanes, $t_S = 3.77$ min (minor), $t_R = 4.50$ min (major). ^{1}H NMR (500 MHz, CDCl$_3$): δ 7.98 (br s, 1H), 7.50 (d, $J = 7.8$ Hz, 1H), 7.35 (d, $J = 8.3$ Hz, 1H), 7.20 (td, $J = 7.6, 1.5$ Hz, 1H), 7.13 (td, $J = 7.3, 1.0$ Hz, 1H), 4.95 (m, 1H), 4.54 (ddd, $J = 13.2, 5.4, 2.0$ Hz, 2H), 3.05 (m, 1H), 2.81–2.91 (m, 2H), 2.49–2.67 (m, 3H), 1.96 (m, 1H). ^{13}C NMR (100 MHz, CDCl$_3$): δ 173.32, 136.29, 133.23, 126.70, 122.05, 119.67, 118.32, 111.00, 107.89, 54.35, 37.61, 31.64, 25.65, 21.01; LRMS (ApCI): 227.0 (100%) [M + H]$^+$. FTIR (CH$_2$Cl$_2$ thin film): 3400, 3258, 2979, 2919, 1667, 1437, 1421, 1308, 1265, 907, 733 cm^{-1}.

Reduction of the amide: A solution of compound (*R*)-**34** (28 mg, 0.124 mmol) in anhydrous THF (2 mL) was treated with LAH (28 mg, 0.744 mmol). The reaction was stirred at rt for 16 h, cooled to 0 °C, and slowly quenched with H$_2$O (200 μL), 4 N NaOH (400 μL), and H$_2$O (800 μL). The reaction was stirred at 0 °C for 30 min, and then filtered through a pad of Celite, eluting with THF (50 mL), MeOH (50 mL), and CH$_2$Cl$_2$ (50 mL). The combined organic filtrates were dried over Na$_2$SO$_4$, filtered, and concentrated *in vacuo*. The crude residue was purified by preparative thin layer chromatography (1% TEA in 12% CH$_2$Cl$_2$:MeOH = 88:12) to afford 25 mg (95%) of the natural product (*R*)-**35**. Off-white solid. $[\alpha]^{25}_D = +101.9$ (*c* 0.48 CHCl$_3$). *ee* = 97%, HPLC: Chiralcel OD-H, 13/87 IPA/hexanes, 1.3 mL min^{-1}, $t_S = 10.08$ min (minor), $t_R = 12.42$ min (major). ^{1}H NMR (500 MHz, CDCl$_3$): δ 8.09 (br s, 1H), 7.47 (d, $J = 7.6$ Hz, 1H), 7.33 (d, $J = 7.6$ Hz, 1H), 7.15 (td, $J = 7.6, 1.0$ Hz, 1H), 7.10 (td, $J = 7.6, 1.0$ Hz, 1H), 4.25–4.28 (m, 1H), 3.33 (ddd, $J = 13.2, 5.7, 2.2$ Hz, 1H), 3.09 (ddd, $J = 13.2, 10.7, 4.9$ Hz, 1H), 2.86–2.99 (m, 3H), 2.66 (ddt, $J = 15.6, 4.4, 2.0$ Hz, 1H), 2.24–2.32 (m, 1H), 1.81–2.00 (m, 3H). ^{13}C NMR (100 MHz, CDCl$_3$): δ 135.95, 135.22, 127.31, 121.44, 119.40, 118.10, 110.69, 107.80, 56.97, 49.27, 45.94, 29.39, 23.39, 17.75; LRMS: 212.9 (100%) [M + H]$^+$. FTIR (CH$_2$Cl$_2$ thin film): 3404, 3053, 2962, 2849, 1451, 1327, 1122, 907, 737 cm^{-1}.

Synthesis of the THGC 45 (Scheme 7.7b)

A solution of [Pd$_2$dba$_3$]·CHCl$_3$ (3.6 μmol) and the chiral ligand **44** (7.9 μmol, 11 mol%) in anhydrous CH$_2$Cl$_2$ (1.0 mL) was stirred until the solution color turned from deep red to orange (about 30 min). Then, the carbonate **43** (0.07 mmol) dissolved in 0.5 mL of CH$_2$Cl$_2$ and Li$_2$CO$_3$ (2 equiv.) was added. The resulting reaction mixture slowly turned yellow. The reaction was stirred overnight and then quenched with water (4 mL) and extracted with EtOAc. The combined organics were dried over Na$_2$SO$_4$ and concentrated under reduced pressure. The crude was purified by passage through a pad of silica gel to give **45** (93%). Yellow solid, m.p. = 109–113 °C. [α]$_D$ = +22.5 (c 0.7, CHCl$_3$). ee = 93%, Chiralcel OD, 90:10 Hexane:IPA, 0.5 mL min^{-1}, t = 26.5 min (major), t = 27.9 min (minor). ^{1}H NMR (300 MHz, CDCl$_3$): δ 2.70 (dd, J = 6.6, 11.4 Hz, 1H), 3.03 (dd, J = 5.2, 11.4 Hz, 1H), 3.69–3.71 (m, 2H), 3.72–3.74 (m, 1H), 3.80 (d, J = 2.8 Hz, 2H), 5.17 (dd, J = 1.5, 9.9 Hz, 1H), 5.29 (dd, J = 1.5, 17.1 Hz, 1H), 5.91–6.03 (m, 1H), 7.05–7.16 (m, 2H), 7.29–7.43 (m, 6H), 7.55 (d, J = 7.0 Hz, 1H), 7.72 (br, 1H). ^{13}C NMR (75 MHz, CDCl$_3$): δ 38.7, 50.2, 57.3, 62.0, 110.1, 111.0, 115.9, 119.4, 119.6, 121.6, 117.6, 128.2(2C), 128.7(2C), 129.4, 132.0, 136.3, 138.1, 140.4. LC-ESI-MS: 289 (M + 1). IR (neat): 3399, 3256, 3058, 2920, 2810, 1615, 1494, 1455, 1257, 1098 cm^{-1}.

Synthesis of aminomethylindole 52 (Scheme 7.9)

In a dry Schlenk tube, N-sulfonyl imine **50** (0.20 mmol) and chiral phosphoric acid **12** (17.3 mg, 0.020 mmol) were dissolved in toluene (1 mL) under argon. The solution was stirred for 10 min at rt and then for another 5 min at −40 °C. Subsequently, 4,7-dihydroindole **49** (0.22 mmol) was added in one portion at −40 °C. After the reaction was complete (monitored by TLC), saturated aqueous NaHCO$_3$ (3 mL) was added to quench the reaction. The mixture was extracted with EtOAc (10 mL). The organic layer was washed with brine (5 mL) and dried over anhydrous Na$_2$SO$_4$. The solvents were removed under reduced pressure. The residue was dissolved in CH$_2$Cl$_2$ (10 mL) and p-benzoquinone (0.66 mmol) was added to the vial. After the reaction was complete (monitored by ^{1}H NMR), the reaction was diluted with CH$_2$Cl$_2$ (20 mL) and the organic layer was washed with 2 N NaOH (2 × 20 mL), brine (20 mL), and dried over anhydrous Na$_2$SO$_4$. Solvent removal and column chromatography (petroleum ether:EtOAc = 5:1) afforded compound (R)-**51** (88% yield). [α]$^{20}_D$ = +21.9 (c 1.0, Acetone). ee = 99%, Chiralcel OD-H, 90/10 hexanes/IPA, 0.7 mL min^{-1}, t_R = 31.04 min (major), t_S = 27.14 min (minor). ^{1}H NMR (300 MHz, CDCl$_3$): δ 2.31 (s, 3H), 5.56 (d, J = 8.7 Hz, 1H), 5.69 (d, J = 8.7 Hz, 1H), 5.91–5.92 (m, 1H), 7.01–7.26 (m, 10H), 7.42 (d, J = 7.5 Hz, 1H), 7.53 (d, J = 8.1 Hz, 2H), 8.51 (br, 1H). ^{13}C NMR 75 MHz, CDCl$_3$): δ 21.4, 56.1, 102.5, 111.1, 119.8, 120.5, 122.3, 127.1, 127.3, 127.7, 128.2, 128.6, 129.4, 136.4, 136.6, 137.2, 138.0, 143.6. HRMS (EI): m/z C$_{22}$H$_{20}$N$_2$O$_2$S (M$^+$) requires 376.1245, found 376.1237. IR (film) 3360, 3271, 3046, 1598, 1493, 1485, 1425, 1317, 1298, 1162, 1085, 1058, 913, 830, 805, 784, 702, 680, 566, 539 cm^{-1}.

Removal of the tosyl group: To a solution of compound (+)-(R)-**51** (95 mg, 0.252 mmol) in anhydrous THF (1.0 mL) at −78 °C was added liquid ammonia (∼5 mL), and then sodium (excess, ~50 mg) was added in portions until a deep blue coloration persisted for 5 min. The reaction was quenched at −78 °C with slow addition of solid NH$_4$Cl (~0.5 g). Ammonia was evaporated and the residue was partitioned between water (5 mL) and CH$_2$Cl$_2$ (5 mL). The aqueous layer was extracted with CH$_2$Cl$_2$ (4 × 5 mL). The combined organic phase was dried over Na$_2$SO$_4$ and concentrated. The residue was dissolved in THF (3 mL) and water (1.5 mL) was added followed by Na$_2$CO$_3$ (60 mg, 0.566 mmol). The resulting suspension was stirred at 0 °C for 5 min, and benzyl chloroformate (60 µL, 0.42 mmol) was introduced to this mixture via a syringe. The reaction mixture was stirred at rt for an additional 10 min, and then diluted with EtOAc (10 mL). The organic layer was collected, washed with brine and dried over Na$_2$SO$_4$. After removal of the solvent, the residue was flash chromatographed (petroleum ether: EtOAc = 5 : 1 to 3 : 1) to give compound (R)-**52** (88% yield). $[\alpha]^{20}_D = +19.6$ (c 1.0, acetone). ee = 97%, HPLC: Chiralcel AD-H, 80/20 hexanes/IPA, 1.0 mL min^{-1}, t_R 14.64 min (major), t_S = 17.37 min (minor). ^{1}H NMR (300 MHz, CDCl$_3$): δ 5.16 (s, 2H), 5.79–5.81 (m, 1H), 6.14 (d, J = 7.8 Hz, 1H), 6.20 (br, 1H), 7.11–7.39 (m, 13H), 7.57 (d, J = 7.5 Hz, 1H), 8.55 (br, 1H). ^{13}C NMR (75 MHz, CDCl$_3$): δ 53.6, 67.2, 101.5, 110.9, 119.8, 120.4, 122.0, 127.2, 127.7, 128.1, 128.2, 128.5, 128.8, 135.9, 136.2, 138.8, 139.1, 156.2. HRMS (ESI): m/z C$_{23}$H$_{21}$N$_2$O$_2$ [M + H]$^+$ requires 357.1597, found 357.1599. IR (film) 3318, 3035, 1695, 1496, 1455, 1341, 1287, 1231, 1122, 1040, 792, 750, 738, 698 cm^{-1}.

7.3
Total Synthesis of Pyrrole-Containing Compounds

Pyrroles are important heterocycles that are often found in natural products, medicinal agents and intermediates in multi-step syntheses. Many of these compounds may be accessed via FC reactions. However, probably because of the relatively instability of pyrroles toward acidic environments, methodologies for the catalytic enantioselective FC reaction of pyrroles are somewhat limited. Nevertheless, this methodology has already been used in a number of total syntheses, especially of pyrrole alkaloids.

7.3.1
Synthesis of (+)-Heliotridane

The FC alkylation of pyrroles with α,β-unsaturated 2-acyl imidazoles in the presence of py-BOX-Sc(III) complexes developed by Evans was applied to the synthesis of (+)-heliotridane **58** [23], an hexahydro-1H-pyrrolizine alkaloid isolated from *Heliotropium lasiocarpum* [27]. The reaction of pyrrole **53** with the acyl imidazole **54**, prepared via Wittig olefination, with catalyst **7** gave the alkylated pyrrole **55** in 96% yield and with ee = 94% (Scheme 7.10). Cleavage of the imidazole in the absence of an external nucleophile afforded a 2,3-dihydropyrrolizone **56**, upon internal acylation of the pyrrole nitrogen. The conditions usually employed for the cleavage of the imidazole (excess MeI in DMF) failed in this case. However, the use of

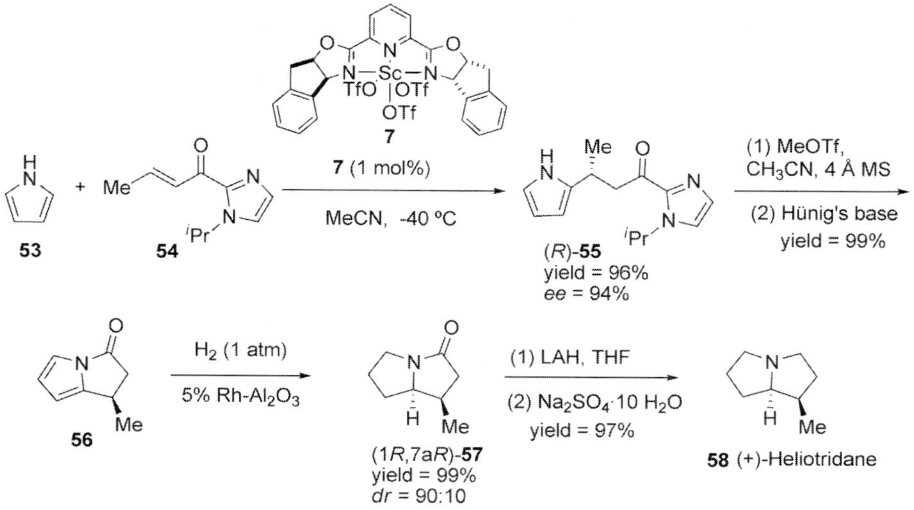

Scheme 7.10 Synthesis of (+)-heliotridane **58**.

a slight excess of methyl triflate in acetonitrile for the methylation of the imidazole and the use of DMAP or Hünig's base to promote cyclization afforded compound **56** in quantitative yield in a one-pot operation. The hydrogenation of **56** afforded **57** in quantitative yield (90:10 dr), and the subsequent amide reduction with LAH provided (+)-heliotridane in 97% yield.

7.3.2
Synthesis of the Indolizidine Alkaloids Tashiromine, epi-Tashiromine, Razhinal, Rhazinilam, Leuconolam and epi-Leuconalam

The intramolecular catalytic enantioselective FC alkylation of a pyrrole incorporating an N-tethered Michael acceptor has been used for the synthesis of several indolizidine alkaloids [28]. Reaction of the N-acyl-2-oxazolidinone **61** with the (R,R)-[PhBOXCu(SbF$_6$)$_2$] [**62**-Cu(SbF$_6$)$_2$] complex afforded compound **63** in 95% yield and 88% ee (Scheme 7.11) [28a]. The enolate of **63** was reacted with the Davis oxaziridine reagent to give the α-hydroxylated derivative **64** as a 1:1 diastereomeric mixture, reduction of this mixture with lithium borohydride afforded diol **65** which was immediately cleaved with NaIO$_4$ in aqueous buffer. The resulting highly sensitive aldehyde was immediately reduced with NaBH$_4$ to alcohol **66**. Hydrogenation of compound **66** using rhodium on alumina as catalyst in hexafluoroisopropanol afforded a chromatographically separable mixture of (−)-tashiromine **67** (85%) and (−)-epi-thashiromine **68** (5%) [29].

A similar approach was used by the same authors [28b] in the synthesis of the alkaloids [30] razhinal **79**, rhazinilam **80**, leuconolam **81** and epi-leuconalam **82**. In this case the intramolecular FC reaction was carried out with MacMillan organocatalysis on enal **69** to give the bicyclic compound **71**, with 74% ee, that

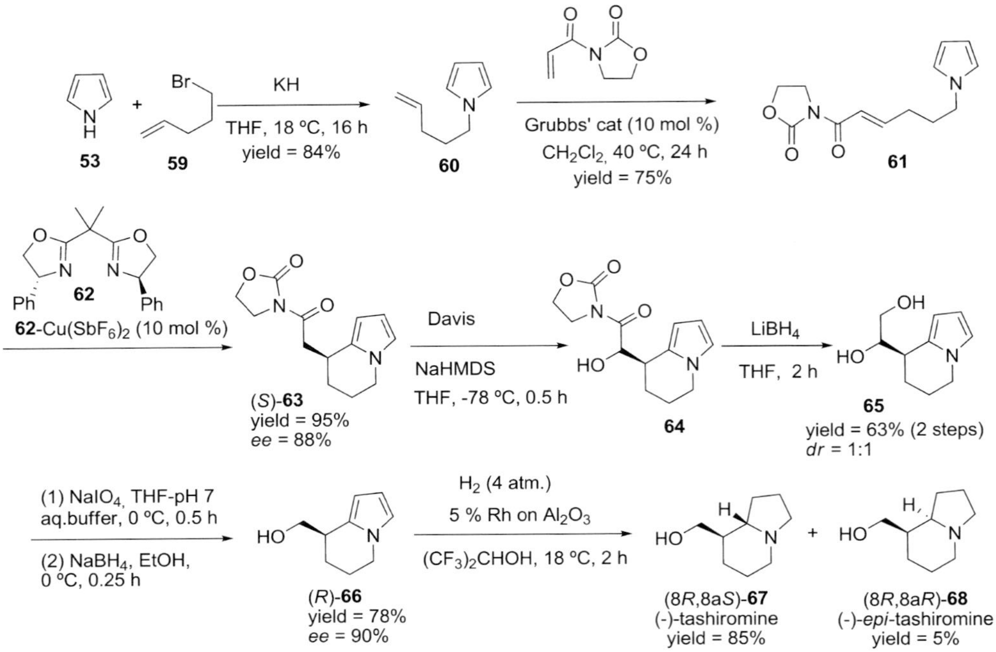

Scheme 7.11 Synthesis of (−)-tashiromine **67** and (−)-epi-tashiromine **68**.

was immediately reduced to alcohol **72**. This compound was transformed into ester **75** in three steps involving mesylation, substitution with cyanide and hydrolysis of the nitrile group. Vilsmeier–Haack formylation of the pyrrole moiety afforded aldehyde **76** which was subjected to iodination affording iodide **77** in a completely regioselective manner. Suzuki–Miyaura cross-coupling of **77** with o-aminophenylboronic ester afforded the arylated pyrrole **78** (64%) which engaged in a simple two-step lactamization procedure to deliver (−)-razhinal **79**. Decarbonylation of (−)-razhinal with Wilkinson's catalyst gave (−)-razhinilam **80** in 89% yield. Finally, treatment of **80** with an excess of PCC afforded a chromatographically separable mixture of (−)-leuconalam **81** and (+)-epi-leuconalam **82** in 28% and 46% yield, respectively (Scheme 7.12).

7.3.3
Synthesis of Pyrrolo[1,2-a]pyrazines

Antilla *et al.* [31] have developed an organocatalytic enantioselective FC reaction with imines as electrophiles to give chiral 2-(2-aminoalkyl)pyrroles. The reaction is best catalyzed by chiral phosphoric acid **12** and allows a variation of substituents on the N and C atoms of the heterocycle. This reaction has been applied to the enantioselective synthesis of pyrrolo[1,2-a]pyrazine derivatives, which are medicinally important com-

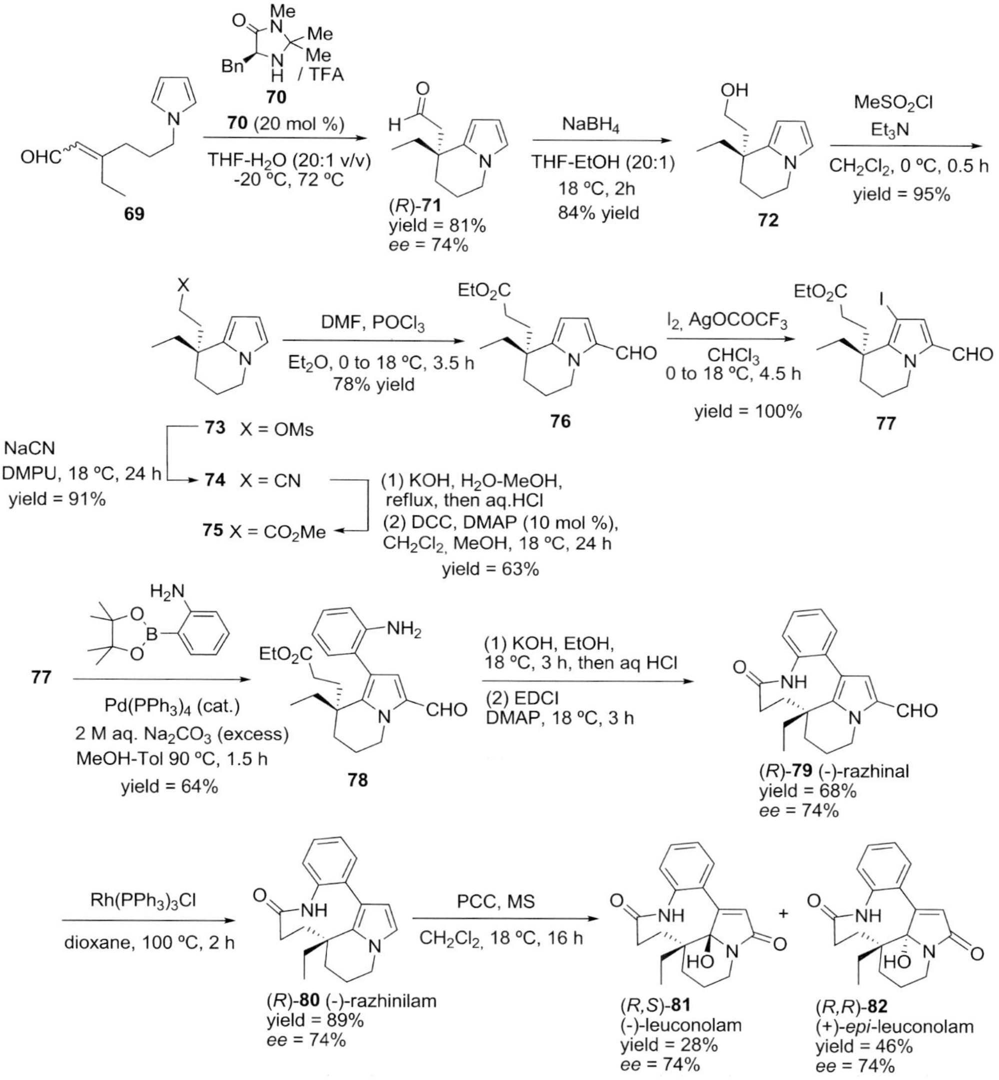

Scheme 7.12 Synthesis of (−)-razhinal **79**, (−)-razhinilam **80**, (−)-leuconolam **81** and (+)-epi-leuconolam **82**.

pounds with reported antiarrhythmic, antiamnesic, psychotropic, antihypertensive and other activities. Alkylation of 1-(2-bromoethyl)pyrrole **84** with N-acyl imine **83** gave an 87% yield of **85** with 92% ee, which was transformed into the antihypertensive drug precursor **86** in just one step in the presence of NaH, without racemization and in good yield (Scheme 7.13) [32].

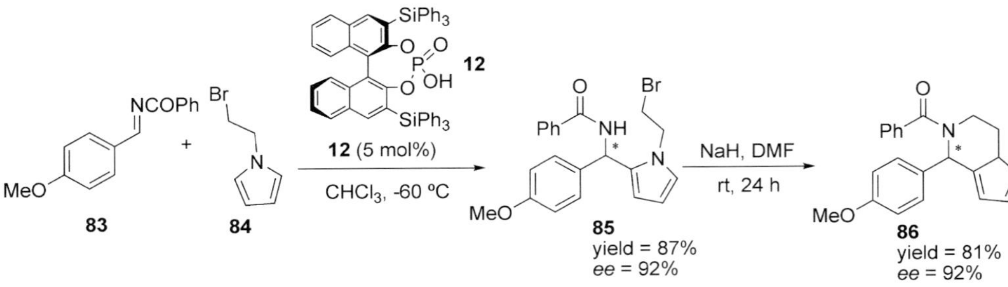

Scheme 7.13 Synthesis of antihypertensive drug precursor pyrrolo[1,2-a]pyrazine **86**.

7.3.4
Experiments: Selected Procedures

Synthesis of (+)-heliotridane 58 (Scheme 7.10)

To a solution of catalyst complex **7** (0.078 mmol) in acetonitrile (30 mL) under N_2 at −40 °C was added **54** (713 µL, 697 mg, 3.9 mmol) and pyrrole (**53**, 1.95 mL,1.87 g, 31.2 mmol). After 16 h of moderate stirring, the mixture was purified by flash silica chromatography to afford 870 mg (93% yield) of (*R*)-**55**. R_f= 0.55, (Hexanes:EtOAc 40 : 60). $[α]^{25}_D = -69.0$ (*c* 1.0, CH_2Cl_2). *ee* = 94%, HPLC: Chiralcel OD-H, 5/95 IPA/ hexanes, 0.8 mL min^{-1}, t_R = 10.6 min (major), t_S = 11.5 min (minor). ^{1}H NMR (500 MHz, $CDCl_3$): δ 8.80 (br s, 1H), 7.25 (s, 1H), 7.16 (s, 1H), 6.60 (d, *J* = 0.98 Hz, 1H), 6.09 (d, *J* = 2.9 Hz, 1H), 5.95 (s, 1H), 5.53 (sept., *J* = 6.8 Hz, 1H), 3.56–3.48 (m, 2H), 3.39–3.30 (m, 1H), 1.44 (t, *J* = 5.9 Hz, 6H), 1.38 (d, *J* = 6.8 Hz, 3H). ^{13}C NMR (100 MHz, $CDCl_3$): δ 192.8, 143.5, 136.9, 129.5, 121.1, 116.2, 107.6, 103.4, 49.2, 47.7, 28.2, 23.6, 23.5, 20.4. HRMS (ES$^+$): $C_{14}H_{19}N_3O$ [M + H]$^+$ requires 246.1609, found 246.1605. IR (film): 3389.2, 3104.9, 2967.0, 2932.2, 1672.4, 1619.8, 1463.8, 1453.6, 1395.1, 1254.7, 1199.2, 1150.5, 1111.1, 1088.3, 1027.8, 995.6, 948.0, 916.3, 784.5, 717.7 cm^{-1}.

Cyclization to compound 56: To a 0.5-dram oven-dried vial capped with a septum and purged with dry N_2, was added 0.5 mL (0.159 M in acetonitrile,0.0795 mmol) of (*R*)-**55**. To the magnetically stirred solution at rt was added methyl triflate (10.7 µL, 15.6 mg, 0.095 mmol). The solution was stirred for 2 h, until only a trace amount of **55** remained, as shown by TLC (hexanes:EtOAc = 40 : 60). Hünig's base (41.1 µL, 0.24 mmol) was added to the solution. The reaction was stirred until the appearance of a significant amount of product as shown by TLC (∼15 min). The reaction was directly purified by flash chromatography (CH_2Cl_2) to afford 10.7 mg (yield = 99%) of (*R*)-**56**. $[α]^{25}_D = +92.2$ (*c* 0.76, CH_2Cl_2). ^{1}H NMR (400 MHz, $CDCl_3$): δ 6.99 (d, *J* = 2.8 Hz, 1H), 6.44 (t, *J* = 3.1 Hz, 1H), 5.97–5.95 (m, 1H), 3.43–3.34 (m, 1H), 3.23 (dd, *J* = 8.0, 18.4 Hz, 1H), 2.58 (dd, *J* = 3.8, 18.4 Hz, 1H), 1.36 (d, *J* = 6.7 Hz, 3H). ^{13}C NMR (100 MHz, $CDCl_3$): δ 171.5, 145.5, 118.9, 110.6, 103.6, 43.3, 27.3, 20.8. HRMS (ES$^+$): C_8H_9NO [M + H]$^+$ requires 136.0762, found: 136.0764. IR (film):

2966.2, 2931.0, 2872.2, 1755.0, 1732.5, 1682.6, 1633.8, 1567.9, 1537.1, 1519.6, 1466.6, 1455.0, 1396.0, 1280.3, 1272.6, 1137.5, 1096.4 cm^{-1}.

Hydrogenation of the pyrrole ring: A sealed flask containing (R)-**56** (230 mg, 1.7 mmol) and 5% Rh-Al$_2$O$_3$ (40 mg) in EtOH (12 mL) was purged with H$_2$ (balloon). The solution was vigorously stirred at rt under H$_2$ for 19 h. The catalyst was removed via filtration through a plug of Celite eluting with EtOAc. The reaction mixture was concentrated *in vacuo* to afford 238 mg (yield = 99%) of (1R,7aR)-**57** as a 90:10 mixture of C-7 diastereomers. R_f = 0.34 (EtOAc:Hexanes:MeOH = 45:45:10). $[\alpha]^{25}_D$ = +23.9 (c 1.8, CH$_2$Cl$_2$). ^{1}H NMR (400 MHz, CDCl$_3$): δ 3.91 (dt, J = 6.4, 9.7 Hz, 1H), 3.52–3.40 (m, 1H), 3.04–2.94 (m, 1H), 2.84 (dd, J = 8.0, 16.4 Hz, 1H), 2.55–2.45 (m, 1H), 2.11–1.88 (m, 3H), 1.70–1.60 (m, 1H), 1.54–1.43 (m, 1H), 0.92 (d, J = 7.2 Hz, 3H). ^{13}C NMR (100 MHz, CDCl$_3$): δ 174.4, 64.9, 42.9, 41.0, 29.5, 26.8, 24.9, 15.8. HRMS (ES+): C$_8$H$_{13}$NO [M + H]$^+$ requires 140.1075, found: 140.1079. IR (film): 2963.2, 2877.1, 1694.9, 1460.7, 1425.4, 1407.8, 1378.0, 1349.0, 1296.1, 908.2, 896.6 cm^{-1}.

Reduction of the amide: To a solution of (1R,7aR)-**57** (35 mg, 0.251 mmol) in THF (2.5 mL) was added LAH (38.2 mg, 1.01 mmol). The vial was sealed with a cap and placed in an oil bath and stirred at 65 °C for 6.5 h. The solution was cooled to rt before Na$_2$SO$_4$·10 H$_2$O (324 mg, 1.01 mmol) was slowly added to the vial. The resulting solution was stirred at rt for 12.5 h. The solid material in the vial was removed by filtration with Et$_2$O. The solution was concentrated *in vacuo* to yield 30.4 mg (yield 97%) of (+)-heliotridane (**58**) as a 90:10 mixture of C-7 diastereomers favoring the desired product. $[\alpha]^{25}_D$ = +53.9 (c 1.6, Et$_2$O). ^{1}H NMR (400 MHz, CDCl$_3$) δ 3.44 (dt, J = 7.1, 9.3 Hz, 1H), 3.18 (ddd, J = 3.1, 7.1, 10.1 Hz, 1H), 3.00 (ddd, J = 6.2, 9.4, 11.0 Hz, 1H), 2.55 (ddd, J = 3.5, 7.4, 10.9 Hz, 1H), 2.45 (dt, J = 6.2, 9.7 Hz, 1H), 2.50–2.20 (m, 1H), 1.88–1.78 (m, 1H), 1.76–1.66 (m, 2H), 1.56–1.64 (m, 1H), 1.30–1.46 (m, 2H), 0.97 (d, J = 7.0 Hz, 3H). ^{13}C NMR (100 MHz, CDCl$_3$) δ 68.4, 56.4, 54.3, 35.5, 31.9, 26.6, 26.5, 15.0. HRMS (ES+): C$_8$H$_{15}$N [M + ACN]$^+$ requires 167.1548, found: 167.1552. IR (film): 2952.0, 2617.5, 2843.3, 1502.5, 1476.5, 1462.5, 1444.9, 1393.1, 1375.8 cm^{-1}.

Synthesis of compound 72 (Scheme 7.12)

A magnetically stirred mixture of (5S)-2,2,3-trimethyl-5-phenylmethyl-4-imidazolidinone monotrifluoroacetate (**70**, 347 mg) and water (1.8 mL) in THF (30 mL) was cooled to −20 °C and treated, dropwise, with a solution of compound **69** (1.00 g, 5.23 mmol) in THF (8.0 mL). The reaction mixture was stirred at −20 °C for 72 h, quenched by dropwise addition of satd aq sodium bicarbonate (20 mL) and then warmed to 18 °C and treated with Et$_2$O (40 mL) and water (10 mL). The separated organic phase was washed with brine (1 × 10 mL) and dried over MgSO$_4$, filtered and concentrated under reduced pressure. Flash chromatography (hexane:EtOAc = 95:5) afforded 812 mg (81%) of aldehyde (R)-**71**. Colorless unstable oil. ^{1}H NMR (300 MHz, CDCl$_3$): δ 9.64 (t, J = 3.3 Hz, 1H), 6.52 (m, 1H), 6.14 (m, 1H), 5.94

(m, 1H), 3.92 (m, 2H), 2.60 (m, 2H), 2.05–1.80 (m, 6H), 0.91 (t, $J = 7.5$ Hz, 3H). ^{13}C NMR (75 MHz, CDCl$_3$): δ 203.8, 133.6, 119.0, 107.5, 104.3, 53.2, 45.1, 36.9, 34.1, 31.2, 19.9, 8.5. HRMS (EI) C$_{12}$H$_{17}$NO [M]$^+$ requires 191.1310, found 191.1316. MS (EI, 70 eV): m/z 191 (M$^+$, 60%), 163 (48), 162 (68) 149 (40), 148 (100), 134 (85), 133 (52), 118 (40), 80 (32), 41 (25). IR (KBr, neat) 3098, 2944, 2878, 2737, 1710, 1486, 1462, 1326, 1270, 1204, 1079, 954, 774, 710, 610 cm^{-1}. *Reduction of the aldehyde*: Sodium borohydride (238 mg, 6.3 mmol) was added in portions to a magnetically stirred solution of aldehyde **71** (0.8 g, 4.19 mmol) and EtOH (0.5 mL) in THF (10 mL) maintained at 18 °C under an atmosphere of nitrogen. After 2 h the reaction mixture was treated sequentially with satd aq potassium hydrogen sulfate (1.0 mL), water (2.0 mL) and EtOAc (20 mL). The separated organic phase was dried (MgSO$_4$), filtered and concentrated under reduced pressure. Flash chromatography (hexane:EtOAc 80 : 20) afforded 679 mg (84%) of compound (*R*)-**72**. Colorless oil, [α]$_D$ = −11 (*c* 0.9, CHCl$_3$). *ee* = 74%, HPLC: Chiralpak AS-H, 95/5 hexane/IPA, 0.9 mL min^{-1}, t_R = 20.2 (major), t_S = 33.4 min (minor).

*Synthesis of pyrrolo[1,2-a]pyrazine **86** (Scheme 7.13)*

To a flame-dried reaction tube was added *N*-acyl imine **83** (0.2 mmol), catalyst **12** (8.8 mg, 0.01 mmol) and anhydrous chloroform (1 mL) under argon. The resulting solution was stirred at −60 °C for about 10 min. Then, pyrrole **84** (0.2 mmol) was added via a syringe and the mixture was stirred at −60 °C for 46 h. Direct column chromatography (hexane: EtOAc = 4 : 1 to 2 : 1) gave 34 mg (87% yield) of compound **85**. White solid, m.p. = 139–141 °C. [α]$^{20}_D$ = −2.5 (*c* 1.70, CHCl$_3$). *ee* = 92%, HPLC: Chiralcel AD-H, 90/10 hexane/IPA, 1.0 mL min^{-1}, t = 23.47 min (minor), t = 29.08 min (major). ^{1}H NMR (250 MHz, CDCl$_3$): δ 3.39–3.53 (m, 2H), 3.80 (s, 3H), 4.21–4.29 (m, 2H), 5.92 (d, J = 1.3 Hz, 1H), 6.13 (t, J = 3.2 Hz, 1H), 6.45 (d, J = 8.3 Hz, 1H), 6.68–6.73 (m, 2H), 6.86 (d, J = 8.8 Hz, 2H), 7.26 (d, J = 8.8 Hz, 2H), 7.40–7.53 (m, 3H), 7.78 (d, J = 7.0 Hz, 2H). ^{13}C NMR (62.5 MHz, CDCl$_3$): δ 30.8, 48.1, 49.4, 55.4, 107.9, 108.7, 114.1, 122.1, 127.1, 128.4, 128.7, 131.8, 132.3, 132.5, 134.0, 159.2, 166.1. HRMS (ESI): C$_{21}$H$_{21}$BrN$_2$O$_2$ ([M + H]$^+$ requires 413.0859, found 413.0847. IR (neat): 3334, 2960, 2925, 1628, 1513, 1488, 1296, 1245 cm^{-1}.

*Cyclization of compound **85***: To a solution of compound **85** (0.05 mmol, 20.5 mg) in anhydrous DMF (0.5 mL) was added NaH (3.0 mg, 50%, 1.2eq.) at 0 °C. The resulting mixture was stirred at rt for 24 h. The reaction was quenched with satd aq NH$_4$Cl solution, and extracted with CH$_2$Cl$_2$ (3 × 6 mL), dried over Na$_2$SO$_4$. Flash chromatography (hexane: EtOAc = 4 : 1 to 2 : 1) afforded 13.4 mg (81%) of compound **86**. White solid. [α]$^{20}_D$ = +134.6 (*c* 0.73, CHCl$_3$). *ee* = 92%, HPLC: Chiralcel AD-H, 85/15 hexane/IPA, 1.0 mL min^{-1}, t = 25.49 min (major), t = 38.87 min (minor). ^{1}H NMR (250 MHz, CDCl$_3$): δ 3.45–4.07 (m, 7H), 5.98 (s, 1H), 6.25 (s, 1H), 6.66 (d, J = 1.0 Hz, 1H), 6.81 (d, J = 12.8 Hz, 1H), 7.17–7.26 (m, 3H), 7.42 (m, 5H). ^{13}C NMR (62.5 MHz, CDCl$_3$): δ 44.6, 55.3, 106.2, 108.7, 113.8, 119.3, 126.5, 126.8, 128.7, 129.9, 135.7, 159.1. HRMS (ESI): C$_{21}$H$_{20}$N$_2$O$_2$ ([M + H]$^+$ requires 333.1597, found 333.1599. IR (neat): 2964, 2927, 1630, 1508, 1416, 1245 cm^{-1}.

7.4
Friedel–Crafts Alkylation of Furan Derivatives in Total Synthesis

Compared to other heteroaromatic compounds, the catalytic enantioselective FC reaction with furans has been less developed. Nevertheless, furans are present in many natural and bioactive compounds and are valuable intermediates in organic synthesis, for example, for the preparation of butenolides or dihydropyrones. Jørgensen [33] first and Jurczak [34] later have described the FC alkylation of 2-substituted furans with glyoxylates to give 2-furyl-hydroxyacetates **90**. In the latter approach, the authors anticipate that, after reduction to diols, these compounds could be transformed into linear and cyclic compounds such as polyhydroxylated compounds, sugars and aza-sugars, under oxidizing conditions (Scheme 7.14). However, to the best of our knowledge, none of the diols currently available via catalytic enatioselective FC alkylation of furans have been used for this purpose.

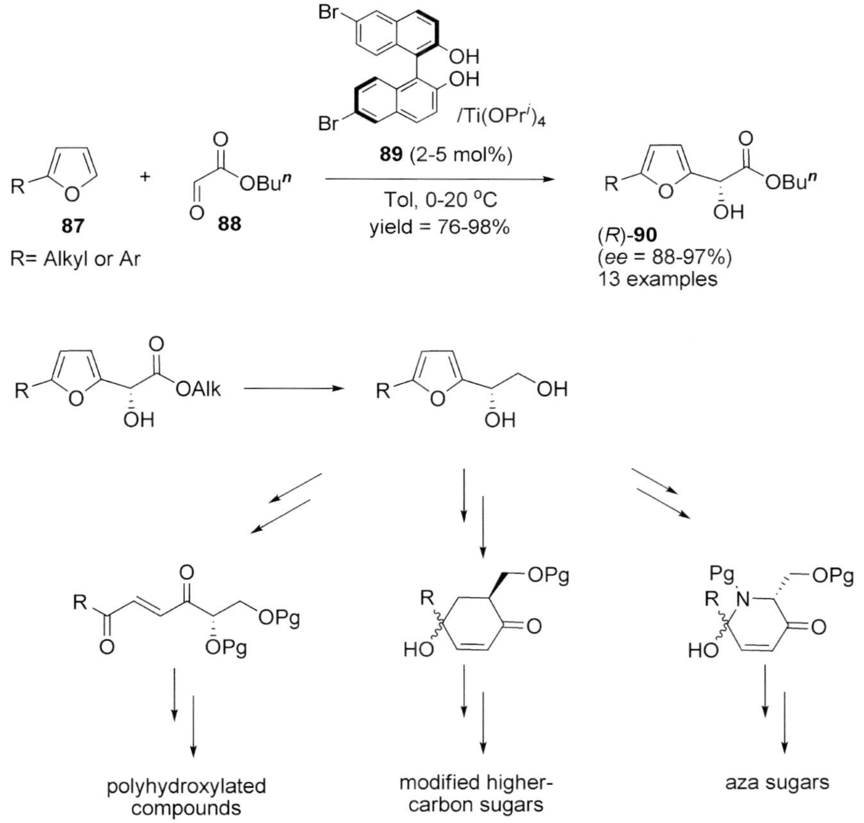

Scheme 7.14 Synthesis of 2-(2-furanyl)-1,2-ethanediols and potential synthetic transformations.

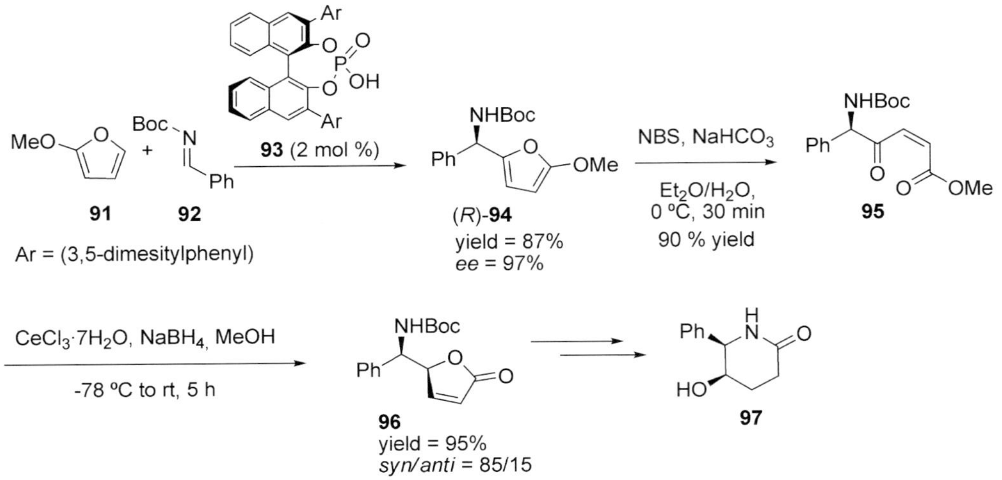

Scheme 7.15 Synthesis of 2-aminobutenolides.

7.4.1
Synthesis of Aminobutenolides

Terada et al. have reported an organocatalytic asymmetric aza-Friedel–Crafts alkylation of 2-metoxhyfuran **91** with N-acyl imines using a BINOL-derived phosphoric acid in dichloroethane as the solvent at −35 °C [35]. The reaction with the imine **92**, derived from benzaldehyde, in the presence of **93** takes place with good yield (87%) providing compound **94** with 97% ee (Scheme 7.15). The aza-Achmatowitcz reaction of **94** cleaved the furan ring cleanly to form the 1,4-dicarbonyl compound **95** in 96% ee. Subsequent reductive cyclization of **95** under Luche conditions led to butenolide **96** in 95% yield, without loss of enantiomeric excess. These kinds of compounds can be transformed [36] into piperidones such as **97**, which are synthetic precursors for piperidine-ether based hNK_1 antagonists [37].

7.4.2
Experiments: Selected Procedures

Synthesis of butenolide 96 (Scheme 7.15)

To a solution of (R)-**93** (1.95 mg, 0.002 mmol) in DCE (1 mL) at −35 °C under N_2, was added N-Boc protected imine **92** (20.5 mg, 0.1 mmol) and 2-methoxyfuran (**91**, 11.1 μL, 0.12 mmol). The resulting solution was stirred for 24 h at this temperature. The reaction mixture was purified by column chromatography (hexane:EtOAc = 12 : 1 to 8 : 1) to give 2-furyl amine (R)-**94** in 87% yield. White solid. R_f = 0.40 (hexane: EtOAc = 1 : 4). ee = 97%, HPLC: Chiralpak AD-H, 95/5 hexane/IPA, 1.0 mL min^{-1}), t_R = 14.9 (major), t_S = 18.0 min (minor). ^{1}H NMR (270 MHz, CDCl$_3$): δ 1.43 (brs, 9H), 3.80 (s, 3H), 5.04 (d, J = 3.1 Hz, 1H), 5.24 (br s, 1H), 5.79 (br s, 1H),

5.94 (d, $J = 3.1$ Hz, 1H), 7.23–7.38 (m, 5H). ^{13}C NMR (67.8 MHz, CDCl$_3$): δ 28.3, 52.6, 57.7, 79.7, 79.8, 108.7, 126.9, 127.5, 128.5, 139.9, 143.6, 154.8, 161.4. HRMS (ESI): m/z C$_{17}$H$_{21}$NaNO$_4$ [M + Na]$^+$ requires 326.1363, found 326.1364. IR (KBr): 3354, 2984, 2943, 1678, 1614, 1585, 1518, 1367, 1319, 1256, 1163, 1043, 1009, 947, 880, 746 cm^{-1}.

Oxidative cleavage of the furyl ring: To a stirred solution of (R)-**94** (151.7 mg, 0.5 mmol) in Et$_2$O (5 mL) was added satd aq NaHCO$_3$ (5 mL). To the resulting biphasic solution was added NBS (89.0 mg, 0.5 mmol) over 15 min as solid at 0 °C. The resulting clear solution was diluted with water and extracted with CH$_2$Cl$_2$, dried over Na$_2$SO$_4$ and filtrated. Column chromatography (EtOAc/Hexane 1 : 10 to 1 : 4) gave 1,4-dicarbonyl compound (R)-**95** (90% yield). Oil, $R_f = 0.17$ (hexane:EtOAc = 1 : 4). ee = 96%, HPLC: Chiralpak AD-H, 90/10 hexane/IPA, 1.0 mL min^{-1}, t_S = 16.5 min (minor), t_R = 21.0 min (major). ^{1}H NMR (270 MHz, CDCl$_3$): δ 1.41 (brs, 9H), 3.78 (s, 3H), 5.55 (brd, $J = 6.2$ Hz, 1H), 6.03 (br, 1H), 6.05 (d, $J = 12.2$ Hz, 1H), 6.32 (d, $J = 12.2$ Hz, 1H), 7.29–7.39 (m, 5H). ^{13}C NMR (67.8 MHz, CDCl$_3$): δ 28.3, 52.3, 64.2, 79.9, 127.2, 128.0, 128.6, 129.1, 136.2, 137.8, 154.7, 165.6, 197.4. HRMS (ESI)C$_{18}$H$_{23}$NaNO$_4$ ([M + Na]$^+$ requires 342.1312, found 342.1312. IR (KBr): 3389, 2976, 2943, 1738, 1724, 1693, 1624, 1501, 1439, 1354, 1290, 1244, 1167, 1051, 891, 700 cm^{-1}.

Reductive cyclization to γ-butenolide: To a solution of (R)-**95** (0.06 mmol) and CeCl$_3$·7H$_2$O (0.06 mmol) in MeOH (1 mL) at −78 °C was added portion-wise NaBH$_4$ (0.06 mmol) and the resulting solution was allowed to warm to rt for 4 h. The mixture was diluted with saturated NH$_4$Cl and extracted with CH$_2$Cl$_2$. The organic phase was dried over Na$_2$SO$_4$ and the usual purification method gave *tert*-butyl (R)-[(S)-2,5-dihydro-5-oxofuran-2-yl)]phenylmethylcarbamate **96** as a 85:15 diastereomeric mixture (95%). *syn*-**96**, (major isomer). $R_f = 0.25$ (hexane:EtOAc = 1 : 2). ee = 96%, HPLC: AD-H, 90/10 hexane/IPA 90/10, 1.0 mL min^{-1}, t = 11.3 min (minor), t = 25.3 min (major). ^{1}H NMR (270 MHz, CDCl$_3$): δ 1.43 (brs, 9H), 5.06 (br, 1H), 5.28 (br, 1H), 5.42 (br, 1H), 5.96 (brd, $J = 4.3$ Hz, 1H), 7.14–7.40 (m, 6H). ^{13}C NMR (67.8 MHz, CDCl$_3$): δ 28.3, 56.4, 80.5, 84.5, 123.1, 127.1, 128.5, 128.8, 135.5, 152.9, 155.9, 172.3. HRMS (ESI): m/z C$_{18}$H$_{23}$NaNO$_4$ ([M + Na]$^+$ requires 312.1206, found 312.1207. IR (KBr): 3371, 2984, 2943, 1778, 1763, 1697, 1599, 1514, 1499, 1369, 1250, 1157, 1094, 926, 829, 702 cm^{-1}.

7.5
Friedel–Crafts Alkylation of Arenes in Total Synthesis

Arenes are constituents of many natural and bioactive compounds. In general, benzene derivatives are less reactive toward electrophiles than most of the five-membered ring heterocycles. Therefore, most of the examples of catalytic enantioselective reactions with arenes have been carried out with aromatic compounds substituted with strongly electron-donating groups, that is, phenols, phenyl ethers and anilines.

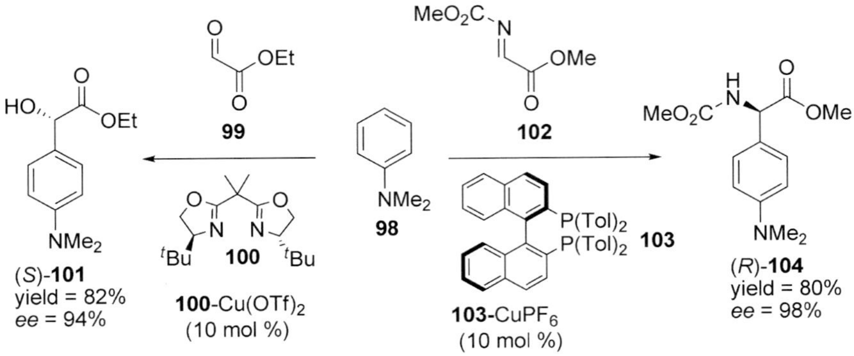

Scheme 7.16 Synthesis of mandelic acid derivatives and α-amino acids.

7.5.1
Synthesis of Optically Active Mandelic Acid Derivatives and Aromatic α-Amino Acids

Jørgensen et al. have reported the FC reaction between substituted anilines **98** and ethyl glyoxylate **99** in the presence of the (S)-tert-BuBOX-Cu(OTf)$_2$ complex **100**-Cu(OTf)$_2$ to provide p-aminomandelic acid derivatives **101** which are starting materials for the synthesis of many biologically active compounds [33]. The reaction takes place with high enantioselectivity and allows use of a variety of substituted anilines as well as other aromatic amines and heterocyclic compounds such as 2-substituted furans (Scheme 7.16). The same group has developed a related aza-FC reaction [38] that uses N-acyl imines of alkyl glyoxylates **102** as electrophiles to give the corresponding aromatic and heteroaromatic α-amino acids **104** with very high enantiomeric excesses (Scheme 7.16). The reaction is catalyzed by a chiral BINAP-Cu(I) complex **103** and can also be performed with substituted furans, thiophenes, pyrroles and aromatic amines.

7.5.2
Synthesis of Optically Active Chromanes

The chromane (dihydrobenzopyran) core is present in a large number of naturally occurring compounds, for instance vitamin E, and in many other bioactive compounds and drugs. The chromane system can be built up via intramolecular ring-closing FC alkylation type reactions of aromatic ethers. For instance, Xiao et al. have describe the cyclization of the ether-tethered enal **105** in the presence of the imidazolidinone **3** and 3,5-dinitrobenzoic acid to give compound **106** in 88% yield and with 90% ee (Scheme 7.17) [19].

An interesting approach for the synthesis of chromanes starting from phenols **107** and unsaturated α-keto esters **108** has been reported by the group of Jørgensen [39]. These researchers have developed a tandem reaction involving an oxa-Michael addition followed by an intramolecular FC alkylation. The reaction proceeds under the influence of BOX-Mg catalyst **109**-Mg^{2+} to give diastereomerically pure products

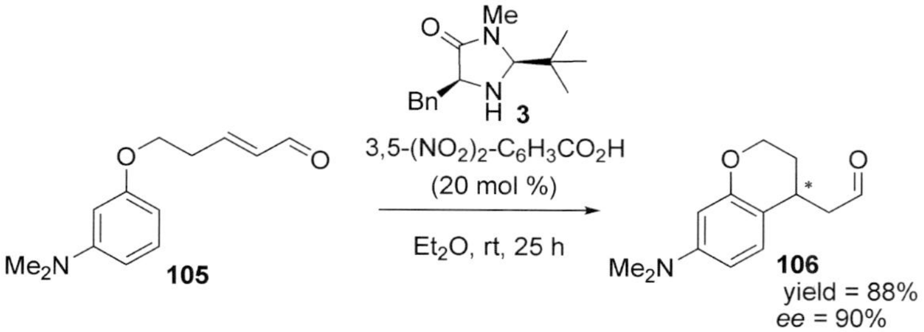

Scheme 7.17 Synthesis of chromanes via intramolecular FC alkylation of ether-tethered enals.

110 with enantioselectivities up to 73% *ee* (Scheme 7.18). However, studies carried out with this reaction revealed that the second step of the reaction (the FC reaction) is totally diastereoselective. The enantioselectivity of the reaction is determined in the oxa-Michael addition step.

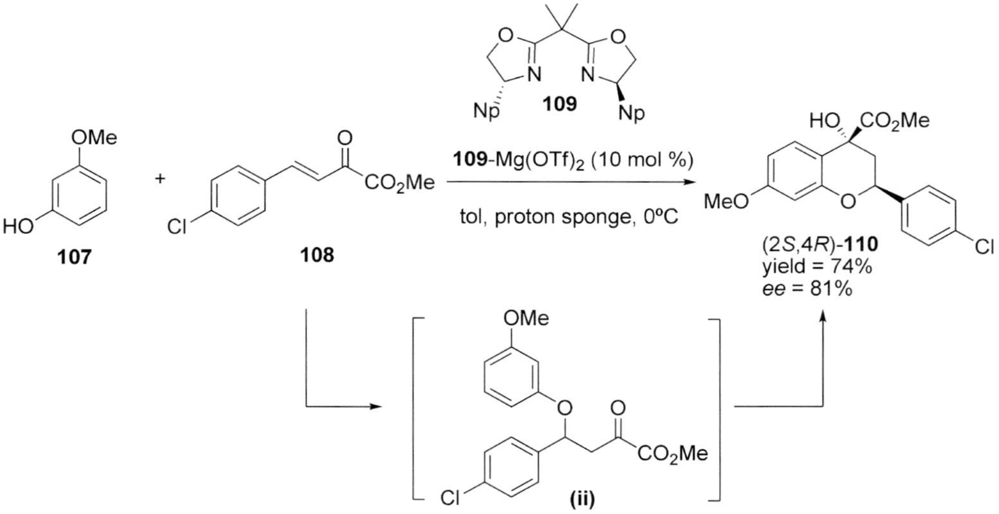

Scheme 7.18 Synthesis of chromanes via tandem oxa-Michael addition and FC alkylation.

7.5.3
Experiments: Selected Procedures

Synthesis of α-hydroxy acid 101 (Scheme 7.16)

To a flame-dried Schlenk tube was added Cu(OTf)$_2$ (18.1 mg, 0.05 mmol) and (S)-*tert*-BuBOX **100** (15.5 mg, 0.055 mmol). The mixture was dried under vacuum for 1–2 h, anhydrous THF (2 mL) was added under N$_2$. After stirring for 0.5 h, freshly distilled ethyl glyoxylate (255 mg, 2.5 mmol) and N,N-dimethylaniline (0.5 mmol) were added. After stirring at 0 °C for 36 h, the reaction mixture was filtered through a short pad of silica gel with Et$_2$O, concentrated and purified by column chromatography (Et$_2$O:hexanes 70 : 30) to give 91 mg (82%) of compound (S)-**101**. White solid, m.p. = 101–103 °C. $[\alpha]_D = +98.7$ (c 0.23, CHCl$_3$). ee = 94%, GC: Astec B-PM, T_{in} 80 °C, 20 °C min^{-1} to 135 °C, 1 °C min^{-1} to 145 °C, 0.2 °C min^{-1} to 158 °C, t_R = 38.5 min (minor), t_S = 39.2 min (major). ^{1}H NMR (400 MHz, CDCl$_3$): δ 7.28–7.23 (m, 2H), 6.73–6.68 (m, 2H), 5.07 (d, J = 6.0 Hz, 1H), 4.27 (dq, J = 10.5, 7.2 Hz, 1H), 4.15 (dq, J = 10.5, 7.2 Hz, 1H), 3.33 (d, J = 6.0 Hz, 1H), 2.96 (s, 6H), 1.23 (t, J = 7.2 Hz, 3H). ^{13}C NMR (100 MHz, CDCl$_3$): δ 174.4, 150.8, 127.8, 126.4, 112.6, 73.0, 62.2, 40.7, 14.3. HRMS C$_{12}$H$_{17}$NO$_3$ (M$^+$) requires 223.1208, found 223.1207.

Synthesis of α-amino acid 104 (Scheme 7.16)

To a flame-dried Schlenk tube equipped with a magnetic stirring bar is added the aza-ylide reagent (140 mg, 0.40 mmol) which is dried under vacuum at rt for 1 h before toluene (0.66 mL) is added by syringe. After stirring for 10 min, a 0.42 M solution of methyl glyoxylate in toluene (1.14 mL, 0.48 mmol) is added to the suspension and stirring at rt continued for another 15 min. The white milky slurry containing **102** is cooled to −78 °C. (Complete imine formation should be checked by ^{1}H NMR, as full conversion of the aza-ylide is crucial). A catalyst solution prepared by drying CuPF$_6$·4MeCN (15 mg, 0.040 mmol) and (R)-Tol-BINAP (**103**, 30 mg, 0.044 mmol) in a flame-dried Schlenk tube under vacuum for 1 h and subsequently dissolved in a mixture of CH$_2$Cl$_2$ (0.20 mL) and toluene (2.0 mL) is added by syringe and, after stirring for 30 s, N,N-dimethylaniline (0.80 mmol) is added in one portion. After 44 h, the reaction mixture is filtered through a plug of silica and eluted with Et$_2$O (50 mL). Flash chromatography gave compound (R)-**104** in 80% yield. $[\alpha]_D = -146.3$ (c 0.8, CHCl$_3$). ee = 98%, HPLC: Chiralcel OD, 90/10 hexane/IPA, 1 mL min^{-1}, t_S = 17.2 min (minor), t_R = 22.8 min (major). ^{1}H NMR (400 MHz, CDCl$_3$): δ 7.18 (d, J = 8.6 Hz, 2H), 6.65 (d, J = 8.6 Hz, 2H), 5.61 (bd, J = 5.8 Hz, 1H), 5.22 (d, J = 7.0 Hz, 1H), 3.68 (s, 3H), 3.64 (s, 3H), 2.91 (s, 6H). ^{13}C NMR (100 MHz, CDCl$_3$): δ 171.9, 155.9, 150.4, 127.9, 123.6, 112.3, 57.3, 52.5, 52.2, 40.3. HRMS C$_{13}$H$_{18}$N$_2$O$_4$ [M + Na]$^+$ requires 289.1164, found 289.1168.

Synthesis of chromane 106 (Scheme 7.17)

To a solution of the α, β-unsaturated aldehyde **105** (85 mg, 0.371 mmol) in ether (3.7 mL) was added (2S,5S)-**3** (18.3 mg, 0.0741 mmol) and 3,5-dinitrobenzoic acid (15.7 mg, 0.0741 mmol). The resulting mixture was stirred at rt for 25 h and passed through a cold silica gel plug eluting with Et$_2$O and concentrated. The resulting residue was chromatographed (petroleum ether:EtOAc) to give 75 mg (88%) of compound **106** (tentatively assigned R-configuration). Colorless oil. $[\alpha]^{25}_D = -16.0$ (c 1.0, CHCl$_3$). ee = 90%, HPLC: Chiralpak AS-H, 90/10 hexanes/IPA, 1 mL min^{-1}, t = 11.9 min (major), t = 13.3 min (min). ^{1}H NMR (400 MHz, CDCl$_3$): δ 9.84(s, 1H), 6.93 (d, J = 7.6 Hz, 1H), 6.33 (dd, J = 8.6, 2.6 Hz, 1H), 6.17 (d, J = 2.4 Hz, 1H), 4.15 (t, J = 5.2 Hz, 2H), 3.39 (dd, J = 4.6 Hz, 1H), 2.89 (s, 6H), 2.64–2.71 (m, 1H), 2.13–2.18 (m, 1H), 1.71–1.77 (m, 1H). ^{13}C NMR (100 MHz, CDCl$_3$) δ 201.9, 155.3, 150.7, 128.9,

112.5, 106.2, 100.5, 63.6, 50.8, 40.6, 28.3, 27.5. MS: m/z 229.0. Anal $C_{13}H_{17}NO_2$ requires C, 71.21; H, 7.81; N, 6.39, found C, 71.10; H, 7.76; N, 6.38.

Synthesis of chromane 110 (Scheme 7.18)

To a flame-dried Schlenk tube was added Mg(OTf)$_2$ (8.0 mg, 0.025 mmol) and (S)-**109** (12 mg, 0.028 mmol). The mixture was dried under vacuum for 0.5 h and distilled anhydrous toluene (0.5 mL) was added. After stirring for 0.5 h, unsaturated ester **108** (58 mg, 0.25 mmol) and p-methyl-N,N-dimethylaniline (3.6 μL, 0.025 mmol) were added and the mixture was cooled to 0 °C. After addition of **107** (55 μL, 0.50 mmol), the mixture was stirred overnight at 0 °C. Flash chromatography (Et$_2$O:pentane 20:80) afforded 77 mg of (2S,4R)-**110** (89%). Oil. $[\alpha]_D = +0.28$ (c 0.63, CHCl$_3$), ee = 73%, HPLC: Chiralcel OD, 95/5 hexane/IPA, $t = 12.9$ min (major), $t = 14.4$ min (minor). ^{1}H NMR (400 MHz, CDCl$_3$): δ 7.31 (dt, $J = 2.0, 8.4$ Hz, 2H), 7.19 (dt, $J = 2.0$, 8.4 Hz, 2H), 6.60 (dd, $J = 0.8, 8.8$ Hz, 1H), 6.45–6.41 (m, 2H), 4.43 (d, $J = 2.0$ Hz, 1H), 4.27 (dd, $J = 6.0, 13.2$ Hz, 1H), 3.90 (s, 1H), 3.74 (s, 3H), 2.38 (td, $J = 2.0$, 13.2 Hz, 1H), 2.24 (dd, $J = 5.6, 13.2$ Hz, 1H). ^{13}C NMR (100 MHz, CDCl$_3$) δ 170.3, 159.8, 152.4, 142.3, 132.9, 130.4, 130.1, 129.1, 117.4, 108.8, 102.1, 101.9, 94.6, 55.6, 55.5, 53.8, 36.9, 36.8. HRMS [M$^+$Na] requires 371.0662, found 371.0665.

7.6 Asymmetric Synthesis of Natural Products Based on Diastereoselective Friedel–Crafts Reactions

In this last section we will present several targeted syntheses of enantioenriched natural products, in which diastereoselective FC alkylations were used as the key step. In these reactions, a new stereogenic center is stereoselectively formed under the influence of another stereogenic center present in a chiral reaction partner or chiral auxiliary.

7.6.1 Synthesis of Hapalindole Alkaloids

The group of Baran has developed a procedure for the direct coupling of indoles with carbonyl compounds [40]. The coupling is achieved by reaction of deprotonated indole and the carbonyl enolate in the presence of an oxidant ion. This type of transformation shows high levels of chemoselectivity, regioselectivity, stereoselectivity and practicality (easy scalable). It can be used to construct quaternary carbon centers and is amenable to asymmetric synthesis. As limitations, the reaction is not amenable to a wide range of heterocyclic scaffolds (pyridine, pyrimidine, furan, thiophene, N-protected indoles and pyrroles, pyrazole and indazole do not react), electron-deficient indoles do not couple well, and methyl ketones are prone to homodimerization. Mechanistic evidence suggests two plausible mechanistic interpretations (pathways A and B) as outlined in Scheme 7.19 for the reaction between indole **10** and (R)-carvone **111**, the radical anion pathway B being preferred [40c].

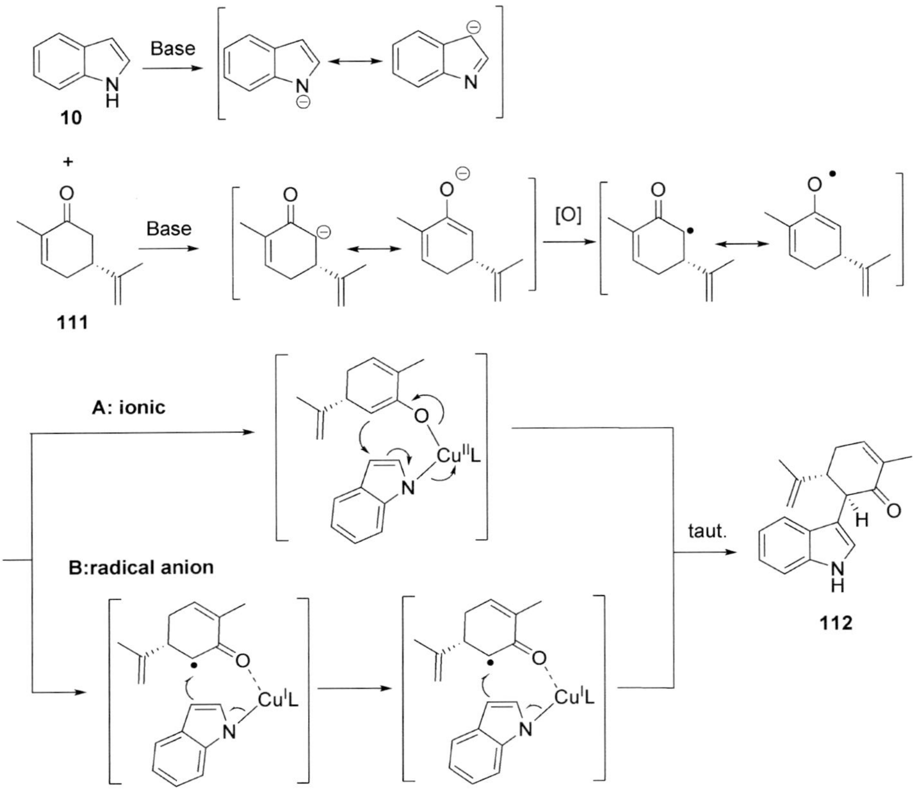

Scheme 7.19 Direct indole and (R)-carvone coupling. Possible mechanistic pathways.

The reaction outlined in Scheme 7.19 has been applied to the synthesis of several alkaloids of the hapalindole family (Scheme 7.20) [40a], such as (+)-hapalindole Q **116** [41] and (−)-12-epi-fischerindole U isothiocyanate **119** [42]. Coupling between indole and (R)-carvone was accomplished upon treatment with LHMDS followed by addition of copper(II) ethylhexanoate to furnish **112** in 53% isolated yield as a single diastereoisomer. Deprotonation of the indole NH of **112**, conjugate reduction and stereoselective quenching of the resulting enolate with acetaldehyde gave alcohol **113** which was dehydrated with Martin sulfurane to give compound **114** in 75% overall yield. Microwave-enhanced reductive amination with NaBH$_3$CN/ NH$_4$OAc furnished the amine **115** as a 6 : 1 mixture of diastereomers. Finally, conversion of the amine into isothiocyanate with CS(imid)$_2$ gave (+)-hapalindole Q **116** in 63% yield. On the other hand, treatment of **114** with TMSOTf brought about a biomimetic acid-catalyzed ring closure to afford ketone **117** in 75% yield (based on recovered starting material). Reductive amination furnished amine **118** as a 10 : 1 mixture of diastereomers (60% yield). This was followed by isothiocyanate formation to give (−)-**119** that has the opposite configuration to the natural product [42].

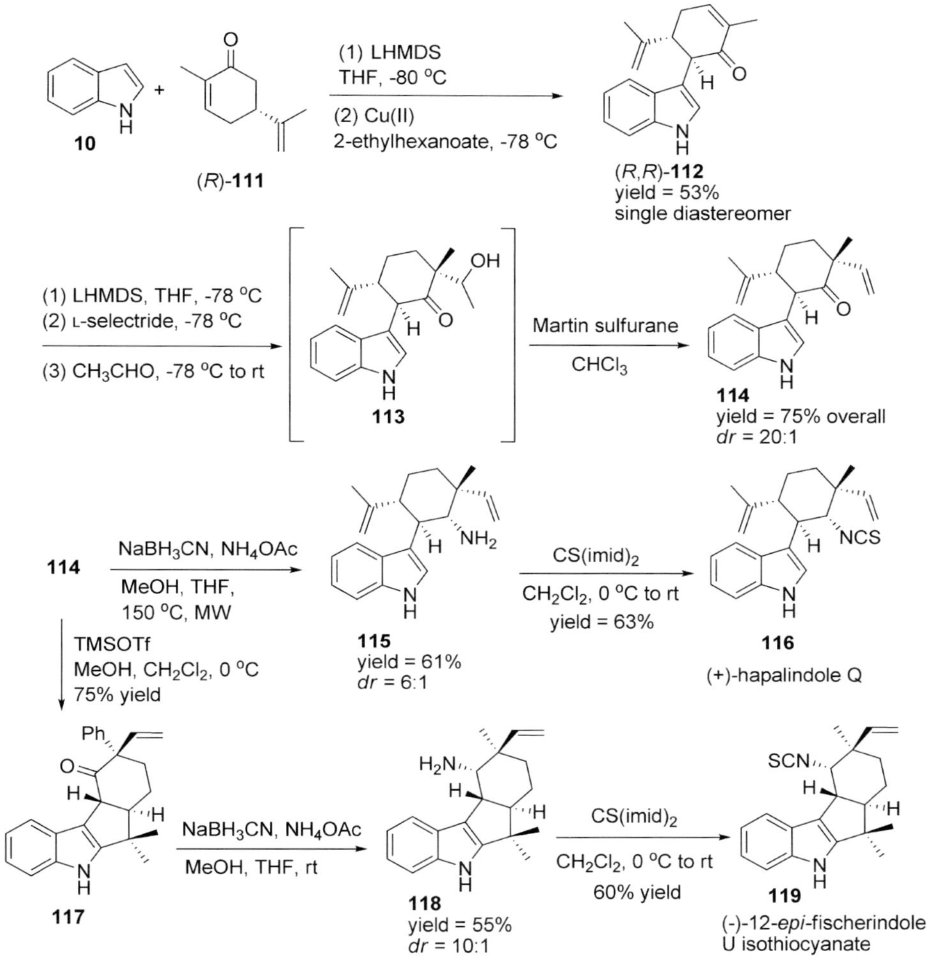

Scheme 7.20 Synthesis of (+)-hapalindole Q **116** and (−)-12-epi-fischerindole U isothiocyanate **119**.

The same authors have reported the synthesis of other alkaloids of the same family, namely fischerindoles I **121** and G **122** and welwitindolinone A **123**, starting from carvone oxide **120** using a similar coupling strategy (Scheme 7.21) [40b].

7.6.2
Synthesis of Acremoauxin A and Oxazinin 3

The above coupling reaction can be accomplished with carbonyl compounds other than ketones, as shown in the synthesis of acremoauxin A **129** and oxazinin 3 **133** reported by Baran et al. (Scheme 7.22) [40c]. Acremoauxin A was isolated from *Acremonium roseum* and exhibits plant-growth inhibition [43]. Its synthesis

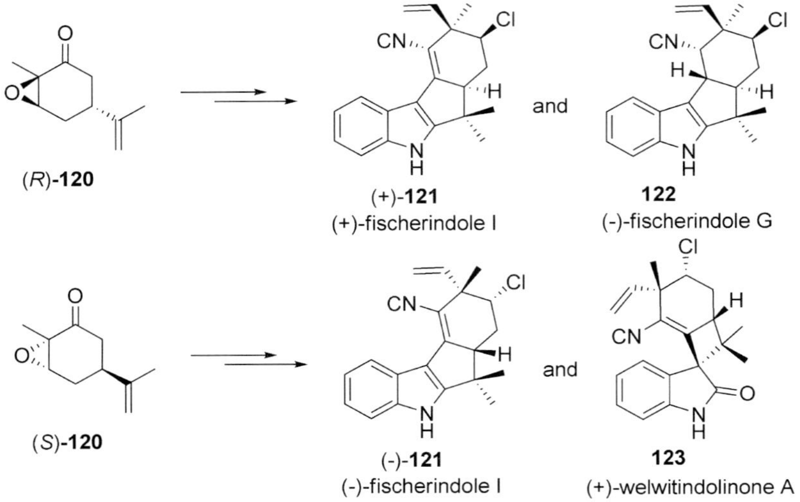

Scheme 7.21 Synthetic approaches to fischerindole I **121**, fischerindole G **122** and welwitindolinone A **123**.

commenced with the union of indole **10** and camphorsultam propionate **124** to provide a 49% yield of the coupled product **125** as a single diastereoisomer ($dr > 20:1$). Hydrolysis of the chiral auxiliary provided **126** (83%), which was coupled with the known arabinitol derivative **127** (prepared from mannitol) to give compound **128** in 69% yield. Deprotection of the acetal with AcOH gave a 62% yield of acremoauxin A **129**, which was spectroscopically identical to the natural product [43]. The synthesis of oxazinin 3 began with the known compound **130** (derived in six steps from tyrosine), which was protected as the pivaloyl amide **131**. Direct indole coupling on this substrate provided compound **132** in moderate yield (40%) and good diastereoselectivity ($dr\ 8:1$). Deprotection of the pivaloyl and benzyl groups furnished oxazinin 3 **133** as a single enantiomer bearing features consistent with those of the natural product [44].

7.6.3
Synthesis of (S)-Ketoralac

The above direct coupling can be extended to pyrrole under slightly modified conditions. The utility of this reaction was showcased in a synthesis (Scheme 7.23) [40c, d] of the nonsteroidal anti-inflammatory drug ketoralac **138**. It is known that the (S)-enantiomer of **138** is significantly more active than the (R)-antipode [45]. The asymmetric synthesis started by installing the appropriate Oppolzer sultam **135** as a chiral auxiliary on the known pyrrole acid **134** (**136**, yield = 100%). This was followed by intramolecular coupling to forge the bicyclic core of ketoralac. Surprisingly, the usual conditions for this reaction failed; success could only be attained when ferrocenium hexaflurophosphate was used as the oxidant providing the extremely unstable product **137** (yield = 65%, $dr = 4.5:1$, determined by ^{1}H NMR) that needed

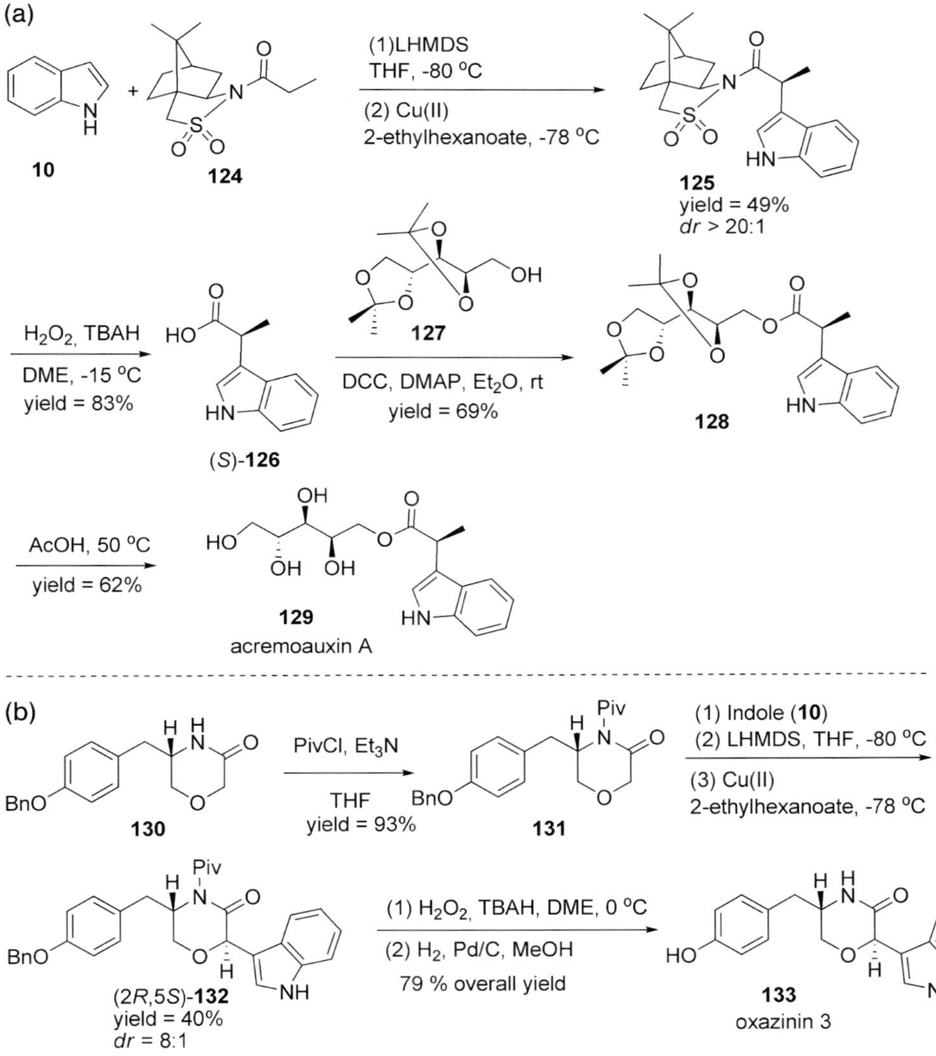

7.6 Asymmetric Synthesis of Natural Products Based on Diastereoselective Friedel–Crafts Reactions

Scheme 7.22 Synthesis of acremoauxin A **129** and oxazinin 3 **133**.

to be immediately benzoylated. Hydrolysis of the chiral auxiliary gave (S)-ketorolac **138** in good yield (38% over two steps) and enantiopurity ($ee = 90\%$).

7.6.4
Synthesis of (−)-Lintetralin

Aryltetralins are a major subgroup of lignans that display an array of interesting pharmacological properties. (+)-Lintetralin was isolated from *Phyllanthus niruri*

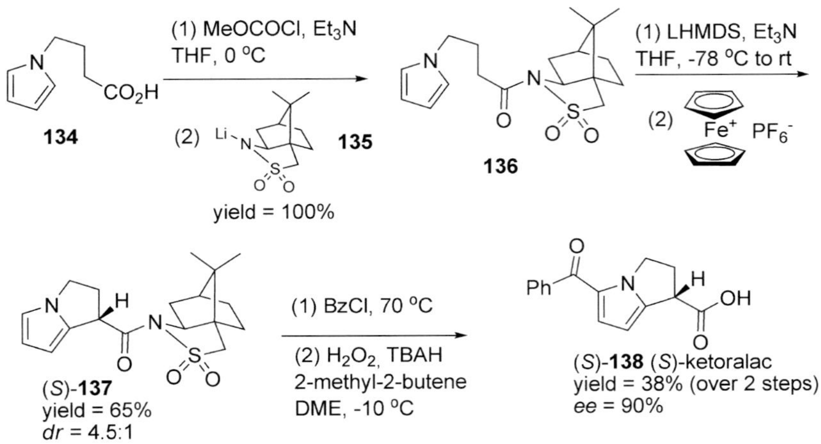

Scheme 7.23 Synthesis of (S)-ketoralac **138**.

Linn, a herbal plant with liver-protecting effects [46]. Enders et al. have reported the first asymmetric synthesis of (−)-lintetralin **146** [47]. The synthesis started with an asymmetric Strecker reaction using piperonal **140** and enantiopure amine **139**, followed by Michael reaction of lithiated **141** to 5H-furan-2-one to afford the 1,4-adduct **142** in 55% yield (Scheme 7.24). The diastereoselectivity of the Michael addition was 88% which could be increased to de > 98% upon chromatography. Alkylation of **142** with 3,4-dimethoxy benzyl bromide and cleavage of the chiral auxiliary took place with excellent diastereoselectivity providing **143** (de, ee > 98%). Reduction of **143** with LAH gave triol **144** as a 2 : 1 mixture of two diastereomers that were not separated. FC cyclization of this key intermediate was accomplished by $BF_3 \cdot Et_2O$ affording diol **145** in 85% yield with complete diastereoselectivity. Subsequent deprotonation of the diols with NaH and methylation with MeI afforded (−)-lintetralin **146**. The spectroscopic data of the synthetic material were in accordance with those of the natural product, although the optical rotation signs differed, establishing for the first time the absolute configuration of natural (+)-lintetralin.

7.6.5
Synthesis of (+)-Erogorgiaene

Erogorgiaene **152**, isolated from the sea whip *Pseudopterogorgia elisabethae*, displays activity against *Mycobacterium tuberculosis* and other mycobacteria [48]. Yadav et al. [49] have reported a total synthesis of **152** which relies on an unprecedented highly diastereoselective FC reaction of an oxetane (Scheme 7.25). The chiral oxetane **148** was prepared from tolylacetic acid **147** in a nine-step synthetic sequence involving diastereoselective chiral auxiliary-guided alkylation and aldol reactions. The crucial FC reaction was carried out with $BF_3 \cdot Et_2O$ to afford tetralin **149** (81% yield) as a single diastereomer, with all stereogenic centers introduced. Alcohol **149** was oxidized under Swern conditions and a Wittig olefination was

7.6 Asymmetric Synthesis of Natural Products Based on Diastereoselective Friedel–Crafts Reactions

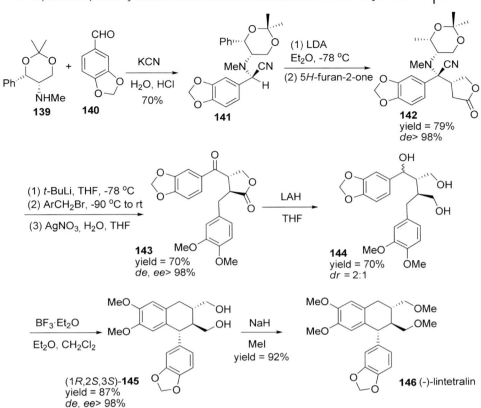

Scheme 7.24 Synthesis of (−)-lintetralin **146**.

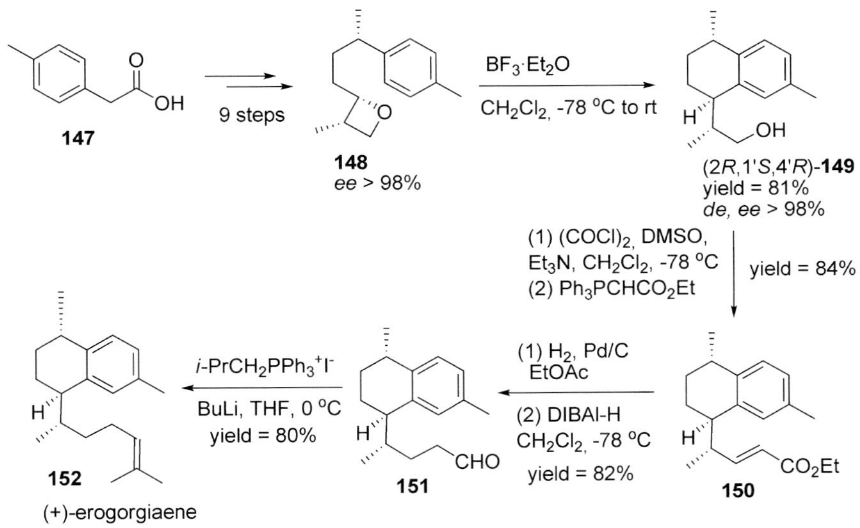

Scheme 7.25 Synthesis of (+)-erogorgiaene **152**.

subsequently carried out to obtain unsaturated ester **150** in 84% yield. Hydrogenation of the double bond and reduction of the ester with DIBAL-H furnished 82% of aldehyde **151**, which was then treated with isopropyltriphenylphsphorane to give (+)-erogorgiaene **152** in 80% yield.

7.6.6
Synthesis of (+)-Bruguierol C

Jennings *et al.* [50] have reported the total synthesis of (+)-bruguierol C, isolated from the stem of the *Brugiera gymmorrhiza* mangrove tree, which exhibits antimicrobial activity [51]. The key step is a Marson-type FC alkylation via an oxocarbenium ion that allows stereoselective formation of a quaternary stereogenic center (Scheme 7.26). Asymmetric allylboration of aldehyde **153** with Brown's allylborane reagent furnished alcohol **154** in 70% yield and with $ee = 90\%$. Hydroboration of the olefinic portion with dicyclohexylborane followed by basic oxidation provided the diol **155** in 91% yield. Other hydroboration procedures did not lead to a sufficient amount of **155**. Selective oxidation of the primary alcohol with TPAP-NMO gave lactone **156** in 68% yield. This was followed by treatment with MeLi to give lactol **157** quantitatively, which was treated with $BF_3 \cdot Et_2O$ to give the tricyclic compound **159** (58% yield) with total diastereoselectivity through oxocarbenium **158b**. Finally, treatment of **159** with TBAF furnished (+)-bruguierol C **160** with 85% yield. The synthetic product showed spectral data in good agreement with the natural sample.

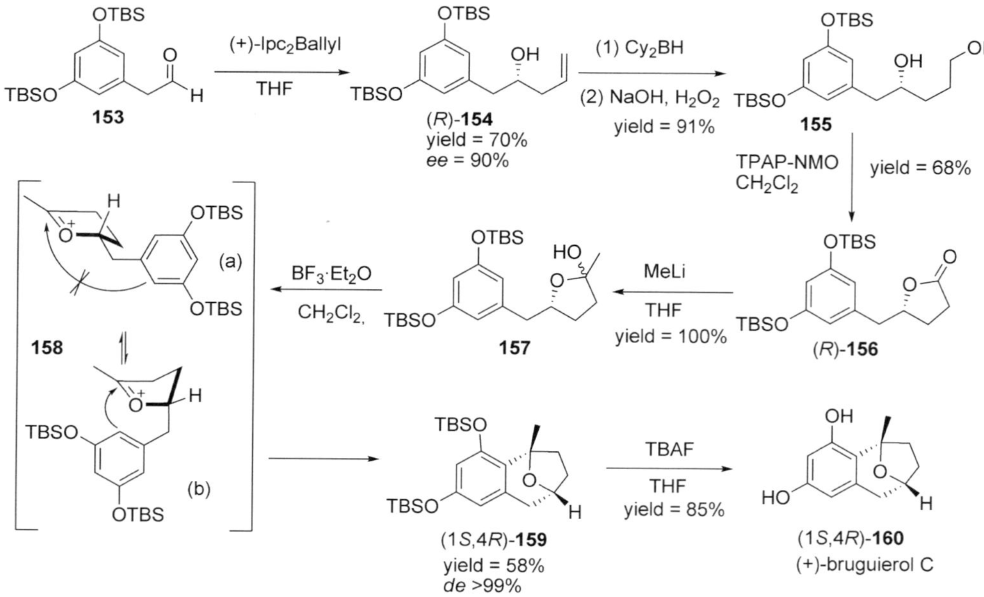

Scheme 7.26 Synthesis of (+)-bruguierol C **160**.

7.6.7
Synthesis of (−)-Talaumidin

An intermolecular FC alkylation on an oxocarbenium ion is the key step in an enantioselective synthesis of (−)-talaumidin **169** [52], a neurotrophic diaryltetrahydrofuran-type lignan isolated from *Aristolocha arcuata* [53] (Scheme 7.27). The oxocarbenium precursor **165** was prepared from alcohol **162** (obtained in 8 steps from aldehyde **161** with total diastereo- and enantioselectivity). Oxidation of the primary alcohol **162** with PDC and then $NaClO_4$ yielded a carboxylic acid, which was converted to the γ-lactone **163** by deprotection of the TBS group in 72% yield over three steps. The chirality of the newly generated chiral center C-4 (*R*) could be inverted to the desired natural *S*-configuration upon treatment of **163** with MeONa to give **164**. Subsequent DIBAL-H reduction, followed by treatment with methylorthoformate yielded acetal **165** as an anomeric mixture. The crucial FC step was achieved upon treatment of **165** with 1,2-methylenedioxybenzene **166** and $SnCl_4$, which only afforded the desired (5*S*)-**168** in 89% yield along with 2% of talumidin, through cation **167**. Finally, debenzylation of **168** with $Pd(OH)_2$ furnished (−)-(2*S*,3*S*,4*S*,5*S*)-talaumidin **169** in 77% yield.

Besides these targeted syntheses, a number of interesting diastereoselective FC reactions have been described, which allow the construction of complex structures and the synthesis of potential precursors for natural products or analogs [54]. Unfortunately, a detailed treatment of these reactions is outside the scope of this chapter.

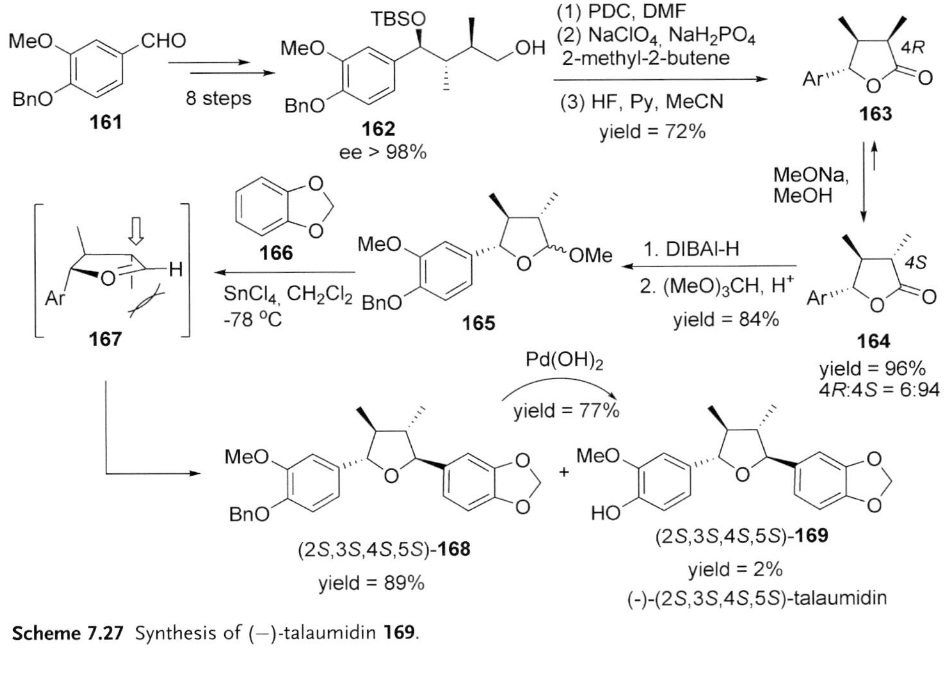

Scheme 7.27 Synthesis of (−)-talaumidin **169**.

7.6.8
Experiments: Selected Procedures

Synthesis of (+)-hapalindole 116 (Scheme 7.20)

(R)-Carvone (**111**, 5.0 g, 33.3 mmol) and indole (**10**, 7.8 g, 66.6 mmol) were dissolved in benzene (50 mL) and the solvent was removed *in vacuo*. The mixture was cooled to −78 °C and a 0.1 M solution of LHMDS in THF (99.9 mmol) was then added over 1 min. After 30 min stirring at −78 °C, the rubber septum was removed and solid copper(II) 2-ethylhexanoate (17.5 g, 49.9 mmol) was rapidly added to the solution and allowed to stir for 12 h at −78 °C after replacing the rubber septum. The solution was then allowed to warm gradually to rt and immediately quenched by pouring into 1 N HCl (500 mL) and partitioned with EtOAc (500 mL). The organic layer was further washed with 1 N HCl (500 mL), 1 N NaOH (500 mL), water (500 mL), and brine (500 mL), and then dried (MgSO$_4$). Flash chromatography (hexanes:EtOAc = 5 : 1 to 2 : 1) afforded the coupled product (R,R)-**112** (yield = 53%) along with recovered carvone and indole. R_f = 0.25 (hexanes:EtOAc = 3 : 1). M.p. = 129–130 °C (cyclohexane:Et$_2$O:benzene). [α]$_D$ = +55 (*c* 3.6, CH$_2$Cl$_2$). ^{1}H NMR (500 MHz, CDCl$_3$): δ 8.15 (s, 1H), 7.44 (d, *J* = 7.5 Hz, 1H), 7.25 (d, *J* = 8.0 Hz, 1H), 7.15 (t, *J* = 7.5 Hz, 1H), 7.07 (t, *J* = 7.5 Hz, 1H), 6.82 (s, 1H), 6.71 (d, 2.0 Hz, 1H), 4.67 (s, 1 H), 4.63 (s, 1 H), 3.91 (d, *J* = 10.5 Hz, 1H), 3.25–3.30 (m, 1H), 2.44–2.60 (m, 2H), 1.90 (s, 3H), 1.62 (s, 3H). ^{13}C NMR (125 MHz, CDCl$_3$) δ 199.7, 146.1, 143.9, 136.5, 135.6, 127.2, 123.0, 121.8, 119.4, 119.3, 113.0, 112.7, 111.6, 49.3, 48.5, 31.1, 19.6, 16.5. HRMS (MALDI) C$_{18}$H$_{20}$NO [M + H]$^+$ requires 266.1545, found 266.1532. IR (film): 3338, 2919, 1655, 1458, 1365, 1248, 1098, 907, 736 cm^{-1}.

Introduction of the vinyl group: Compound (R,R)-**112** (100 mg, 0.38 mmol) was azeotroped with benzene (2 mL) and cooled to −78 °C. 0.1 M LHMDS (0.57 mmol) was then added to the flask over 1 min at −78 °C. After 20 min, 1 M L-Selectride (0.396 mmol) was added and stirred for 1 h. Acetaldehyde (0.14 mL, 2.26 mmol) was then added and stirred for 15 min at that temperature before a 30 wt.% solution of H$_2$O$_2$ (1.3 mL) and 2 N NaOH (1.8 mL) was added dropwise. The reaction was allowed to warm to rt and stirred vigorously for 12 h. The reaction was extracted with EtOAc (3 × 3 mL). The organic extract was washed with water (2 × 4 mL), 5% NaHSO$_4$ (4 mL), brine (4 mL), and dried (MgSO$_4$). The solvent was removed *in vacuo* and the residue dissolved in benzene (2 mL) and then concentrated *in vacuo* and then dissolved in CHCl$_3$ (3.8 mL) and Martin Sulfurane (280 mg, 0.415 mmol) was added. After 10 min, the solvent was removed and the residue chromatographed (hexanes: EtOAc = 5 : 1 to 2 : 1) to give 83 mg, (75% yield over two steps) of pure **114**. R_f = 0.36 (CH$_2$Cl$_2$). M.p. = 178–179 °C (CH$_2$Cl$_2$). [α]$_D$ = +119 (*c* 1.3, CH$_2$Cl$_2$). ^{1}H NMR (500 MHz, CDCl$_3$) δ 8.05 (br s, 1H), 7.33 (d, *J* = 8.0 Hz, 1H), 7.29 (d, *J* = 8.0 Hz, 1H), 7.13 (t, *J* = 8.0 Hz, 1H), 7.05 (t, *J* = 8.0 Hz, 1H), 6.89 (d, *J* = 2.5 Hz, 1H), 6.16 (dd, *J* = 17.5, 11.0 Hz, 1H), 5.34 (d, *J* = 11 Hz, 1H), 5.18 (d, *J* = 17.5 Hz, 1H), 4.62 (s, 1H), 4.54 (s, 1H), 4.29 (d, *J* = 12.5 Hz, 1H), 2.92 (td, *J* = 12.0, 3.5 Hz, 1H), 2.19 (td, *J* = 14.5, 2.5 Hz, 2H), 1.86–1.77 (m, 2H), 1.54 (s, 3H), 1.21 (s, 3H). ^{13}C NMR (125 MHz,

CDCl$_3$) δ 211.0, 147.0, 143.2, 136.2, 127.8, 123.4, 121.8, 119.4, 119.1, 116.3, 112.2, 111.9, 111.4, 53.6, 52.5, 48.9, 39.4, 28.8, 25.2, 18.6. HRMS (ESI): C$_{20}$H$_{24}$NO [M + H]$^+$ requires 294.1852, found 294.1850. IR (film) 3369, 2928, 1702, 1458, 1371, 1099, 917, 737 cm^{-1}.

Introduction of the isothiocyanate moiety: Ammonium acetate (606 mg, 7.87 mmol) and NaCNBH$_3$ (124 mg, 1.97 mmol) were dissolved in MeOH (1.97 mL). Indole **114** (57.7 mg, 0.20 mmol) was dissolved in THF (0.4 mL), added to the methanol solution, stirred for seven days at rt and then quenched with 5% NaHCO$_3$ (3 mL) and extracted with diethyl ether (3 × 5 mL). The organic layers were washed with 1 N HCl (5 mL) and the aqueous layer was brought to pH 8.0 with 2 N NaOH (10 mL). The aqueous layers were extracted with ethyl acetate, washed with brine, dried (MgSO$_4$), and the solvent removed *in vacuo* to give the crude amine **115** (35.2 mg, 61%, $dr = 3:1$) and recovered indole **114** (5.0 mg). Amine **115** (9.0 mg, 0.031 mmol) was then dissolved in CH$_2$Cl$_2$ (0.5 mL) and CS(imid)$_2$ (6.0 mg, 0.034 mmol) added. The solution was allowed to stir at rt for 4 h to afford, after flash chromatography (hexane:CH$_2$Cl$_2$ = 3 : 2), 6.5 mg (63% yield) of (+)-hapalindole Q (**116**) and 2.0 mg (19%) of its C-11 epimer 11-epi-**116**. (Performing the same scale reductive amination in a Biotage/Personal Chemistry microwave vessel at 150 °C for 2 min led to a similar yield but with a $dr = 6:1$). Hapalindole Q (**116**). $R_f = 0.29$ (hexanes:CH$_2$Cl$_2$ = 3 : 2). Clear oil. [α]$_D$ = + 28 (c 1.2, CH$_2$Cl$_2$). ^{1}H NMR (500 MHz, CHCl$_3$) δ 8.00 (br s, 1H), 7.65 (d, J = 8.0 Hz, 1H), 7.35 (d, J = 8.0 Hz, 1H), 7.18 (t, J = 8.0 Hz, 1H), 7.11 (t, J = 8.0 Hz, 1H), 7.00 (d, J = 2.0 Hz, 1H), 6.23 (dd, J = 17.5, 11.0 Hz, 1H), 5.38 (d, J = 11.0 Hz, 1H), 5.28 (d, J = 17.5 Hz, 1H), 4.52 (br s, 1H), 4.46 (s, 1H), 3.88 (br s, 1H), 3.14 (br s, 1H), 2.76 (br s, 1H), 2.00 (dt, J = 14.0, 2.0 Hz, 1H), 1.82 (qd, J = 13.0, 2.0 Hz, 1H), 1.63–1.47 (m, 2H), 1.47 (s, 3H), 1.24 (s, 3H). ^{13}C NMR (125 MHz, CDCl$_3$) δ 147.3, 139.3, 131.5, 130.1, 127.1, 122.4, 122.3, 119.9, 119.7, 116.8, 112.3, 112.1, 111.9, 50.3, 42.6, 36.8, 30.2, 27.8, 25.7, 19.4, 14.6. HRMS (ESI) C$_{21}$H$_{25}$N$_2$S [M + H]$^+$ requires 337.1733, found 337.1732. IR (film) 3423, 2931, 2095, 1458, 1334, 1096, 893, 740 cm^{-1}. 11-epi-hapalindole Q (11-epi-**116**, identical to the known natural product 12-epi-hapalindole D isothiocyanate). $R_f = 0.37$ (hexanes:CH$_2$Cl$_2$ = 3 : 2). Clear oil. [α]$_D$ = −111 (c 0.2, CH$_2$Cl$_2$). ^{1}H NMR (500 MHz, CDCl$_3$) δ 8.10 (br s, 1H), 7.44 (d, J = 8.0 Hz,1H), 7.38 (d, J = 8.0 Hz, 1H), 7.19 (t, J = 8.0 Hz, 1H), 7.12 (t, J = 8.0 Hz, 1H), 7.10 (d, J = 2.0 Hz, 1H), 6.01 (dd, J = 18.0, 11.0 Hz, 1H), 5.38 (d, J = 11.0 Hz, 1H), 5.29 (dd, J = 18.0 Hz, 1H), 4.80 (s, 1H), 4.64 (s, 1H), 3.95 (d, J = 2.0 Hz, 1H), 3.51 (dd, J = 12.0, 2.0 Hz, 1H), 2.84–2.80 (m, 1H), 1.83–1.79 (m, 3H), 1.71–1.67 (m, 1H), 1.47 (s, 3H), 1.16 (s, 3H). ^{13}C NMR (125 MHz, CDCl$_3$): δ 147.8, 143.0, 136.0, 126.9, 124.0, 122.2, 119.7, 117.8, 115.2, 114.7, 112.5, 111.6, 66.9, 43.9, 42.0, 37.9, 31.6, 28.9, 28.7, 18.9. HRMS (ESI) C$_{21}$H$_{25}$N2S [M + H]$^+$ requires 337.1733, found 337.1725. IR (film) 3415, 2928, 2097, 1456, 1338, 1097, 918, 741 cm^{-1}.

Synthesis of (S)-ketoralac 138 (Scheme 7.23)

Compound **134** (460 mg, 3.00 mmol) was dissolved in benzene (3 mL) and the solvent removed *in vacuo*. The compound was then dissolved in anhydrous THF (15 mL) and cooled to 0 °C. Et$_3$N (433 μL, 3.25 mmol) was then added, followed by methyl chloroformate (232 μL, 3.0 mmol), and stirring was continued for 60 min. In a separate flask, (S)-2,10-camphorsultam (538 mg, 2.5 mmol) was dried *in vacuo* for 30 min, dissolved in THF (12.5 mL) and cooled to −78 °C. A solution of 2.5 M BuLi (1.05 mL, 2.63 mmol) was then added and the solution stirred for 20 min. The anhydride solution was then filtered through a medium glass frit under a blanket of nitrogen to remove the triethylamine hydrochloride salt. The filtrate was then cannulated into a solution of the lithiate **135** at −78 °C. The reaction was allowed to immediately warm to rt and quenched with 20% K$_2$CO$_3$ solution (20 mL) and partitioned with EtOAc (30 mL). The organic layer was washed with water (20 mL) and brine (20 mL) and dried (MgSO4). Removal of the solvent *in vacuo* gave 876 mg of pure sultam **136** (100% yield). R_f = 0.32 (hexane:EtOAc = 3 : 1). Brown oil. $[\alpha]_D$ = −49 (c 10.6, CH$_2$Cl$_2$). ^{1}H NMR (400 MHz, CDCl$_3$): δ 6.65 (t, J = 2 Hz, 2H), 6.13 (t, J = 2.4 Hz, 2H), 3.90–3.99 (m, 2H), 3.84–3.87 (m, 2H), 2.68–2.74 (m, 2H), 2.04–2.15 (m, 4H), 1.87–1.92 (m, 2H), 1.35–1.45 (m, 2H), 1.14 (s, 3H), 0.97 (s, 3H). ^{13}C NMR (100 MHz, CDCl$_3$): δ 170.8, 120.5 (2C), 108.1 (2C), 65.1, 52.8, 48.4 (2C), 47.7, 44.6, 38.4, 32.8, 32.3, 26.4, 26.3, 20.8, 19.8. LRMS (ESI): [M + H]$^+$ 351, [M + Na]$^+$ 373.

Diastereoselective coupling: Compound **136** (39.5 mg, 0.11 mmol) was dried *in vacuo* for 30 min, dissolved in dry THF (8.0 mL), and cooled to −78 °C. Triethylamine (30 μL, 0.23 mmol) was added to the reaction mixture, followed by 1 M LHMDS (135 μL). Stirring was continued for 20 min, after which time the reaction was warmed to 12 °C. After 15 min the septum was removed and solid ferrocenium hexafluorophosphate (28.0 mg, 0.085 mmol) was added rapidly, after which the septum was replaced. The reaction was vigorously stirred for 5 min, until the reaction mixture was yellow and all the ferrocenium salt was consumed. After this time, the reaction was diluted with 3 : 1 hexane:EtOAc (15 mL) and filtered through a short plug of silica gel. The solvent was removed *in vacuo* to give 54.8 mg (65% yield) of the crude reaction mixture **137** (dr = 4.5 : 1). This compound was quite unstable and was reacted immediately after preparation. The crude mixture was dissolved in benzoyl chloride (200 μL) and stirred at 70 °C for 4 h. The reaction was then cooled to rt, diluted with CH$_2$Cl$_2$ (10 mL) and washed with 2 N NaOH (3 × 10 mL), water (10 mL) and brine (10 mL). The organic layer was dried (MgSO$_4$) and the solvent removed *in vacuo*. Preparative TLC purification (CH$_2$Cl$_2$), gave 5.6 mg (27% yield brsm) of pure benzoylated product and recovered **136** (23.7 mg). At this point, the two diastereomers were also successfully separated (3.9 mg major, 1.7 mg minor). The major diastereoisomer (3.9 mg, 0.0086 mmol) was dissolved in DME (35 μL) and cooled to −10 °C. Isobutylene (2.7 μL, 0.0258 mmol) was then added to the reaction mixture, followed by 30% H$_2$O$_2$ (1.9 μL, 0.0172 mmol) then a 40% TBAH solution (11.2 μL, 0.0172 mmol). Stirring was continued at −10 °C for 3 h, then quenched with four drops of 1.5 M Na$_2$SO$_3$, followed by stirring for 1 h. The reaction was then acidified with 1 N HCl (5 mL) and extracted with EtOAc (3 × 5 mL). The organic layers were combined, dried (MgSO$_4$), and the solvent removed *in vacuo*. Preparative TLC (EtOAc) gave 1.2 mg (yield = 58%) of (S)-**138**. R_f = 0.17 (EtOAc). Off-white solid, m.p. >250 °C. ee = 90%, HPLC: Chiralpak AD, 90/10/0.1 hexane/IPA/TFA, 0.8 mL min^{-1}, t_R = 12.6 min (minor), t_S = 13.9 min (major). ^{1}H NMR (400 MHz, CDCl$_3$): δ 7.79 (d, J = 7.2 Hz, 2H), 7.51 (t, J = 7.2 Hz, 1H), 7.43 (t, J = 7.6 Hz, 2H), 6.81 (d, J = 4 Hz, 1H), 6.13 (d, J = 4 Hz, 1H), 4.53–4.58 (m, 1H), 4.43–4.48 (m, 1H), 4.08–4.12 (m, 1H), 2.79–2.92 (m, 2 H). ^{13}C NMR (150 MHz, CDCl$_3$): δ 186.0, 176.6, 142.4, 140.0, 132.3, 129.8 (2C), 129.0 (2C), 128.2, 125.9, 104.2, 48.4, 43.1, 31.9. HRMS (ESI): C$_{15}$H$_{14}$NO$_3$ [M + H]$^+$ requires 256.0968, found 256.0975. IR (film) 2917, 2848, 1735, 1713, 1621, 1596, 1569, 1467, 1431, 1403, 1270, 1205.

Synthesis of (+)-lintetralin 146 (Scheme 7.24)

To a stirred suspension of LAH (80 mg, 2.1 mmol) in THF (10 mL) was added at 0 °C lactone **143** (200 mg, 1.0 mmol) dissolved in dry THF (5 mL). The ice bath was removed and the mixture was stirred at rt for 5 h. EtOAc was added dropwise, after which aqueous NH$_4$Cl (5 mL) and H$_2$O (5 mL) were added. The mixture was filtered and extracted with EtOAc. The organic layer was dried with MgSO$_4$ and evaporated *in vacuo*. The crude triol **144** (192 mg, 95% crude yield) was then dissolved in CH$_2$Cl$_2$. BF$_3$·Et$_2$O (3.0 mmol) was slowly added and the solution was stirred at rt for 1 h. Then the reaction was quenched with satd aq NaHCO$_3$ (5 mL), extracted with CH$_2$Cl$_2$, and dried over MgSO$_4$. Removal of the solvent and column chromatography (EtOAc) afforded 160 mg (yield = 87%) of (1*R*,2*S*,3*S*)-**145** as a single diastereomer. M.p. 72–75 °C. [α]$_D^{27}$ = −18.3 (*c* 0.4, CHCl$_3$). ^{1}H NMR (400 MHz, CDCl$_3$): δ 1.65–1.72 (m, 1H) 1.89–1.94 (m, 1H), 2.57–2.62 (m, 2H), 2.86 (s, 2H), 3.40 (dd, J = 5.2, 11.4 Hz, 1H), 3.76 (s, 3H), 3.57–3.71 (m, 4H), 3.75 (s, 3H), 5.83 (s, 2H), 6.13, (s, 1H), 6.47 (s, 1H), 6.50 (s, 1H), 6.57 (d, J = 7.7 Hz, 1H), 6.65 (d, J = 7.7 Hz, 1H). ^{13}C NMR (75 MHz, CDCl$_3$): δ 33.5, 40.1, 48.3, 48.8, 56.1, 56.1, 62.8, 66.6, 101.1, 108.2, 109.4, 111.1, 113.2, 123.0, 128.6, 131.9, 139.6, 146.3, 147.4, 147.5, 148.1. MS (ESI): m/z 373 (M$^+$ + 1, 23), 372 (M$^+$, 100), 353 (7), 323 (46%), 296 (14%), 283 (9%). Anal. C$_{21}$H$_{24}$O$_6$ requires C, 67.73; H, 6.50; found: C, 67.70; H, 6.64.

Methylation of compound 145: MeI (200 mg, 1.4 mmol) was added to a solution of the diol (1*R*,2*S*,3*S*)-**145** (100 mg, 0.27 mmol) in dry THF. Sodium hydride (0.6 g, 60% dispersion in oil) was added portion-wise to the solution over 20 min, followed by more MeI (110 mg 0.77 mmol). After being stirred at rt for 2.5 h, the mixture was cooled to 0 °C and methanol (5 ml) was added. The organic phase was evaporated and the residue was chromatographed (pentane:Et$_2$O = 1 : 1) to give 100 mg (92%) of (−)-lintetralin **146**. Colorless liquid. [α]$_D^{24}$ = −4.0 (*c* 0.1, CHCl$_3$). ^{1}H NMR (400 MHz, CDCl$_3$): δ 1.68–1.75 (m, 1H), 2.04–2.09 (m, 1H), 2.75 (d, J = 7.7 Hz, 2H), 3.04 (dd, J = 3.3, 9.6 Hz, 2H), 3.20 (s, 3H), 3.28 (s, 3H), 3.30 (dd, J = 3.0, 9.6 Hz, 1H), 3.35 (dd, J = 6.3, 15.6 Hz, 1H), 3.41 (dd, J = 3.8, 9.3 Hz, 1H), 3.54 (s, 3H), 3.76 (s, 3H), 3.91 (d, J = 10.4 Hz, 1H), 5.85 (dd, J = 1.4, 3.8 Hz, 1H), 6.16 (s, 1H), 6.49 (d, J = 1.6 Hz, 1H), 6.52 (s, 1H), 6.57 (dd, J = 1.6, 8.0 Hz, 1H), 6.66 (d, J = 8.0 Hz, 1H). ^{13}C NMR (75 MHz, CDCl$_3$): δ 33.4, 36.5, 45.3, 47.4, 56.0, 56.1, 59.1, 59.2, 71.3, 75.5, 101.0, 108.0, 109.6, 111.2, 113.1, 122.9, 129.1, 132.1, 139.9, 146.1, 147.2, 147.4, 147.9. MS (EI, 70 eV): m/z 401 (M$^+$ + 1, 24), 400 (M$^+$, 100), 368 (14), 323 (67). Anal. C$_{23}$H$_{28}$O$_6$ requires C, 68.98; H, 7.05; found C, 68.70; H, 6.83.

Synthesis of (+)-bruguierol C 160 (Scheme 7.23)

To a solution of lactone (R)-**156** (250 mg, 0.57 mmol) dissolved in dry Et$_2$O (5 mL) was added 1.6 M MeLi (0.48 mL, 0.77 mmol) dropwise under argon at $-78\,^\circ$C. The reaction was left stirring for 1.5 h and quenched with std. aq. NH$_4$C. The aqueous layer was then extracted (3 × 20 mL) with Et$_2$O. The combined organic extracts were dried (MgSO$_4$), filtered, and concentrated under reduced pressure, which afforded the crude lactol product **157**. To a solution of crude lactol dissolved in CH$_2$Cl$_2$ (5 mL) was added BF$_3$·OEt$_2$ (0.14 mL, 1.14 mmol) dropwise under argon at $-20\,^\circ$C. The solution was stirred for 2 h and quenched with std. aq. NH$_4$Cl. The aqueous layer was extracted with Et$_2$O (3 × 20 mL). The usual procedure followed by flash chromatography (hexanes:Et$_2$O) afforded 143 mg (58% yield) of compound (1S,4R)-**159**. $R_f = 0.4$ (hexane:EtOAc = 7 : 3). Colorless oil. $[\alpha]_D^{25} = +16.2$ (c 0.03, CH$_2$Cl$_2$). ^{1}H NMR (360 MHz, CDCl$_3$): δ 6.17 (d, J = 2.5 Hz, 1H), 6.12 (d, J = 2.2 Hz, 1H), 4.61 (m, 1H), 3.31 (dd, J = 16.2, 5.0 Hz, 1H), 2.34 (d, J = 16.2 Hz, 1H), 2.21 (m, 1H), 2.09 (m, 1H), 1.84 (s, 3H), 1.76 (m, 1H), 1.63 (m, 1H), 1.01 (s, 9H), 0.96 (s, 9H), 0.30 (s, 3H), 0.24 (s, 3H), 0.17 (s, 6H). ^{13}C NMR (90 MHz, CDCl$_3$): δ 154.1, 152.2, 135.1, 126.8, 113.4, 108.5, 80.4, 73.1, 42.0, 37.8, 26.0, 25.6, 24.2, 18.5, 18.1, -3.5, -3.9, -4.4, -4.4. HRMS (EI): m/z C$_{24}$H$_{42}$O$_3$Si$_2$ (M$^+$) requires 434.2673, found 434.2674. IR (CHCl$_3$) 2954, 2930, 2897, 2355, 1601, 1571, 1424, 1372, 1279, 1190, 1082, 894, 830, 778 cm^{-1}.

Deprotection of the hydroxy groups: To a solution of (1S,4R)-**159** (90 mg, 0.21 mmol) dissolved in THF (5 mL) was added TBAF (0.63 mL, 0.63 mmol) at rt. The reaction was stirred for 1.5 h and quenched with std. aq. NH$_4$Cl. The aqueous layer was extracted with EtOAc (3 × 20 mL). The combined organic extracts were dried with MgSO$_4$, filtered, and concentrated under reduced pressure. Flash chromatography (hexanes:EtOAc = 60 : 40) afforded 37 mg (85% yield) of (+)-(1S,4R)-bruguierol C (**160**). $R_f = 0.4$ (Hexane:EtOAc = 1 : 1). White solid. $[\alpha]_D^{25} = +4.2$ (c 0.005, MeOH). ^{1}H NMR (360 MHz, CD$_3$OD): δ 6.08 (d, J = 2.2 Hz, 1H), 6.02 (d, J = 2.5 Hz, 1H), 4.58 (m, 1H), 3.20 (dd, J = 16.2, 5.0 Hz, 1H), 2.35 (d, J = 16.2 Hz, 1H), 2.20 (m, 1H), 2.10 (m, 1H), 1.81 (s, 3H), 1.74 (m, 1H), 1.63 (m, 1H). ^{13}C NMR (125 MHz, CD$_3$OD): δ 157.5, 155.5, 136.0, 122.3, 108.1, 102.0, 82.3, 75.0, 42.9, 38.8, 31.0, 24.4. HRMS (EI): m/z C$_{12}$H$_{14}$O$_3$ (M$^+$) requires 206.0943, found 206.0947. IR (CHCl$_3$) 3735, 3633, 3326, 2923, 2854, 2360, 1608, 1464, 1348, 1296, 1161, 1028, 997, 836 cm^{-1}.

Acknowledgments

Financial support from the Ministerio de Educación y Ciencia and FEDER (CTQ 2006-14199/BQU) is gratefully acknowledged. C.V. thanks the Generalitat Valenciana for a pre-doctoral grant.

Abbreviations

Ac	acetyl
Bn	benzyl
BOX	bis(oxazoline)
CBz	carbobenzoxy
COX	cyclooxygenase
Cy	cyclohexyl
DBU	diazabicycloundecene
DCA	dichloroacetic acid
DCC	dicyclohexylcarbodiimide
DCE	1,2-dichloroethane
de	diastereomeric excess
DIBAL-H	diisobutylaluminum hydride
DMAP	4-dimethylaminopyridine
DME	dimethoxyethane
DMF	*N,N*-dimethylformamide
DMSO	dimethyl sulfoxide
DMPU	1,3-dimethyl-3,4,5,6-tetrahydro-2(1*H*)-pyrimidone
dr	diastereomeric ratio
EDCl	1-(3-dimethylaminopropyl)-3-ethyl-carboimide hydrochloride
ee	enantiomeric excess
FC	Friedel–Crafts
GnRH	gonadotropin-releasing hormone
HMDS	hexamethyldisilane
HT	hydroxytriptamine
IPA	isopropyl alcohol
LAH	lithium aluminum hydride
LHMDS	lithium hexamethyldisilane
NBS	*N*-Bromosuccinimide
OTf	triflate
PCC	pyridinium chlorochromate
PDC	pyridinium dichromate
Pg	protecting group
Piv	pivaloyl
PMP	*p*-methoxyphenyl
PS	Pictet–Spengler
PyBOX	bis(oxazolinyl)pyridine
TBAH	tetrabutylammonium hydroxide
TBME	*tert*-butylmethyl ether
TBS	*tert*-butyldimethylsilyl
TEA	triethylamine
TFA	trifluoroacetic acid
THF	tetrahydrofuran
TMS	trimethylsilyl

THBC	tetrahydro-β-carboline
THPI	tetrahydropyrano[3,4-b]indole
THGC	tetrahydro-γ-carboline
TOX	tris(oxazoline)
Ts	tosyl
TTCE	1,1,2,2-tetrachloroethane

References

1 Erker, G. and van der Zeijden, A.A.H. (1990) *Angewandte Chemie – International Edition*, **29**, 512–514.

2 Austin, J.F. and MacMillan, D.W.C. (2002) *Journal of the American Chemical Society*, **124**, 1172–1173.

3 Black, W.C., Bayly, C., Belley, M., Chart, C.-C., Charleson, S., Denis, D., Gauthier, J.Y., Gordon, R., Guay, D., Kargman, S., Lau, C.K., Leblanc, Y., Mancini, J., Ouellet, M., Percival, D., Roy, P., Skorey, K., Tagari, P., Vickers, P., Wong, E., Xu, L. and Prasit, P. (1996) *Bioorganic & Medicinal Chemistry Letters*, **6**, 725–730.

4 (a) Evans, D.A., Fandrick, K.R. and Song, H.-J. (2005) *Journal of the American Chemical Society*, **127**, 8942–8943. (b) Evans, D.A., Fandrick, K.R., Song, H.-J., Scheidt, K.A. and Xu, R. (2007) *Journal of the American Chemical Society*, **129**, 10029–10041.

5 (a) Riggle, B. (2002) U.S. Patent, US 6399646 B1. 20020604, CAN 136:397348, AN 2002: 425421; (b) Kato, K., Tanaka, S., Gong, Y.-F., Katayama, M. and Kimoto, H. (1999) *World Journal of Microbiology & Biotechnology*, **15**, 631–633; (c) Matsuda, K., Toyoda, H., Nishio, H., Nishida, T., Dohgo, M., Bingo, M., Matsuda, Y., Yoshida, S., Harada, S., Tanaka, H., Komai, K. and Ouchi, S. (1998) *Journal of Agricultural and Food Chemistry*, **46**, 4416–4419.

6 Farr, R.N., Alabaster, R.J., Chung, J.Y.L., Craig, B., Edwards, J.S., Gibson, A.W., Ho, G.-J., Humphrey, G.R., Johnson, S.A. and Grabowski, E.J.J. (2003) *Tetrahedron: Asymmetry*, **14**, 3503–3515.

7 Itoh, J., Fuchibe, K. and Akiyama, T. (2008) *Angewandte Chemie – International Edition*, **47**, 4016–4018.

8 Bandini, M., Garelli, A., Rovinetti, M., Tommasi, S. and Umani-Ronchi, A. (2005) *Chirality*, **17**, 522–529.

9 (a) Hendricks, R.T., Sherman, D., Strulovici, B. and Broka, C.A. (1995) *Bioorganic & Medicinal Chemistry Letters*, **5**, 67–72; (b) Chu, L., Fisher, M.H., Goulet, M.T. and Wyvratt, M.J. (1997) *Tetrahedron Letters*, **38**, 3871–3874; (c) Stevenson, G.I., Smith, A.L., Lewis, S., Michie, S.G., Neduvelil, J.C., Patel, S., Marwood, R., Patel, S. and Castro, J.L. (2000) *Bioorganic & Medicinal Chemistry Letters*, **10**, 2696–2697; (d) Glennon, R.A., Lee, M., Rangisetty, J.B., Dukat, M., Roth, B.L., Savage, J.E., McBride, A., Rauser, L., Hufeisen, S. and Lee, D.K.H. (2000) *Journal of Medicinal Chemistry*, **43**, 1011–1018; (e) Smith, A.L., Stevenson, G.I., Lewis, S., Patel, S. and Castro, J.L. (2000) *Bioorganic & Medicinal Chemistry Letters*, **10**, 2693–2696.

10 Sui, Y., Liu, L., Zhao, J.-L., Wang, D. and Chen, Y.-J. (2007) *Tetrahedron*, **63**, 5173–5183.

11 Orme, M.W. Sawyer, J.S. and Schultze, L.M. (2002) PCT Int. Appl. WO2002098428 A1. 20021212, CAN 138:24726, AN 2002:946116.

12 Zhou, J., Ye, M.-C., Huang, Z.-Z. and Tang, Y. (2004) *The Journal of Organic Chemistry*, **69**, 1309–1320.

13 (a) Jeannin, L., Nagy, T., Vassileva, E., Sapi, J. and Laronze, J.-Y. (1995) *Tetrahedron Letters*, **36**, 2057–2058; (b) Nemes, C.,

Jeannin, L., Sapi, J., Laronze, M., Seghir, H., Auge, F. and Laronze, J.-Y. (2000) *Tetrahedron*, **56**, 5479–5492.

14 King, H.D., Meng, Z., Denhart, D., Mattson, R., Kimura, R., Wu, D., Gao, Q. and Macor, J.E. (2005) *Organic Letters*, **7**, 3437–3440.

15 Taber, M.T., Wright, R.N., Molski, T.F., Clarke, W.J., Brassil, P.J., Denhart, D.J., Mattson, G.K. and Lodge, N.J. (2005) *Journal of Pharmacology, Biochemistry, and Behavior*, **80**, 521–528.

16 Rueping, M., Nachtshein, B.J., Moreth, S.A. and Bolte, M. (2008) *Angewandte Chemie – International Edition*, **47**, 593–596.

17 (a) Yamada, H., Kawate, T., Matsumizu, M., Nishida, A., Yamaguchi, K. and Nagakawa, M. (1998) *The Journal of Organic Chemistry*, **63**, 6348–6354; (b) Seayad, J., Seayad, A.M. and List, B. (2006) *Journal of the American Chemical Society*, **128**, 1086–1087; (c) Wanner, M.J., van der Haas, R.N.S., de Cuba, K.R., van Maarseveen, J.H. and Hiemstra, H. (2007) *Angewandte Chemie – International Edition*, **46**, 7485–7487; (d) Taylor, M.S. and Jacobsen, E.N. (2004) *Journal of the American Chemical Society*, **126**, 10558–10559; (e) Raheem, I.T., Thiara, P.S., Peterson, E.A. and Jacobsen, E.N. (2007) *Journal of the American Chemical Society*, **129**, 13404–13405.

18 Kam, T.-S. and Sim, K.-M. (1998) *Phytochemistry*, **47**, 145–147.

19 Li, C.-F., Liu, H., Liao, J., Cao, Y.-J., Liu, X.-P. and Xiao, W.-J. (2007) *Organic Letters*, **9**, 1847–1850.

20 Bandini, M., Melloni, A., Piccinelli, F., Sinisi, R., Tommasi, S. and Umani-Ronchi, A. (2006) *Journal of the American Chemical Society*, **128**, 1424–1425.

21 (a) Han, X. and Widenhoefer, R.A. (2006) *Organic Letters*, **8**, 3801–3804; (b) Liu, C. and Widenhoefer, R.A. (2007) *Organic Letters*, **9**, 1935–1938.

22 (a) Çavdar, H. and Saraçoglu, N. (2005) *Tetrahedron*, **61**, 2401–2405; (b) Çavdar, H. and Saraçoglu, N. (2006) *The Journal of Organic Chemistry*, **71**, 7793–7799.

23 Evans, D.A. and Fandrick, K.R. (2006) *Organic Letters*, **8**, 2249–2252.

24 Blay, G., Fernández, I., Pedro, J.R. and Vila, C. (2007) *Tetrahedron Letters*, **38**, 6731–6734.

25 Kang, Q., Zheng, X.-J. and You, S.-L. (2008) *Chemistry – A European Journal*, **14**, 3539–3542.

26 Varney, M.D., Appelt, K., Kalish, V., Reddy, M.R., Tatlock, J., Palmer, C.L., Romines, W.H., Wu, B.-W. and Musick, L. (1994) *Journal of Medicinal Chemistry*, **37**, 2274–2284.

27 (a) Menshikov, G. (1933) *Berichte*, **66B**, 875–878; (b) Honda, T., Yamane, S., Naito, K. and Suzuki, Y. (1995) *Heterocycles*, **40**, 301–310.

28 (a) Banwell, M.G., Beck, D.A.S. and Smith, J.A. (2004) *Organic and Biomolecular Chemistry*, **2**, 157–159; (b) Banwell, M.G., Beck, D.A.S. and Willis, A.C. (2006) *Arkivoc*, 163–174.

29 (a) Gage, J.L. and Branchaud, B.P. (1997) *Tetrahedron Letters*, **38**, 7007–7010; (b) Ha, D.-C., Park, S.-H., Choi, K.-S. and Yun, C.-S. (1998) *Bulletin of the Korean Chemical Society*, **19**, 728–730; (c) David, O., Blot, J., Bellec, C., Fargeau–Bellassoued, M.-C., Haviari, G., Célérier, J.-P., Lhommet, G., Gramain, J.-C. and Gardette, D. (1999) *The Journal of Organic Chemistry*, **64**, 3122–3131; (d) Kim, S.-H., Kim, S.-I., Lai, S. and Cha, J.K. (1999) *The Journal of Organic Chemistry*, **64**, 6771–6775.

30 (a) Kam, T.-S., Tee, Y.-M. and Subramaniam, G. (1998) *Natural Product Letters*, **12**, 307–310; (b) Goh, S.H., Razak Mohd Ali, A. and Wong, W.H. (1989) *Tetrahedron*, **45**, 7899–7902.

31 Li, G., Rowland, G.B., Rowland, E.B. and Antilla, J.C. (2007) *Organic Letters*, **9**, 4065–4068.

32 Peresada, V.P., Medvedev, O.S., Likhosherstov, A.M. and Skoldinov, A.P. (1987) *Khimiko-Farmatsevticheskii Zhurnal*, **21**, 1054–1059.

33 Gatherhood, N., Zhuang, W. and Jørgensen, K.A. (2000) *Journal of the American Chemical Society*, **122**, 12517–12522.

34 Majer, J., Kwiatkowski, P. and Jurczak, J. (2008) *Organic Letters*, **10**, 2955–2958.

35 Uraguchi, D., Sorimachi, K. and Terada, M. (2004) *Journal of the American Chemical Society*, **126**, 11804–11805.

36 Carswell, E.L., Snapper, M.L. and Hoveyda, A.H. (2006) *Angewandte Chemie – International Edition*, **45**, 4230–7233.

37 (a) Yoon, Y.-J., Joo, J.-E., Lee, K.-Y., Kim, Y.-H., Oh, C.-Y. and Ham, W.-H. (2005) *Tetrahedron Letters*, **46**, 739–741; (b) Harrison, T., Williams, B.J., Swain, C.J. and Ball, R.G. (1994) *Bioorganic & Medicinal Chemistry Letters*, **21**, 2545–2550.

38 Saaby, S., Bayon, P., Aburel, P.S. and Jørgensen, K.A. (2002) *The Journal of Organic Chemistry*, **67**, 4352–4361.

39 (a) van Lingen, H.L., Zhuang, W., Hansen, T., Rutjes, F.P.J.T. and Jørgensen, K.A. (2003) *Organic and Biomolecular Chemistry*, **1**, 1953–1958; (b) for another example see also: Lyle, M.P.A., Draper, N.D. and Wilson, P.D. (2005) *Organic Letters*, **7**, 901–904.

40 (a) Baran, P.S. and Richter, J.M. (2004) *Journal of the American Chemical Society*, **126**, 7450–7451; (b) Baran, P.S. and Richter, J.M. (2005) *Journal of the American Chemical Society*, **127**, 15394–15396; (c) Richter, J.M., Whitefield, B.W., Maimone, T.J., Lin, D.W., Castroviejo, M.P. and Baran, P.S. (2007) *Journal of the American Chemical Society*, **129**, 12857–12869; (d) Baran, P.S., Richter, J.M. and Lin, D.W. (2005) *Angewandte Chemie – International Edition*, **44**, 609–612.

41 Moore, R.E., Cheuk, C., Yang, X.-Q.G., Patterson, G.M.L., Bonjouklian, R., Smitka, T.A., Mynderse, J.S., Foster, R.S., Jones, N.D., Swartzendruber, J.K. and Deeter, J.B. (1987) *The Journal of Organic Chemistry*, **52**, 1036–1043.

42 Isolation of (+)-**115**: Stratmann, K., Moore, R.E., Bonjouklian, R., Deeter, J.B., Patterson, G.M.L., Shaffer, S., Smith, C.D. and Smitka, T.A. (1994) *Journal of the American Chemical Society*, **116**, 9935–9942.

43 Sassa, T., Yoshida, N. and Haruki, E. (1989) *Agricultural and Biological Chemistry*, **53**, 3105–3107.

44 Ciminiello, P., Dell'Aversano, C., Fattorusso, E., Forino, M., Magno, S., Lanaro, A. and Di Rosa, M. (2001) *European Journal of Organic Chemistry*, 49–53.

45 Guzman, A., Yuste, F., Toscano, R.A., Young, J.M., Van Horn, A.R. and Muchowski, J.M. (1986) *Journal of Medicinal Chemistry*, **29**, 589–591.

46 (a) Satyanarayana, P., Subrahamanyam, P., Viswanatham, K.N. and Ward, R.S. (1988) *Journal of Natural Products*, **51**, 44–49; (b) Thyagarajan, S.P., Subramaniam, S., Thirursalasundari, T. and Blumgerg, B.S. (1988) *Lancet*, **2**, 764–766.

47 Enders, D., Del Signore, G. and Berner, O.M. (2003) *Chirality*, **15**, 510–513.

48 Rodriguez, A.D. and Ramirez, C. (2001) *Journal of Natural Products*, **64**, 100–102.

49 Yadav, J.S., Basak, A.K. and Srihari, P. (2007) *Tetrahedron Letters*, **48**, 2841–2843.

50 Martinez Solorio, D. and Jennings, M.P. (2007) *The Journal of Organic Chemistry*, **72**, 6621–6623.

51 Han, L., Huang, X., Sattler, I., Moellmann, U., Fu, H., Lin, W. and Grabley, S. (2005) *Planta Medica*, **71**, 160–164.

52 Esumi, T., Hojyo, D., Zhai, H. and Fukuyama, Y. (2006) *Tetrahedron Letters*, **47**, 3979–3983.

53 Zhai, H., Nakatsukasa, M., Mitsumoto, Y. and Fukuyama, Y. (2004) *Planta Medica*, **70**, 598–602.

54 For some additional examples of diastereoselective FC reactions: (a) Cabral, L.M., Costa, P.R.R., Vasconcellos, M.L.A.A., Barreiro, E.J. and Castro, R.N. (1996) *Synthetic Communications*, **26**, 3671–3676; (b) Wipf, P., Hopkins, C.R. (2001) *The Journal of Organic Chemistry*, **66**, 3133–3139; (c) Harmata, M., Hong, X. and Barnes, C.L. (2004) *Organic Letters*, **6**, 2201–2203; (d) Fei, Z. and McDonald, F.E. (2007) *Organic Letters*, **9**, 3547–3550.

8
Industrial Friedel–Crafts Chemistry

Duncan J. Macquarrie

Summary

This chapter outlines the scope of industrial Friedel–Crafts chemistry. The reaction is a key transformation of bulk raw materials into a range of products, ranging from large volume products such as ethylbenzene to more complex, smaller volume, higher value molecules. One clear trend is the shift away from aluminum chloride catalysis, with its associated hazards, towards HF catalysis and beyond HF to solid acid catalysts. Of the solid acids, zeolites are amongst the leading candidates, and have found real applications in industrial Friedel–Crafts chemistry. The impact of the principles of green chemistry on the Friedel–Crafts reaction is discussed.

8.1
Introduction

The Friedel–Crafts (FC) reaction has always had a central role in the chemical industry. It is ideally suited to the basic transformations of hydrocarbons that have allowed the production of a range of bulk chemicals from which essentially everything else can be made. The power and scope of the reaction is such that many of the largest processes, in terms of production volume, involve FC chemistry, often alkylation but also many acylations. In many ways, these processes lie at the opposite end of the chemical spectrum from the asymmetric variants that are the primary focus of this book, but this underlines the enormous importance and versatility of the reaction. While the products of the majority of these alkylations result in the generation of a chiral center (of the relatively simple alkylating agents, only methylation, ethylation and isopropylation cannot provide this feature – higher chain lengths will always give some rearrangement and therefore some chiral centers will arise), this feature has never been relevant for the processes, and the racemate is the desired product.

In recent years, the drive towards cleaner and safer production methods and higher selectivities has given industry the impetus to change some of the FC processes in favor of catalysts featuring higher selectivities and easier handling. As will be seen, this can broadly be represented as a move away from $AlCl_3$ towards (and to a lesser

Catalytic Asymmetric Friedel–Crafts Alkylations. Edited by M. Bandini and A. Umani-Ronchi
Copyright © 2009 WILEY-VCH Verlag GmbH & Co. KGaA, Weinheim
ISBN: 978-3-527-32380-7

extent away from) HF, continuing towards solid acids which have the benefits of relatively simple handling, recovery and reuse, as well as providing the potential for shape selective reactions. In particular, this feature can improve positional selectivity and reduce the extent of double alkylation, both of which are typically serious problems in FC alkylation processes.

8.2
Green Chemistry and the Friedel–Crafts Reaction

The twelve principles of green chemistry were first published 10 years ago, and have formed the basis of a new direction in chemistry over the last decade [1, 2]. These principles outline the requirements for a green process, and include key requirements such as the avoidance of toxic solvents, the use of catalysts, high atom economy (i.e., addition reactions, where all the atoms in the reagents appear in the product, are better than those where by-products appear), low energy usage, and process safety. High selectivity is also clearly a major benefit. Separation and recovery of catalysts is another aspect of catalytic reactions which has to be designed into processes. In this respect, the use of solid acids, where separation is simpler (and recovery/reuse are potentially also easier), represents a significant step forwards in FC catalysis. A great deal of effort has been expended in the development of novel solid acid catalysts for the FC reaction, in particular alkylation, although real progress has also been made with acylations, where true catalysts have been developed [3–6]. This represents a major step forward in comparison with the greater than stoichiometric quantities of Lewis acids which are traditionally required.

8.2.1
Green Chemical Assessment of Friedel–Crafts Processes

Green chemistry has developed over the last 20 years or so into a powerful tool for analyzing chemical processes and their environmental performance. It is becoming possible (although still very difficult) to analyze processes in terms of their environmental impact, and thus to identify areas for improvements, and whether process A is actually environmentally "better" than process B. A wide range of metrics has been developed to aid in this process, from "reaction design" metrics such as atom economy, through to whole process metrics such as the E factor [7–11]. These help to quantify different aspects of a process in order to provide a measure of environmental impact. For example, atom economy [12] measures the proportion of atoms in reagents which end up in the product (addition reactions such as Michael or Diels–Alder reactions giving 100%, S_N2 reactions being lower due to the formation of e.g., HBr from ROH + RBr). FC alkylations depend on the alkylating agent (Figure 8.1) (it should be borne in mind that the synthesis of the alkylating agent has also to be considered). E factor is very simply the mass of waste produced per unit mass product, and so includes the whole reaction plus isolation, and is very simple to measure for a single process, or over an entire factory [13, 14].

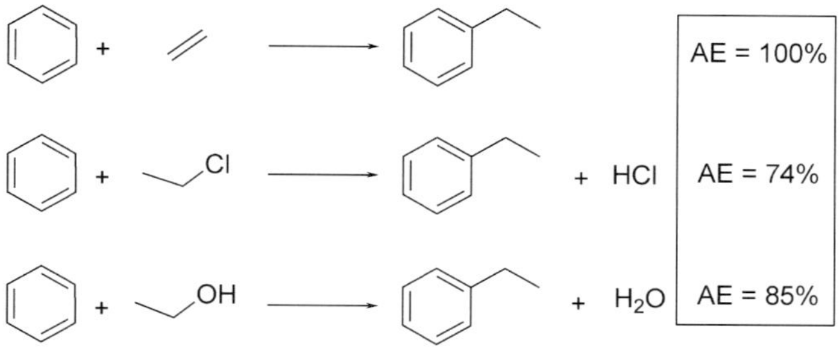

8.2 Green Chemistry and the Friedel–Crafts Reaction | 273

Figure 8.1 Atom economy in FC reactions of benzene as a function of alkylating agent.

For the top reaction, the atom economy is 100%; all atoms in the reagents are included in the products. For the second and third, the atom economies are the molecular weight of the desired product divided by the sum of the molecular weights of the reagents (74% for EtCl and 85% for ethanol, respectively). The nature of the waste is also important, clearly water is less of a difficulty than HCl. However, if we consider, in general, the synthesis of an alkyl halide versus an alcohol, then the picture is less clear, since the alcohol may well be derived from the chloride via hydrolysis, in which case a direct FC reaction with the halide will generally be better, given similar efficiencies of the steps. A knowledge of the synthesis of the starting materials is therefore necessary before a sensible result can be obtained. In order to do this, a more complex metric is required.

The ultimate metric is provided by a life cycle analysis, which looks at all the inputs (energy and chemicals) and outputs (waste) associated with a product throughout its life cycle. This includes contributions from all reagents, solvents and catalysts (and their precursors), transportation of raw materials (and their precursors), intermediates and products, and the fate of the product at the end of its useful life. This makes a full life cycle analysis very complex to carry out, and few have been done rigorously. However, the approach is extremely valuable as an indication of the factors that should be considered, even where all the data is not available. A diagram of a partial life cycle assessment is shown in Figure 8.2 for a catalyst in a FC reaction.

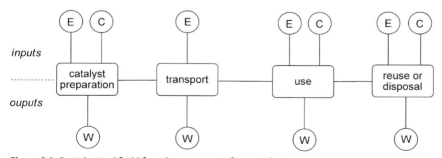

Figure 8.2 Partial, simplified life cycle assessment for a catalyst, indicating the main inputs and outputs which need to be quantified. E = energy, C = chemicals, W = waste.

For the whole process, a similar set of diagrams would be needed for all the components. This is clearly a major undertaking for which key data may not be available. Simplified versions have therefore been developed, which allow a partial answer to be derived [15].

One of the main foci of green chemistry over the last 20 years or so has been the development of new, less polluting catalysts for chemical processes [16]. High on the list to be replaced with cleaner alternatives is aluminum chloride, the traditional catalyst of choice for most industrial FC processes. While the main problems lie with the greater than stoichiometric quantities required in acylation reactions, alkylations are also deserving of attention. The major reasons for this lie in the difficulties in catalyst recovery, along with the toxic and corrosive nature of the waste. A further difficulty lies in the fact that alkylation of the desired product is generally a major problem, meaning that many $AlCl_3$-based processes run at low conversion. Washing out the catalyst not only destroys the catalyst, but means that the recovered starting materials must be thoroughly dried before use. Additionally, the difficulty in recovery and reuse of the catalyst means that fresh catalyst must constantly be used – therefore the process of catalyst production plays a significant part in the environmental footprint of its use. $AlCl_3$ is produced from bauxite via reduction to aluminum and then reaction with either Cl_2 or HCl (producing hydrogen as a by-product) in an energy intensive process.

The exceptional activity of $AlCl_3$ also means that its use is widespread and that replacement catalysts must perform very well to compete. It is fair to say that limited headway has been made in terms of acylation reactions in general (with a few notable exceptions, such as zeolite catalysis, which will be mentioned briefly below). However, more progress has been possible with alkylation reactions, and some promising solid acids have been developed, which display very good levels of activity in some processes, along with the possibility of reuse.

Over the last 50 years, partial replacement of $AlCl_3$ with HF has been successfully carried out. HF has a lower environmental footprint in its preparation (reaction of CaF_2 with sulfuric acid) although the production of $CaSO_4$ in large quantities – 3.7 tonnes per tonne of HF – is a problem. The use of HF has the advantage that it is relatively volatile and condensable, and can be readily recovered and reused. Specialist vessels and equipment must be used, and considerable know-how is required to run plants. The major downside is the risks associated with leaks and emissions as HF is an extremely toxic and hazardous material.

The third major catalyst type, and the most recently implemented, is solid acids, in particular zeolites. The synthesis of these materials requires a silica source (typically sodium silicate, an aluminum source (a salt), a template (typically a quaternary ammonium salt) in an aqueous medium. Calcination burns out the template and leaves a porous solid. The preparation requires energy and results in the complete loss of template. However, the catalysts are readily recovered and reused, meaning that the impact of catalyst preparation per tonne of FC product can be relatively low. In addition, the solids are easily handled, have long lifetimes, and can often deliver significant selectivity benefits, which is a clear benefit in terms of waste minimization and easier separations of the desired product. Other solid acids such as clays and modified silicas are known and have broadly similar benefits to the zeolites, although each must be assessed on its own merits.

While few renewable sources of alkenes exist, renewability has been addressed by Mobil, who have filed a patent which uses pyrolyzed waste plastics as a source of alkenes for alkylation processes [17].

8.3
Heterogeneous Catalysts for the Friedel–Crafts Reaction

The majority of heterogeneous catalysts have large surface areas (typically 100–1000 $m^2 g^{-1}$), and catalytic sites, which may or may not be interconnected, are dispersed on the surface and mainly reside in a network of pores. In order to achieve such large surface areas (a football pitch is about 8000 m^2), these materials are typically highly porous, with pore volumes of about 0.5–1.5 $cm^3 g^{-1}$. As catalysis takes place on the surface, diffusion of reactants to the surface, and of products (normally larger) from the surface, through the pore system, is a key consideration, and the pore size must be large enough to allow this to take place freely, without overly slowing the reaction or causing pore blockage. Most solids used as catalyst supports have random, irregular mesoporous structures (pores between 2–50 nm), which may also contain irregular micropores. Such supports include silicas and aluminas. Clays are lamellar aluminosilicate materials, where the gap between the lamellae is flexible and can be increased or decreased depending on the conditions used (e.g., solvents can expand the layers, thermal treatment can reduce the gap). Highly ordered mesoporous silicas and aluminas are available with pores in the mesoporous range, such as the MCM-41 and SBA-15 materials [18, 19]. These systems have very regular pore systems with a very narrow pore size distribution, which can be readily prepared in the range 2–10 nm by modification of the synthetic procedure. Their larger pore size allows their use in the conversion of larger molecules than is possible with zeolites, and also allows the attachment of larger organic functionality (including catalysts) to the pore walls [20–22], meaning that the relatively large enantioselective catalytic complexes could be readily incorporated into a well-defined solid matrix.

8.3.1
Zeolites

Zeolites represent an extremely important class of solid catalysts – they are used heavily in the petrochemical industry for initial processing such as cracking and isomerization. As such, the vast majority of chemicals available are derived from molecules which were formed within a zeolite. In terms of the FC reaction, most of the major FC processes operated industrially use zeolite catalysis.

Zeolites are prepared by the reaction of a silica precursor (typically sodium silicate), an aluminum precursor (a salt) and a template (usually a small molecule such as a quaternary ammonium salt) under specified conditions of temperature and pH.

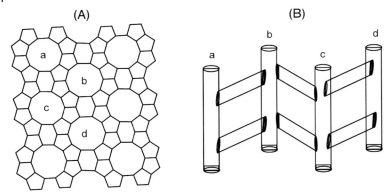

Figure 8.3 Schematic diagrams of zeolite ZSM-5, with one series of channels shown. Vertices represent Si atoms, O atoms are omitted for clarity. The pore channel of relevance is that defined by the 10-silicon ring. (A) is a partial structure looking sideways at (B), with the four pores marked.

Depending on the combination chosen, a series of families of zeolites are formed (examples are ZSM-5 (Figure 8.3), Mordenite, Beta). Each of these has a specific structure and pore system. They are predominantly crystalline materials and, as such, have very well defined and very regular pores. The size of the pores is generally in the range 0.5–1 nm, depending on the zeolite (some have more than one pore system, or have interconnecting pore systems). In addition, zeolites have exceptional thermal stability and very high surface area.

The requisite acidity is provided by the inclusion of Al sites in the structure (in a framework composed predominantly of silicon and oxygen). Each Al in the framework has the consequence of generating a negative charge on the lattice, which is balanced by a counterion. This is usually Na^+ from the synthesis, but exchange of this leads to proton-containing zeolites. The number of acid sites is therefore directly proportional to the number of Al atoms in the zeolite, but the acid strength is inversely proportional. The acid form is usually designated by a prefix H (thus H-ZSM-5, H-Beta), and the Si : Al ratio is often provided.

The presence of strongly acidic sites within highly regular pore structures of the dimensions of small molecules (H-ZSM-5 is widely considered a superacid, and has pores of the same dimension as a benzene molecule) endows them with the remarkable property of shape selectivity, whereby different isomers of a product can form depending on the size/shape of the reactants/transition state/product. A very important example is the formation of *p*-xylene, where extremely high selectivity towards the para-isomer is required. H-ZSM-5, a very strongly acidic zeolite, has a pore dimension 0.53 nm × 0.57 nm, and the para-isomer diffuses through the pores approximately 10 000 times faster than the ortho- or meta- isomers, which have marginally larger effective diameters. In an isomerization reaction (or in the methylation of toluene) the para-isomer is therefore favored due to steric constraints (rather than electronic), and exceptionally high selectivity towards the para-isomer is obtained.

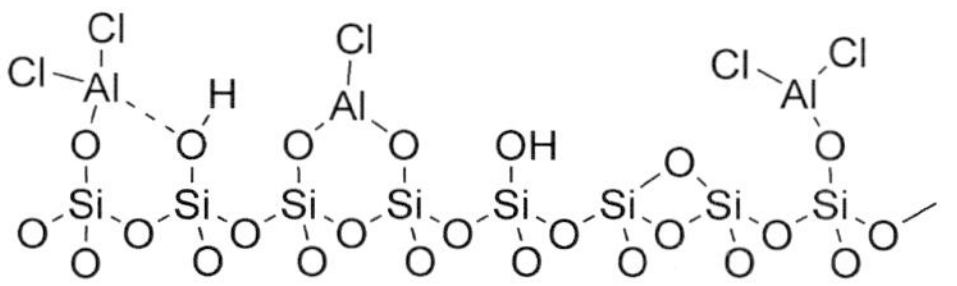

8.3 Heterogeneous Catalysts for the Friedel–Crafts Reaction

[Diagram of surface species: Cl₂Al–O–Si... H coordinated to O–Si... Cl–Al(–O)₂ bridging two Si... surface OH... Cl₂Al–O–Si units on silica surface]

Figure 8.4 Diagrammatic representation of potential surface acidic sites on AlCl₃-treated silica.

8.3.2
Clays and Other Solid Acids [23, 24]

Other solids which are used (typically as supports for Lewis acids such as AlCl$_3$, FeCl$_3$, BF$_3$ or ZnCl$_2$) are clays and silicas. Silicas are themselves very mildly acidic, but insufficiently so to function as a FC catalyst. They can, however, function as a support and attach Lewis acids by adsorption (either by coordination to a surface oxygen with the vacant orbital, or by reaction of e.g., an Al–Cl bond with a surface hydroxy group to give a Si–O–AlCl$_2$ unit (Figure 8.4). This leads to a range of sites, including strong and weak Lewis sites and Brønsted sites (via polarization of surface silanols, as well as via physisorbed HCl) [25].

This is an important feature of these materials, in that the surface sites are often generated by reaction of the Lewis acid with the support surface, and this can generate several different species, with much of the activity being due to Brønsted acidity [26–30].

Clays are largely naturally occurring minerals, which are cheap and readily available. They are layered aluminosilicates, with several variations in structural details from type to type (and from source to source within structural types – e.g., Montmorillonite is subtly different depending on where it is mined (Figure 8.5).

The clays have low inherent acidity, but have exchangeable cations (predominantly sodium) and can therefore be converted to an acidic form. The clays can also be used as a support for Lewis acids in the same way as silicas can. However, the properties of the resultant materials can be significantly different. While clays do not possess the exceptional thermal stability of the zeolites, they have found use recently as catalysts in industrial processes, as they have fewer restrictions in terms of pore size and accessibility and thus can deal with larger molecules.

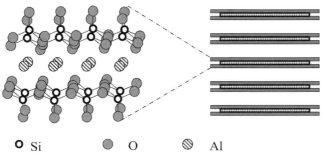

○ Si   ● O   ◍ Al

Figure 8.5 Diagramatic representation of a montmorillonite clay.

Nafion (a perfluorinated polymer with sulfonic acid groups) and its silica supported analog are also materials of interest as very strong solid acids. They have found use in a series of FC acylations [31–35].

8.4
Large Scale Hydrocarbon Processing

8.4.1
Ethylbenzene and Cumene

The production of ethylbenzene from benzene is carried out on a scale of around 15 Mt per year using two processes, one gas phase and one liquid phase (Figure 8.6) [36]. Most of the ethyl benzene produced goes to the production of styrene in a catalytic dehydrogenation process. Recently there has been a move away from the traditional liquid phase FC alkylation method, typically utilizing aluminum chloride, towards the gas phase alternative, which utilizes a solid catalyst [37, 38]. The true catalyst is thought to be a complex of HCl (formed from small amounts of EtCl added early in the reaction), $AlCl_3$, and ethylbenzene. Monoalkylation can be achieved with reasonable selectivity by keeping the benzene in excess, and limiting the extent of conversion. Poly-alkylated material can be recycled through the reactor, where it serves as an ethylating agent for fresh benzene. Operating at higher temperatures, with well-controlled introduction of ethane, and with lower quantities of $AlCl_3$ can result in a very efficient and highly selective monoalkylation. The downside of these processes is the use of aluminum chloride, which requires washing out to give toxic and corrosive waste streams, which in turn means that the recovered benzene must be dried before recycling.

In the Mobil–Badger process, the catalyst is a ZSM-5 zeolite which has pore dimensions not much larger than the product. Very high selectivity to the desired product is obtained, and the benzene to ethene ratio can be closer to 1:1 than with the liquid phase process. Mobil also run a modified process, involving a $H_3PO_4/$

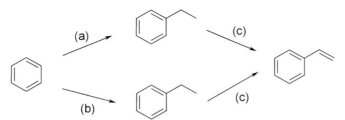

Figure 8.6 Routes to ethylbenzene and styrene production.
(a) Liquid phase route; typically using $AlCl_3$ catalyst (from 0.025 to 1 wt%) at temperatures between 85 and 140 °C. (b) Using solid catalysts (typically ZSM-5 zeolite or similar, temperatures typically 300–450 °C. (c) Catalytic dehydrogenation to styrene uses iron oxide catalysts with promoters.

Mg^{2+}/ZSM-5 catalyst for the shape selective production of p-ethyltoluene which is used in the production of p-methylstyrene.

Cumene processes are broadly similar, although oligomerization is a bigger problem with propene than with ethene. The scale of cumene production is slightly smaller than ethylbenzene, at around 10 Mt per year.

8.4.2
Linear Alkyl Benzenes

Linear alkyl benzenes (LAB, **1**) are precursors of LAS – linear alkyl benzene sulfonates, a major group of surfactants, prepared from benzene and a long chain ($\geq$C12) alkene using a FC catalyst, followed by sulfonation of the alkyl aromatic.

Originally, the alkylation process (Figure 8.7) used either AlCl$_3$ or HF. Problems exist with both processes, AlCl$_3$ generating toxic waste streams and HF generating concerns about containment. With this in mind, several researchers have looked into the potential of zeolitic materials for this process, with mixed results. Small pore zeolites such as H-ZSM-5 (pore dimensions 5.3 Å × 5.6 Å) and H-ZSM-12 (6.2 Å) have little activity, due to pore size constraints, larger pore materials such as mordenite (7 Å) were a little better. Removal of aluminum from the zeolite structure (dealumination) provides a larger pore structure, but rapid deactivation was noted, probably due to lower framework stability. In most cases shape selectivity from the pore structure resulted in high selectivity of the desired 2-phenylalkane product [39–41]. Mesoporous silica treated with immobilized AlCl$_3$ (predominantly $\equiv$Si–O–AlCl$_2$ species) was shown to be very active in the reaction [42]. Reduction in pore size (from 24 to 16 Å) and selective poisoning of external sites led to very high proportions of the desired 2-isomer, along with some reduction in rate. Regeneration of the catalyst after reaction was difficult and after five runs the catalytic activity dropped significantly.

In the mid 1990s, UOP introduced the Detal process [43, 44] which replaced these catalysts with a solid alumino-silicate acid catalyst. Since then, production of LAB has moved increasingly away from AlCl$_3$ (and, to a lesser extent, away from HF) and towards the solid acid route, primarily for ease of handling and recovery of catalyst.

One of the challenges that a catalyst must meet is to deliver a linear product (in this case, this refers to attachment of the benzene moiety as close as possible to the chain end). This aids in biodegradability of the product. The Detal process improves linearity of the product over the HF route (94–95% vs. 92.3%) thereby improving degradability as well as process safety.

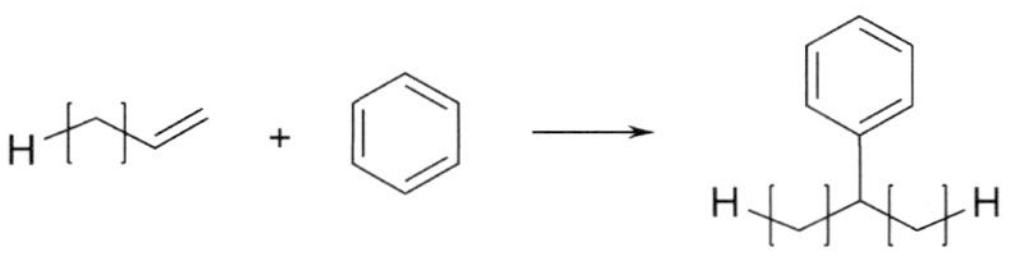

Figure 8.7 The general process for the production of linear alkylbenzenes. The ideal product is where m or $n = 1$.

8.4.3
Dialkylated Biaryls

4,4′-Diisopropyl biphenyl **2** and 2,6-diisopropylnaphthalene **3** are precursors to dicarboxylic acids, important in the manufacture of high performance plastic materials [45]. High selectivity to the desired isomer is of enormous importance (as is the case with *p*-xylene for PET production) and the shape selectivity offered by zeolites is a major factor in the choice of catalyst. Dow Chemical has developed routes to these materials using a modified zeolite that they also use for cumene production. The dealumination of zeolites removes Al atoms from the crystalline matrix of the zeolites and, by so doing, changes the three-dimensional structure of the zeolite. In Dow's case, they dealuminated a mordenite zeolite, and produced a modified material (3-DDM, 3-dimensional dealuminated mordenite) with micropores (0.5–0.7 nm) connected by much larger mesopores (5–10 nm). The presence of the mesopores allows better access to the micropores, reduces blocking, poisoning and diffusional constraints, but does not significantly affect the high selectivity of the zeolite. The effect of small pores is shown by the poor performance of H-ZSM-5, which works very well in the related alkylation of benzene with ethene and propene. With this material, high conversions and very good selectivity to the desired isomer can be obtained, leading to high yields of the desired isomer (Figure 8.8). As can be seen from Figure 8.9, while a few zeolites allow high levels of conversion, 3-DDM is the only zeolite tried which gives high selectivity and yield [46–49].

8.4.4
Alkylanilines and Related Compounds

Diarylamines **4** are well known as antioxidants, and a range of them is produced. The basic diphenylamine nucleus is common to the majority of these structures, and this can have a range of alkyl groups attached to enhance compatibility with the

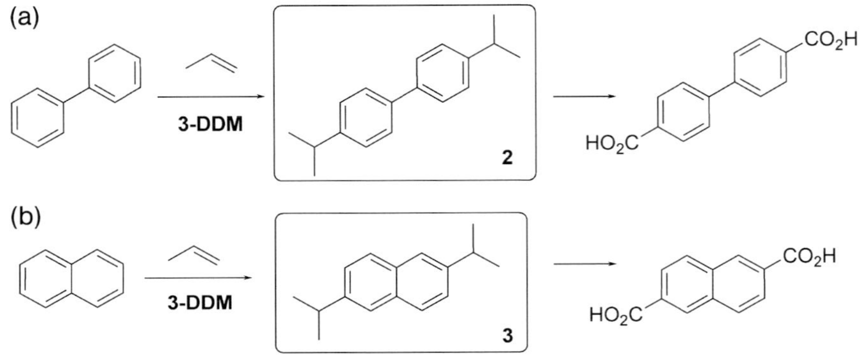

Figure 8.8 Preparation of (a) 4,4′-di-isopropylbiphenyl and (b) 2,6-di-isopropylnaphthalene using 3DDM catalysts. Yields and selectivities shown in Figure 8.9 for **2**. Similar results have been obtained for **3**.

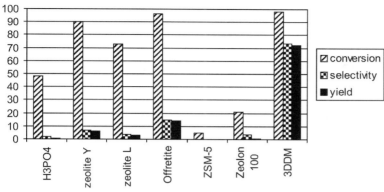

Figure 8.9 Depiction of conversion, selectivity and yield in the synthesis of 4,4′-diisopropyl biphenyl (Figure 8.8a) using a range of acid catalysts. (Selectivity refers to the percentage of dialkylated product).

matrix in which it must work. Ciba Geigy have patented routes to these compounds (and alkylated phenothiazines 5) using clay-based catalysts (e.g., Fulcat) which enable the alkylation to be carried out at reasonable temperatures and pressures (Figure 8.10) [50, 51].

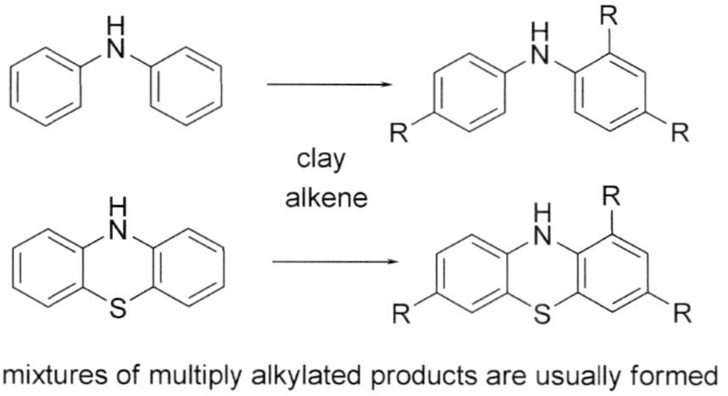

mixtures of multiply alkylated products are usually formed

Figure 8.10 Alkylation of diarylamines and phenothiazines using clay catalysis.

8.4.5
Alkylphenol Production

Alkylphenols are produced at around 0.5 Mt per year using FC alkylation. The products are used in a variety of applications, including antioxidants. As previously seen for alkylbenzenes, production of alkylphenols has relied on liquid phase catalysts such as BF_3 [52–54], $ZnCl_2$ [55], HF [56] and CF_3SO_3H [57], while more recent patents have begun to focus on solid catalysts such as zeolites and solid sulfonic acids such as Deloxan and Nafion-silica composites Continuous processes have also been developed where the feed is flowed over a bed of solid catalyst under supercritical conditions [58, 59].

8.4.6
Diarylmethanes

Diarylmethanes are used for a variety of purposes, and are clearly very well suited to synthesis by FC reaction. The traditional routes are from arene plus (substituted) benzyl chloride, but the benzyl alcohol has also been used. Whether this is necessarily beneficial is unclear, as most benzyl alcohols are prepared from the hydrocarbon via radical chlorination and subsequent displacement of the chlorine with water or hydroxide. Direct oxidation of the hydrocarbon to the alcohol is difficult so, in terms of waste minimization, there may be little advantage in using the alcohol rather than the halide.

The relatively large size of the products has meant that much of the work carried out on diarylmethanes has been done with clays and silicas rather than zeolites and Contract Chemicals have commercialized a catalyst (Envirocat EPZ10) for FC alkylations [60]. Another member of this group of catalysts is clayzic, a catalyst formed from a clay (Montmorillonite K10) and zinc chloride. Upon activation, the clay/$ZnCl_2$ mixture is converted into a very active catalyst. The activation process goes beyond simple drying and it is thought that partial delamination and rearrangement of the clay structure takes place [61, 62]. The resultant material is exceptionally active for the FC alkylation of benzene with benzyl chloride (Figure 8.11) [63]. Some control over double alkylation is possible.

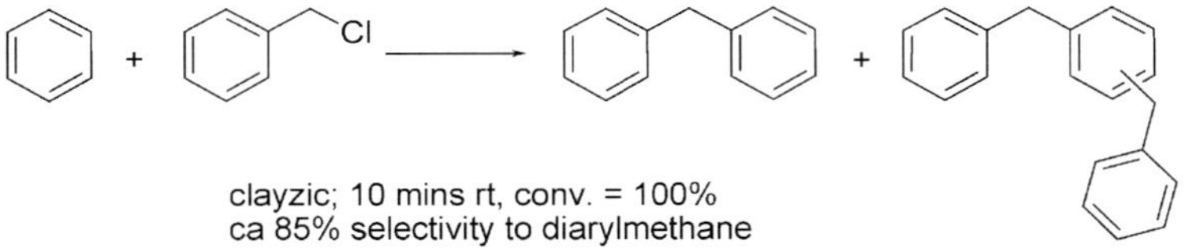

clayzic; 10 mins rt, conv. = 100%
ca 85% selectivity to diarylmethane

Figure 8.11 Formation of diarylmethanes with clay-based catalysts.

8.4.7
Hydroxyalkylation of Aromatics

Hydroxyalkylation of aromatics is a process for producing benzylic alcohols from aldehydes and ketones, or phenylethanols from aromatics and epoxides. (Figure 8.12) Compared to the preceding alkylations, this process allows the introduction of functionality into the product which can then be further utilized in subsequent

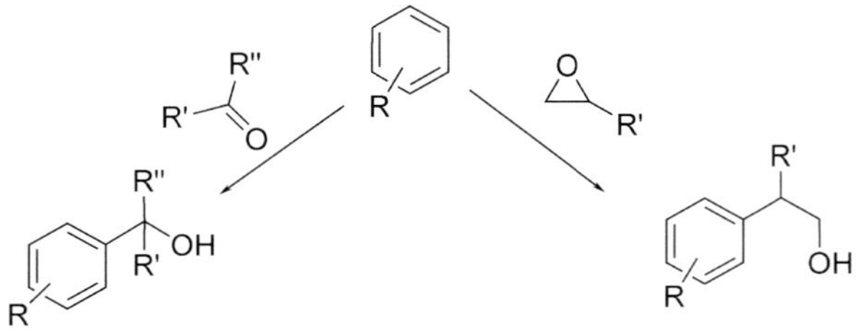

Figure 8.12 Hydroxyalkylation of aromatics.

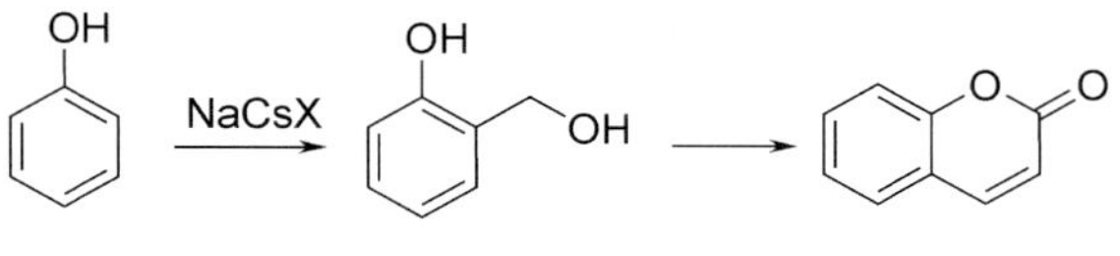

Figure 8.13 Use of saligenol **6** in the synthesis of coumarin, a product of interest in the fragrance industry.

transformations. Less progress has been made in the industrial utilization of this variant, largely due to difficulties of selectivity, and of subsequent reactions, particularly where benzylic alcohols are produced. Other hurdles include the requirement that the aromatic is highly activated, and that bringing together two reactants of very different polarity within the same zeolite pore is often difficult.

The benzylic alcohols are themselves highly active alkylating agents and diarylmethanes are readily formed. In terms of selectivity, zeolites could certainly be used to direct towards para-substitution. However, some important targets (such as saligenol, Figure 8.13) require ortho-selectivity, to which zeolites are less well suited. The best reported alkylation remains that from Asahi, who reported that a zeolite with low acidity (NaCsX) gave a 45% selectivity to the ortho-isomer **6** at 66% conversion under relatively mild conditions.

Despite these difficulties, progress has been made, with the highly active furans being hydroxymethylated using formaldehyde with catalysis by a dealuminated H-mordenite zeolite in good yield and selectivity (Figure 8.14) [64].

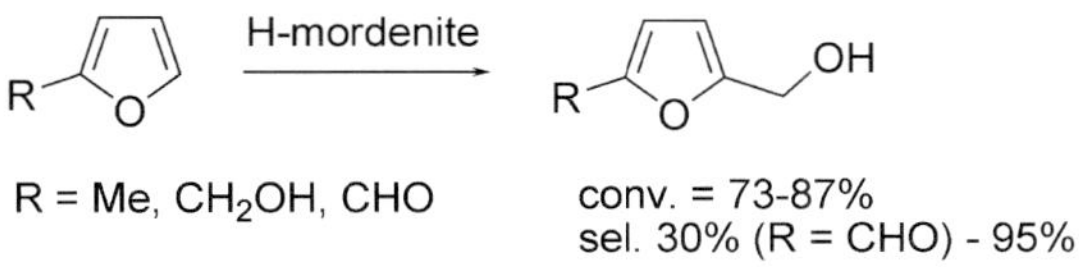

R = Me, CH_2OH, CHO   conv. = 73–87%
                              sel. 30% (R = CHO) – 95%

Figure 8.14 Hydroxymethylation of furans.

This work has been extended to the highly activated guaiacol **7** to prepare a vanillin precursor **8** (Figure 8.15) [65].

A somewhat similar FC oxo-alkylation has been reported by Mitsui chemicals as a stage in the synthesis of herbicides for paddy weeds [66]. They generated a cationic intermediate **9** using BF_3 Et_2O in dichloromethane at 0 °C, and then alkylated intramolecularly to give a fused ring system **10** with retention of stereochemistry (Figure 8.16). Yields were high in most cases, although cleavage of the benzyl group and formation of epoxide **11** (boxed) was sometimes a competing reaction.

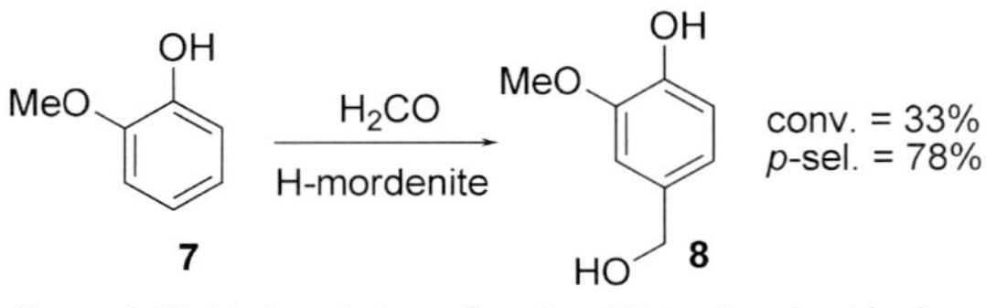

Figure 8.15 Hydroxylation of guaiacol **7** to give the 4-hydroxymethylguaiacol **8**.

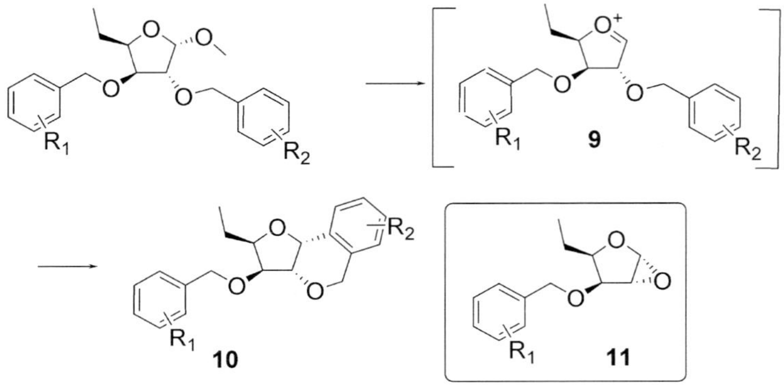

Figure 8.16 Synthesis of tricyclic herbicides using BF_3 catalysis. Conditions: BF_3 Et_2O, CH_2Cl_2, 0 °C to rt, 2 h (82%).

8.4.8
Pechmann Syntheses

The Pechmann synthesis of coumarins also relies on a FC hydroxyalkylation using ketoesters (Figure 8.17). Sulfuric acid has been the traditional catalyst of choice for the reaction but, more recently, zeolites have been shown to be effective catalysts. For example, the synthesis of methylumbelliferone **12**, an important fragrance ingredient and an intermediate in the synthesis of insecticides, can be prepared using H-Beta as catalyst [67, 68].

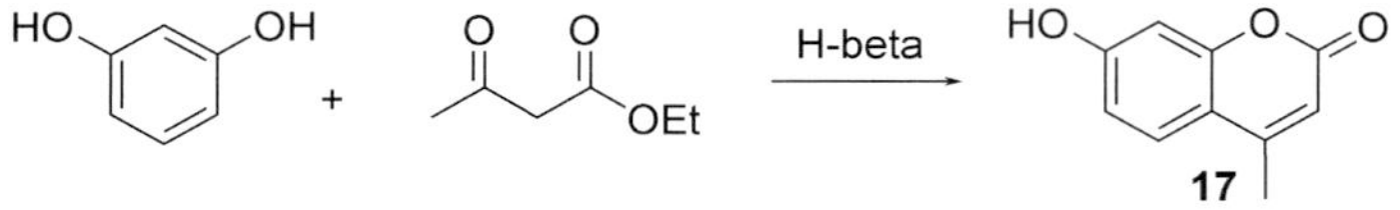

Figure 8.17 Pechmann synthesis of methylumbelliferone **12**.

8.4.9
Use of Epoxides as Alkylating Agents

Alkylations using epoxides are also known. Here, intramolecular alkylations have been found to be much more successful than bimolecular variants. This has been ascribed to the difference in polarity between the two molecules, meaning that adsorption of one is generally favored at the expense of the other. Target molecules include 2-phenylethanol **13**, one of the most important fragrance molecules, produced on a scale of several thousand tonnes per year. Another example of an industrially relevant product is the benzopyran **14** shown in Figure 8.18, again used in the fragrance industry and prepared via a FC reaction.

While most industrial production of this product involves racemic epoxide, the (4S, 7R)-isomer of **14a** is considered to have the most powerful fragrance and routes

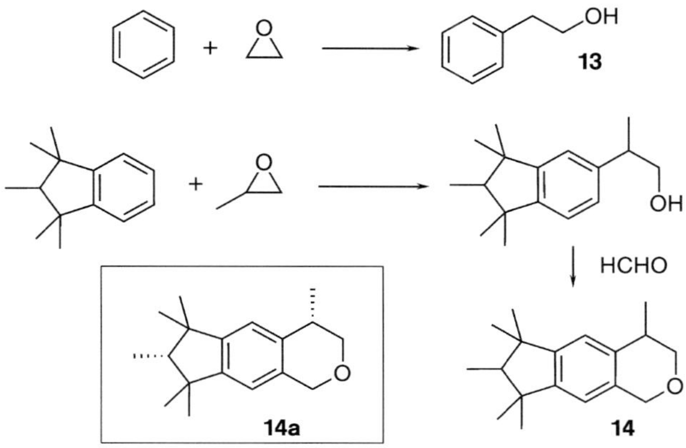

Figure 8.18 Hydroxyethylation in the fragrance industry.

to its formation are known [69]. Traditional approaches to such products have thus relied on the use of AlCl$_3$, often in large amounts, as the complexes formed are good catalysts for epoxide polymerization, which can become a major problem unless large excesses of aromatic and AlCl$_3$ are used. Mechanistically, these reactions involve exclusive substitution at the more substituted end of the epoxide, along with inversion at the reacting C center, meaning that stereochemical information is not lost due to scrambling via planar carbocation intermediates.

Elings et al. have demonstrated that intramolecular alkylations of aromatics with epoxides are relatively efficient with zeolitic catalysts, but much less so where the reaction is bimolecular (Figure 8.19) [70, 71]. Conversions were up to 100% at 80 °C, with selectivity to the alcohol being up to 38%, with dehydration being the major side reaction.

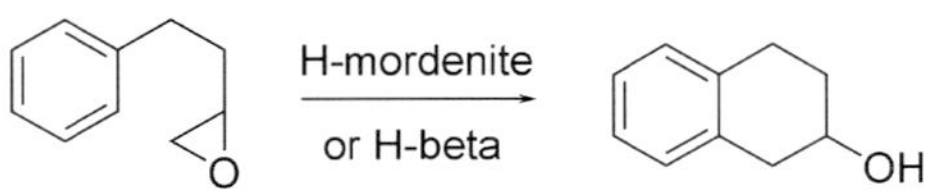

Figure 8.19 Intramolecular FC alkylation with an epoxide.

As can be seen, the intramolecular nature of the reaction inverts the usual positional selectivity, no doubt due to the constraints of the transition state for the expected 5-membered ring.

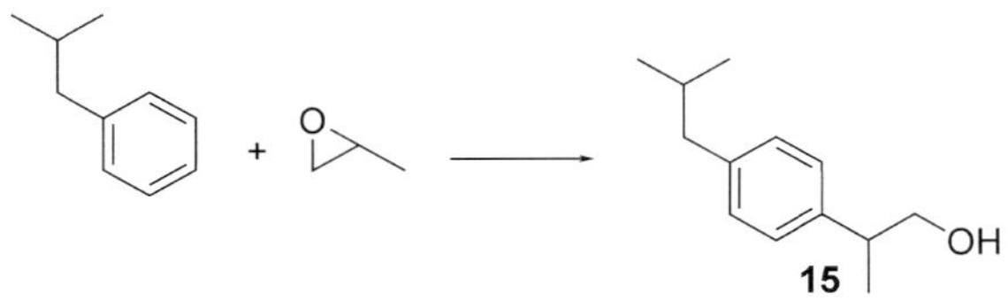

Figure 8.20 FC alkylation of isobutylbenzene with propylene oxide.

Mobil have described the synthesis of 2-(4-isobutylphenyl)propanol, shown in Figure 8.20, via FC reaction of isobutylbenzene with propylene oxide [72], indicating that bimolecular reactions of this type are possible. The level of detail is not sufficient to judge how close this work brings us to an actual commercial process.

8.5
Conclusions and Perspectives

The FC alkylation reaction is ubiquitous in chemistry. It is one of the key transformations in industrial chemistry, especially in the large scale hydrocarbon processing area and will remain of central importance for many years to come. Over the last 25 years or so, industry has moved away from the traditional aluminum chloride-based processes in an effort to reduce waste production. This has manifested itself in a move towards HF, readily separated and recovered for reuse but particularly dangerous if containment is not perfect. Similarly, solid acids, themselves readily recoverable and reusable, have come to prominence. The main group of solid acids which have been used in commercial processes are the zeolites, materials with a long industrial pedigree. They are readily synthesized, available on a large scale, available in a range of structures with controllable acidity (from mild to very strongly acidic) and extremely thermally stable, allowing reactivation and reuse. Other solids have been used with more sporadic success. A relatively small number of processes are known where the FC reaction has been used commercially to prepare more complex structures. However, it must be assumed that much effort is being expended in such endeavors but that little of this has been published in the open literature. It is likely that such work, including enantioselective versions of the FC reaction, will appear over the next few years. Translation of the mostly homogeneous catalytic systems to solid-supported versions is already underway, and significant progress has recently been made with developing effective solid enantioselective catalysts for a range of reactions including FC reactions [73–76].

References

1 Anastas, P.T. and Warner, J.C. (1998) *Green Chemistry, Theory and Practice*, Oxford University, New York.
2 Clark, J.H. and Macquarrie, D.J. (eds) (2002) *Handbook of Green Chemistry and Technology*, Blackwell Science, Oxford.
3 Gunnewegh, E.A., Downing, R.S. and van Bekkum, H. (1995) *Studies in Surface Science and Catalysis*, **97**, 447–453.
4 Ratton, S. (1997) *Chemistry Today*, **3–4**, 33.
5 Vogt, A. and Pfenninger, A. (1987) EP 0701987A1 (to Uetikon AG).
6 Metivier, P. (2001) *Fine Chemicals Through Heterogeneous Catalysis* (eds R.A. Sheldon and H. van Bekkum), Wiley-VCH, Weinheim, pp. 173–204.
7 Constable, D.J.C., Curzons, A.D. and Cunningham, V.L. (2002) *Green Chemistry*, **4**, 521–527.
8 Curzons, A.D. (2001) *Green Chemistry*, **3**, 1–6.
9 Auge, J. (2008) *Green Chemistry*, **10**, 225–231.
10 Andraos, J. (2005) *Organic Process Research & Development*, **9**, 149–163.

11 Gonzalez, M.A. and Smith, R.L. (2003) *Environmental Progress*, **22**, 269–276.
12 Trost, B.M. (1991) *Science*, **254**, 1471–1477.
13 Sheldon, R.A. (1997) *Chemistry & Industry*, 12–15.
14 Sheldon, R.A. (1992) *Chemistry & Industry*, 903–906.
15 Curzons, A.D., Jiminez-Gonzalez, C., Duncan, A.L., Constable, D.J.C. and Cunningham, V.L. (2007) *The International Journal of Life Cycle Assessment*, **12**, 272–280.
16 Sheldon R.A. and van Bekkum H. (eds) (2001) *Fine Chemicals Through Heterogeneous Catalysis*, Wiley-VHC, Weinheim.
17 Collins, N.A., Green, L.A., Gupte, A.A., Marler, D.O. and Tracy, W.J. III (1995) US Patent 5705724. (to Mobil Oil corporation).
18 Beck, J.S., Vartuli, J.C., Roth, W.J., Leonowicz, M.E., Kresge, C.T., Schmitt, K.D., Chu, C.T.W., Olson, D.H., Sheppard, E.W., McCullen, S.B., Higgins, J.B. and Schenker, J.L. (1992) *Journal of the American Chemical Society*, **114**, 10834–10843.
19 Zhao, D.Y., Feng, J.L., Huo, Q.S., Melosh, N., Fredrickson, G.H., Chmelka, B.F. and Stucky, G.D. (1998) *Science*, **279**, 548–552.
20 Hoffmann, F., Cornelius, M., Morell, J. and Fröba, M. (2006) *Angewandte Chemie – International Edition*, **45**, 3216–3251.
21 Taguchi, A. and Schüth, F. (2005) *Microporous and Mesoporous Materials*, **77**, 1–45.
22 De Vos, D.E., Dams, M., Sels, B.F. and Jacobs, P.A. (2002) *Chemical Reviews*, **102**, 3615–3640.
23 Corma, A. and Garcia, H. (2003) *Chemical Reviews*, **103**, 4307–4365.
24 Centi, G. and Perathoner, S. (2008) *Microporous and Mesoporous Materials*, **107**, 3–15.
25 Shorrock, J.K. (2002) Ph.D Thesis, University of York.
26 Algarra, F., Corma, A., Fornés, V., Garcia, H., Martinez, A. and Primo, J. (1993) *Studies in Surface Science and Catalysis*, **78**, 653–659.
27 Corma, A., Fornés, V., Martin-Aranda, R.M., Garcia, H. and Primo, J. (1990) *Applied Catalysis A – General*, **59**, 237–248.
28 Su, B.L. and Barthomeuf, D. (1995) *Applied Catalysis A – General*, **124**, 73–80.
29 Su, B.L. and Barthomeuf, D. (1995) *Applied Catalysis A – General*, **124**, 81–90.
30 Clark, J.H., Rénson, A. and Wilson, K. (1999) *Catalysis Letters*, **61**, 51–55.
31 Kureck, P.R. (1991) US Patent 5126489 (to UOP).
32 Gotto, S., Gotto, M. and Kimura, Y. (1991) *Reaction Kinetics and Catalysis Letters*, **41**, 27–41.
33 Misono, M. (1996) *Advances in Catalysis*, Vol 42, **41**, 113–252.
34 Patil, M.L., Jnanasehwara, G.K., Sabde, D.P., Dongare, M.K., Sudalai, A. and Deshpande, V.H. (1997) *Tetrahedron Letters*, **38**, 2137–2140.
35 Harmer, M., Vega, A.J., Sun, P., Farneth, W.E., Heidekum, A. and Hölderich, W.F. (2000) *Green Chemistry*, **2**, 7–14.
36 Weissermehl, K. and Arpe, H.-J. (1993) *Industrial Organic Chemistry*, VCH, Weinheim.
37 Matar, S., Mirbach, M.J. and Tayim, H.A. (1994) *Catalysis in Petrochemical Processes*, Kluwer Academic Publishers, Dordrecht, p. 27.
38 Degnan, T.F. Jr, Smith, C.M. and Vayat, C.R. (2001) *Applied Catalysis A – General*, **221** (1–2), 283–294.
39 Le, Q.N., Marler, D.O., McWilliams, J.P., Rubin, M.K., Shim, J. and Wong, S.M. (1990) US Patent 4962256 (to Mobil).
40 Hing-Yuan, H., Zhonghui, L. and Enze, M. (1988) *Catalysis Today*, **2**, 321–327.
41 Sivisanker, S. and Thangaraj, A. (1992) *Journal of Catalysis*, **138**, 386–390.
42 Price, P.M., Clark, J.H., Martin, K., Macquarrie, D.J. and Bastock, T.W. (1998) *Organic Process Research & Development*, **2**, 221–225.
43 Stache, H.W. (1995) *Anionic Surfactants: Organic Chemistry*, CRC Press, Cleveland.

44 Kocal, J.A., Vora, B.V. and Imai, T. (2001) *Applied Catalysis A – General*, **221**, 295–301.
45 Schaefgen, J.R. (1975) US Pat 4118372 (to E I du Pont de Namour).
46 Lee, G.S., Maj, J.J., Rocke, S.C. and Garces, J.M. (1989) *Catalysis Letters*, **2**, 243–247.
47 Garces, J., Maj, J.J., Lee, G.J. and Rocke, S.C. (1987) US Pat 4891448 (to Dow).
48 Lee, G.S., Garces, L.M. and Maj, J.J. (1989) US Pat 5015797 (to Dow).
49 Meima, G.R., Lee, G.S. and Garces, J.M. (2001) *Fine Chemicals through Heterogeneous Catalysis* (eds R.A. Sheldon and H. van Bekkum), Wiley-VCH, pp. 151–160.
50 Evans, S. and Allenbach, S. (1994) US Pat 5503759 (to Ciba Geigy).
51 Evans, S., Allenbach, S. and Dubs, P. (2000) US Pat 6407231 (to Ciba Geigy).
52 Motters, H.O. and Peters, T.J. (1953) US Pat 2800451 (to Esso).
53 Tirtiaux, R. and Signouret, J.B. (1957) US Pat 2978423 (to Esso).
54 Lange, A. and Roth, H.P. (2000) DE 100481507 (to Keil and Weinkauf).
55 Teubner, H., Krauer, A., Weuffen, W., Schrötter, E. and Grübel, G. (1982) US Pat 4532367 (to VEB Jenapharm).
56 Suzuki, T. and Naito, S. (1975) US Pat 3976702 (to Mitsubishi).
57 Giolito, S.L. and Mirviss, S.B. (1977) US Pat 4103096 (to Stauffer).
58 Poliakoff, M., Swann, T.M., Tacke, T., Hitzler, M.G., Ross, S.K., Wieland, S. and Smail, F.R. (1999) US Pat 6303840 (to Thomas Swann & Co).
59 Amandi, R., Licence, P., Ross, S.K., Aaltonen, O. and Poliakoff, M. (2005) *Organic Process Research & Development*, **9**, 451–456.
60 Brown, C.M., Clark, J.H., Kybett, A.P. and Macquarrie, D.J. (1989) Eur Pat 89303433.
61 Clark, J.H., Cullen, S.R., Barlow, S.J. and Bastock, T.W. (1994) *Journal of the Chemical Society – Perkin Transactions 2*, 1117–1130.
62 Rhodes, C.N. and Brown, D.R. (1992) *Journal of the Chemical Society – Faraday Transactions*, **88**, 2269–2274.
63 Clark, J.H., Kybett, A.P., Macquarrie, D.J., Barlow, S.J. and Landon, P. (1989) *Chemical Communications*, 1353–1354.
64 Lecomte, J., Finiels, A., Geneste, P. and Moreau, C. (1999) *Journal of Molecular Catalysis A – Chemical*, **140**, 147–163.
65 Moreau, C., Finiels, A., Razigade, S., Gilbert, L. and Fajula, F. (1996) WO Patent 9637452 (to Rhodia).
66 Kakimoto, T., Koizumi, F., Hirase, K., Banba, S., Tanaka, E. and Arai, K. (2004) *Pest Management Science*, **60**, 493–500.
67 Gunnewegh, E.A., Hoefnagel, A.J., Downing, R.S. and van Bekkum, H. (1996) *Recueil des Travaux Chimiques des Pays-Bas*, **115**, 226–232.
68 Gunnewegh, E.A., Hoefnagel, A.J. and van Bekkum, H. (1995) *Journal of Molecular Catalysis A – Chemical*, **100**, 87–92.
69 Fráter, G., Müller, U. and Kraft, P. (1999) *Helvetica Chimica Acta*, **82**, 1656–1665.
70 Elings, J.A., Downing, R.S. and Sheldon, R.A. (1997) *Studies in Surface Science and Catalysis*, **105**, 1125–1133.
71 Elings, J.A. (1997) PhD Thesis, University of Technology, Delft.
72 Altman, L.J., Angevine, P.J. and McWilliams, J.P. (1990) US Pat 5001283 (to Mobil).
73 Thomas, J.M. and Raja, R. (2008) *Accounts of Chemical Research*, **41**, 708–720.
74 Yu, P., He, J. and Gou, C.X. (2008) *Chemical Communications*, 2355–2357.
75 Zhang, Y., Zhao, L., Lee, S.S. and Ying, J.Y. (2006) *Advanced Synthesis and Catalysis*, **348**, 2027–2032.
76 Corma, A. and Garcia, H. (2006) *Advanced Synthesis and Catalysis*, **348**, 1391–1412.

Index

a
achiral rhodium catalysts 211
acremoauxin A/oxazinin 3 255, 257
– synthesis 255–257
2-acyl imidazoles, cleavage 38
adducts elaboration 30
– enantioenriched indole-substituted 30
– – aldehydes 30
– – carboxylic acids 30
– – ketones 30
air-stable chiral P/S ligands 153
Akiyama cyclic transitions 74
alcoholic moiety 70
– bifunctional mode of action 71
alcoholic solvent 24, 25
– direct activation 181
– direct nucleophilic substitution 182
– enantioselectivity 24
– iBuOH 24
– transition state 25
aldehydes/ketones addition 101
– electron-rich aromatic/heteroaromatic compounds 102
– heterogeneous catalysis 118
– organocatalysis 116
– organometallic catalysis 102
– – chiral bisoxazoline-Cu(II) catalysis 108
– – chiral titanium (IV) catalysis 103
aliphatic enones 32
– Friedel–Crafts alkylation 32
alkaloid synthesis 134
alkenes 208
– Lewis/Brønsted acid-catalyzed hydroarylation procedures 204
– Rh-catalyzed hydroarylation 208
– substitution effect 214
– – on rhodium-catalyzed enantioselective alkylation 214

alkenylindoles 214, 216
– enantioselective hydroarylation 216
– Pt(II)-catalyzed enantioselective hydroarylation 214–216
– – ligand/solvent effect 215
alkoxyarenes 104
– Friedel–Crafts reaction 104
– chiral BINOL-Ti complex 104
β-alkyl 2-acyl imidazoles 35
– alkylation products 35
alkylanilines compounds 280, 281
alkylated product conversion 37, 43
– imidazole to carboxylic acid derivatives 37
– tryptophan analog 43
alkylating agent 6–9, 272
alkylation 207
– aromatic aldimines 209–211
– aromatic ketimines 207–209
alkylidene malonates 19–25, 42, 43
– catalytic asymmetric Friedel–Crafts reaction 42, 43
– – experimental procedure 42, 43
– – transition state 42
– enantioselective Friedel–Crafts alkylation reactions 20, 22, 24, 25
N-alkyl indoles 72
– derivatives 72
– Friedel–Crafts alkylation reaction 72
– – bissulfonamide catalyzed with nitroalkenes 72
allenes 146, 160, 161, 163
– Au-catalyzed hydroarylation mechanistic cycle 163
– catalytic cycloarylation reactions 146
– catalytic hydrofunctionalization with indoles 161
– gold(I)-mediated intramolecular hydroarylation with indoles 161

- metallo-catalyzed hydroarylation 160
allenylindoles 217–219
- Au(I)-catalyzed enantioselective hydroarylation 217, 218
- enantioselective hydroarylation 218, 219
allylation reactions 157
- catalytic systems 157
allyl cross-coupling reaction mechanism 148
- Pd-catalyzed 148
allylic alkylation processes 147, 153
- intermolecular approach 147
- - asymmetric allylations 152
- - benzene-like compounds 147
- - indoles 150
- intramolecular approach 156–160
α-amino acid synthesis 250, 252
2-aminobutenolides synthesis 247–249
aminoindane/arene substitution effect 213
- on enantioselective hydroarylation 213
p-aminomandelic acid derivatives 250
aminomethyl indoles synthesis 229–240
annulation processes, effective catalyst 161
antihypertensive drug precursor 243, 244
- pyrrolo[1,2-a]pyrazine 244
- synthesis 244
arenes 8, 102, 157, 158, 162, 163, 176, 186, 203, 204, 249
- catalytic intramolecular alkylation 157, 158
- - via allyl alcohols 158
- enantioselective alkylation 203
- enantioselective functionalizations 8
- Friedel–Crafts alkylation 186, 249–253
- Friedel–Crafts-type reactions 102
- - with α-dicarbonyl compounds 102
- gold-catalyzed hydroarylation 162
- gold-catalyzed regio/diastereoselective alkylation 176
- hydroarylation procedure 163, 204
- intramolecular alkylation 158
- metal-carbene intermediate 204
- thiourea-gold-catalyzed intramolecular diastereoselective cycloalkylation 176
aromatic aldimines 212, 213
- enantioselective cyclization 212
aromatic amines 105, 108, 109, 114
- enantioselective ethyl glyoxylate catalyzed Friedel–Crafts reaction 105, 108, 109, 114
- - by BINOL-titanium complex 105
- - titanium-catalyzed 114
aromatic C–H bonds, asymmetric addition 101

aromatic compounds 1, 26, 101, 102, 110, 119, 148
- 1,3-dimethoxybenzene 26
- asymmetric imines Friedel-Crafts reaction 119
- asymmetric/enantioselective carbonyl compounds alkylation, via 1,2-addition 102
- enantioselective Friedel–Crafts reaction with ethyl trifluoropyruvate 110, 111
- Friedel–Crafts (FC) alkylation 1
- Friedel–Crafts carbonyl compounds reaction 101
- functionalization 148
- π-rich 53
- - asymmetric Friedel–Crafts alkylation by iminium catalysis 53
- reaction conditions 13
aromatic epoxides 173, 176
- gold-catalyzed cycloalkylation 176
- indoles addition 173
aromatic ethers 114, 116
- catalytic enantioselective ethyl trifluoropyruvate Friedel–Crafts reaction 114, 116
- - solvent-free conditions 114, 116
aromatic imines 211
- rhodium-catalyzed enantioselective hydroarylation 211
aromatic ketimines 207, 209, 215
- enantioselective intramolecular hydroarylation 209, 215
- - rhodium-catalyzed 215
aromatic nucleophiles 54
- π-nucleophiles 56
- site-specific alkylation 54
aromatic process 282
- hydroxyalkylation 282
aromatic rings 158
aromatic substrates 49, 50
- enantioselective Michael additions 49, 50
aromatic systems 88, 109, 150
- gold(III) chloride-catalyzed allylic alkylation, with allylic alcohols 150
- Julolidine 109
- N-methylindoline 109
- N-methyltetrahydroquinoline 109
- stereoselective alkylations 88
aryl-alcohols 159
- Friedel–Crafts allylic cycloalkylation 159
aryl aldimines 211
asymmetric aminocatalysis process 23, 50, 55, 61
- application 55

Index

asymmetric electrophilic aromatic substitutions 8
asymmetric Friedel–Crafts reactions 115
– alkylations 12, 167
– – allylic 145
– – chiral organocatalysis 12
– – model 114
– copper(II)-catalyzed 115
asymmetric organometallic catalysis 9
atropisomerization process 206
– atropisomers interconversion 206
– atropselective alkylation 206
– – rhodium(I)-catalyzed protocol 206
– bis-indole preparation 192
aza-Achmatowitcz reaction 248
aza-Friedel–Crafts reaction 120
aza-Wittig reaction 121

b

bench-stable catalysts 116
benzamide 203, 204
– enantioselective ortho-alkylation 203
– enantioselective ortho-hydroarylation 204
benzene 273
– Friedel–Crafts reactions, atom economy 273
benzotriazole 77, 78
– asymmetric addition 77, 78
benzylic stereocenters 145, 146
– synthesis 145
– vinyl units 146
benzylidene malonates 23
– enantioselective Friedel–Crafts alkylation reactions 23
N-benzyltryptamine 138
– asymmetric Picard–Spengler reaction 138
BF_3-mediated reactions 183, 184
biaryls 206
– atropselective alkylation 206, 207
bidentate Michael acceptors 17
bifunctional catalysts, use 67, 75
BINOL ligand 106
BINOL-Ti complexes 104
– catalytic efficiency 104
– catalyzed asymmetric reactions 105
– zirconium adducts 11
biomimetic acid-catalyzed ring closure 254
biosynthetic cyclization process 75
2,2′-bipyridyl copper(II) complex 113
– chiral nonracemic C2-symmetric 113
– enantioselective Friedel–Crafts alkylation reactions 113
bis-N,N'-aryl thiourea catalysts 71
– X-ray structure 72

bis-prophenol ligand 89
bis(oxazoline) ligands 22, 23, 41
– aza-bis(oxazolines) 23
– copper complexes 23
– use 22
bis(oxazolinyl)pyridine (pybox)-scandium(III) triflate complexes 32
bisindole, formation 27
BOX catalytic system 90
– ligands 84, 113
– Mg catalyst 250
– zinc/copper complexes 88
(R)-6,6′-Br_2-BINOL-Ti complex 104
Brønsted–Lowry acids 3, 4, 8, 18, 27, 62, 67, 68, 75
– assisted activation 28
Brønsted base, functionality 78
Brown's allylborane reagent 260
(+)-bruguierol C synthesis 260, 266
butenolide synthesis 248, 249

c

calcination process 274
carbon-carbon bond formation reactions 149, 150, 152
– coupling alkylation reaction 149
– enantioselective coupling reaction 103
– mechanism 150
– ruthenium(IV)-catalyzed 149, 150
– stereoselective formation 49
– – Friedel–Crafts alkylation strategy 49
– tentative mechanistic cycle 149
carbonyl compounds addition, see aldehydes/ketones addition
carbonyl electrophile, role 105
carbonyl-ene reaction 105
– Corey's transition-state model 105
catalyst structure-enantiodifferentiation profile 117
catalytic activation concept, see iminium catalysis
catalytic asymmetric Friedel–Crafts alkylation reactions 1, 19, 223
– bidentate templates 19
– synthesis 223
catalytic Friedel–Crafts reaction 115
– b/w aromatic amines and ethyl glyoxylate 115
– – by chiral bisoxazoline-copper (II) complex 115
catalytic stereoselective reactions 167
catalytic system(s) 17, 31, 35, 62, 121, 152
– chemical efficiencies 17
– chiral entities 62

- features 35
- pyrrole derivatives 35
- stereochemical efficiencies 17
catalytic transformations, importance 22
cationic [(Tol-BINAP)Pd(II)] complexes 39
- binding interaction 39
cationic zirconocene complexes 204
C–C double bonds 145
- hydroarylation 145
- nucleophilic allylic alkylation 145
chalcones 64
- enantioselective Friedel–Crafts alkylation 64
- – by chiral Brønsted acids catalyst 64
chelating compounds 18
Chen's working hypothesis 75
chiral acetate 188
- AuCl$_3$-catalyzed diastereoselective alkylation 188
chiral acid catalyst 63
chiral aldimine derivates 213
- rhodium-catalyzed cyclization 213
chiral aluminium complex 60
- BINOL-based ligand 60
chiral 1-aryl-2,2,2-trifluoroethanol derivatives 104
- synthetic route 104
chiral benzylic alcohols 182
- carbocation, diastereotopic faces 182
chiral BINAP-Cu(I) complex 250
chiral BINOL-based phosphoric acid 63
chiral boron reagents 134
- Pictet–Spengler reaction 134
chiral Brønsted acids 69, 120, 124, 126, 228
- catalyzed reaction 228
chiral carbocations 183
- enantioenriched precursors preparation/separation 183
chiral ferrocenyl compounds 193
- Friedel–Crafts reactions 193
chiral hydrogen bond donors, application 116
chiral imidazoline-aminophenol ligand 86
- CD spectroscopy 86
- combinatorial HTS method 86
chiral ketoester catalyst 113
- coordination geometry 113
chiral Lewis acid 5, 10, 18, 120
- catalysts 81, 130
- catalyzed reactions 18, 77
- organometallic 103
- plausible coordination patterns of complexes 18
chiral metal complexes 18, 40, 80

- chiral ligand 23
- pool 18
chiral organic molecules 123
- thioureas 123
- ureas 123
chiral organometallic catalysts 10
- pictorial representation 10
chiral phosphorus-ligands 11
- BINAP 11
- DPPBA-based ligands 11
- ferrocenyl ligands 11
- phosphoramidites 11
chiral phosphoric acid (CPA) 124, 127, 128, 130, 136, 231, 242
- Pictet–Spengler reaction 136
chiral secondary amine catalysis, see asymmetric aminocatalysis process
chiral silane Lewis acid 121, 124
- asymmetric Friedel–Crafts reaction 124
- – with benzoyl hydrazones 124
chiral titanium (IV) catalysis 103
- catalyst 106
- reaction pathway 103
chromane synthesis 251–253
- via intramolecular Friedel–Crafts alkylation 251
- via tandem oxa-Michael addition 251
chromatographic purification 40
- S-(1,3-benzoxazol-2-yl) thioester conversion 40
- – into amides 40
- – into esters 40
cinchona alkaloids 116
- cinchonidine 116
- cinchonine 116
- derived thioureas 119
cinnamate esters 20
- Friedel–Crafts alkylation 20, 21
clays 277
- properties 277
- solid acids 277, 278
coordinating solvents 24
- CCl$_4$ 24
- CH$_2$Cl$_2$ 24
- toluene 24
- TTCE 24
Corey's transition-state model 106
COX-2 inhibitor 53, 224, 232
- synthesis 224, 232
- – using MacMillan catalyst 224
cyclic ethers 168
- arene-mediated ring-opening 168
- pyrans 168
- THF 168

cyclization reaction 56
– asymmetric organocascade 56
– – aminocatalytic conjugate addition-halogenation sequence 56
– α-chlorination step 56
– domino conjugate addition 56

d

Davis oxaziridine reagent 241
decarbonylation process 211
– rhodium-catalyzed 211
Detal process 279
N,N-dialkylamino aromatics 104
– Friedel–Crafts reaction 104
diarylamines/phenothiazines alkylation reactions 281
diastereoisomers separation 41
– flash chromatography 41
diastereoselective Friedel–Crafts reactions 181, 182, 226, 253
– coupling 264
– gold-catalyzed 187
1,2-dicarbonyl derivatives 108
– p-chlorophenylglyoxal 108
– methyl pyruvate 108
dicarboxylic acids 280
– precursors 280
– – 2,6-diisopropylnaphthalene 280
– – 4,4′-diisopropyl biphenyl 280
Diels–Alder reactions 272
4,7-dihydroindoles 131
– CPA-catalyzed asymmetric Friedel–Crafts reaction 131
– – with N-sulfonyl imines 131
1,3-dimethoxybenzene 119
– enantioselective alkylation 119
– – solid-supported catalyst 119
– – with ethyl 3,3,3-trifluoropyruvate 119
N,N-dimethylaniline 112
– enantioselective hydroxalkylation reaction 112
N-directing group effect 210
– on enantioselective hydroarylation 210
dodecylbenzenesulfonic acid (DBSA) 194
double hydrogen bond interaction 69, 70
– b/w ureas and nitroarenes 69

e

electron-deficient alkenes 68, 69, 76
– 1,2,3-triazoles 76
electron-deficient arenes 158
– intramolecular allylic alkylation 158
– – via In(III)-catalysis 158
electron-deficient heteroaromatics 54

electron-donating groups (EDGs) 3, 146, 249
electron-rich arenes 17, 18, 34, 104, 148, 149, 162, 171
– asymmetric alkylations 18
– catalytic enantioselective fluoral catalyzed Friedel–Crafts reaction 104
– – by BINOL-titanium complex 104
– Friedel–Crafts-type allylations 149
– Friedel–Crafts alkylation 34, 171
– indoles 18
– intramolecular hydroarylation 162
– – with gold(I)-catalysis 162
– Mo-catalyzed allylation 148
– pyrroles 18
electron-rich aromatic compounds 65, 112, 148
– aniline 112
– benzenes cycloalkylation 178
– – Friedel–Crafts mechanism vs. auration-based pathway 178
– enantioselective Friedel–Crafts reactions 120
– – with α-imino ester 120
– Friedel–Crafts alkylation reaction 65
– – with glyoxylate/trifluoropyruvate 112
– indole derivatives 112
– transition-metal-catalyzed alkylation 148
electron-rich furans 88
– lack of reactivity 88
– 2-methylfuran 88
– phenylfuran 88
electron-rich heteroaromatic compounds 121
– catalytic system 121
– furans 121
– pyrroles 121
– thiophenes 121
electron-rich heteroaromatic systems 90
– chiral Lewis acid catalyzed Friedel–Crafts alkylation 90
electron-rich heterocycles 80
– compounds 68
– Friedel–Crafts type alkylation 68
– Michael addition 80
electron-withdrawing groups (EWGs) 3, 158, 161
electrophilic agents 149
– Csp3-based 7
enals 54
– ether-tethered 251
– Friedel–Crafts alkylation 54
– iminium catalyzed asymmetric Friedel–Crafts synthetic utility 54

– intramolecular Friedel–Crafts alkylation 251
enantioenriched allenes 162
– Au(I)-catalyzed annulation with pyrroles 162
enantioenriched phenyl benzyl alcohols reaction 184, 185
– preparation 184
enantioselective acyl-Pictet–Spengler reaction 135, 138
– chiral thiourea-catalyzed 135, 138
– (+)-yohimbine synthesis 135
enantioselective alkylation reactions 203, 207
– intramolecular 228
– transition metal-catalyzed methods 203
enantioselective cyclization 209
– rhodium-catalyzed 209
enantioselective Diels–Alder reactions 72
– Lewis acid catalyzed 72
enantioselective hydroarylation 203, 205–207, 209–211, 213
– gold(I)-catalyzed allene 217
– intramolecular 214
– iridium(I)-catalyzed 206
– phosphine ligands 205
– rhodium-catalyzed 203, 207, 209, 210, 213
enantioselective organocatalytic Friedel–Crafts hydroxyalkylation reaction 117
enantioselective organocatalyzed ring-closing Friedel–Crafts reaction 228
enantioselective Pictet–Spengler reaction 134, 137
– CPA-catalyzed 137
– – via N-benzyltryptamine 137
– – via sulfenyliminium ions 137
enantioselective reactions 124, 187–196, 204, 210, 213, 218, 250
– Brønsted acid catalysis 191
– ruthenium catalysis 187–191
enones 61–63
– asymmetric Friedel–Crafts alkylation 63
– – by chiral primary amine salt catalysts 63
– Friedel–Crafts indole alkylation 62
– – catalytic asymmetric π-rich aromatics 61
– – organocatalytic tools 62
(+)-epi-leuconolam synthesis 243
epoxides 179
– alkylating agents 284
– enantiomerically pure 169–179
– racemic cis/trans-epoxides 180, 181
– – Friedel–Crafts-based kinetic resolution 181
– – Friedel–Crafts reaction 180

– ring-opening mechanism 168–181
– – asymmetric 179–181
– – benzene-based 168
– salen-chromium-catalyzed kinetic resolution 179, 180
(+)-erogorgiaene synthesis 258–260
ethyl glyoxylate 109
– enantioselective Friedel–Crafts reaction 109
– – with Juloindine 109
– – with N-methylindoline 109
– – with N-methyltetrahydroquinoline 109
ethylbenzene/styrene production 278
– using $AlCl_3$ catalyst 278

f

ferrocenyl alcohol 195
– derivatives 193
– reaction 195
fischerindole 256
– synthetic approaches 256
– types 256
Friedel–Crafts adducts conversion 25
– alkylidene malonates into tryptophans 25
– indole into tryptophans 25
Friedel–Crafts alkylation reactions 2–4, 6, 9, 17, 25, 27, 52, 65, 83, 159, 286
– adducts 25
– advantage 20
– allylic 148
– aminocatalytic/asymmetric 52
– catalytic asymmetric Friedel–Crafts alkylation protocols 17
– enantioselective 13, 17
– – transformation 190
– general aspects/historical background 3–5
– intermediate complexes 4
– intramolecular approach 65–67
– Marson-type 260
– Michael-type 7
– pathway 83
– pictorial representation 3
– products 22, 26
– published papers number 2
– silver-based LA-catalysis 159
Friedel–Crafts (FC) reaction 5, 21, 24, 54, 55, 114, 271, 272, 275
– acyl methylenemalonates 21
– addition reactions 18
– – aromatic C–H bonds 18
– – catalytic enantioselective 18
– – heteroaromatic C–H bonds 18
– adduct(s) 29, 32, 36, 76
– arylidene malonates 24

- asymmetric reaction 51
- chemistry 271
- ethenetricarboxylates 21
- fluoral catalyzed 114
- – by BINOL-Ti complexes 114
- green chemical assessment 272–275
- heterogeneous catalysts 275–278
- naphthols 75
- NH-indoles 24
- organocascade strategy 57
- product(s) 27, 76, 86, 105, 131
- – – intramolecular aza-Michael addition 131
- protocol 88
- regioselectivity constraints 53, 54
- role 271
- working model 106
furan 192
- derivatives 247
- Friedel–Crafts alkylation 247–249
- ring, oxidative cleavage 125
- Ru-catalyzed reaction 192
2-(2-furanyl)-1,2-ethanediols 247
- synthesis/synthetic transformations 247

g
galaxolide intermediate 170
- synthesis 170
gold 160, 218
- catalysis mechanism 160, 176, 177, 184–187
- catalyzed enantioselective conversion 221
- (I)-catalyzed cyclization 218
- (I)-catalyzed enantioselective hydroarylation 218
- – – intramolecular 229
gonadotropin-releasing hormone (GnRH) antagonists 225
- tryptamine derivatives 225
green chemistry 274
- Friedel–Crafts reaction 272
- objective 274
- principles 272
Grignard reagent 37

h
hapalindole alkaloids synthesis 253–255, 262, 263
- (−)-12-epi-fischerindole U isothiocyanate synthesis 254, 255
- (+)-hapalindole Q 254, 255
(+)-harmicine synthesis 229, 237, 238
Heck-type direct metal-catalyzed oxidative cycloarylations 147

(+)-heliotridane synthesis 37, 38, 240, 241, 244, 245
- 2,3-dihydro-1H-pyrrolizine precursor preparation 37, 38
Henry reaction, CD analysis 86
heteroaromatic compounds 26, 88, 112, 145, 150
- 2-methyl furan 26
- Friedel–Crafts (FC) reactions, stereoselective protocols 145
- furans 112
- gold(III)chloride-catalyzed allylic alkylation, with allylic alcohols 150
- pyrroles 112
- stereoselective alkylations 88
- thiophene 112
heterobiaryl compounds 203, 206
- atropisomerization 206
- atropselective alkylation 203
N-heterocycles 78
- 1,2,3-triazoles 78
hetero-Diels–Alder (HDA) reactions 179
heterogeneous catalyst synthesis 126, 275
- zeolites 275, 276
2-(4,5-hexadienyl)indole 217
- on gold-catalyzed enantioselective hydroarylation 217
- – – ligand/silver salt/solvent effect 217
hexafluoroisopropanol (HFIP) 24
HIV protease inhibitors 231
Horner–Emmons reaction 32
- with benzaldehyde 32
hydroarylation, enantioselectivity 208
hydrogen-bond 11
- asymmetric catalysis 11
- complex 117
- interaction 78, 108
α-hydroxy acid 251
- synthesis 251, 252
α′-hydroxy enones 28–31, 44
- catalytic asymmetric Friedel–Crafts reaction 44
- – – experimental procedure 44, 45
- Friedel–Crafts alkylation 29
- reaction with N-methylpyrrole 28
- reaction with pyrrole 28
hydroxylactams 134, 136
- enantioselective Pictet–Spengler-type cyclization 134, 136
1-hydroxynaphthalene 103
- enantioselective pyruvic ester addition 103
- – – Zr-dibornacyclopentadienyl complex 103

i

imidazole groups 37
– adducts elaboration 37
– – into indole-substituted aldehydes/ketones 37
imidazolidinone catalyst 51, 53, 55
– evolution 52
– π-facial shielding 51
– iminium geometry control 51
imines 119, 120
– activation modes 125
– catalytic asymmetric Friedel–Crafts reaction 120
– – intermolecular approach 120
– – organocatalysis 123
– – organometallic catalysis 120
– – Pictet–Spengler reaction 132
– directing group effect 212
– – on enantioselective cyclization 212
– Friedel–Crafts reactions 119
– Lewis acid activation 123
iminium activation approach 60
iminium catalysis 50, 54
– benzofurans 54
– N-Boc indoles 54
– 2-formyl furans 54
iminium ion formation 52
iminoarenes 207
– Rh(I)-catalyzed enantioselective hydroarylation 207–214
α-imino esters 119, 120, 122, 123
– asymmetric Friedel–Crafts reactions 119, 120, 122, 123
– – Cu(I)-catalyzed 122, 123
– – with aromatic compounds 122
– – with heteroaromatic compounds 123
indium(III) catalysis mechanism 172–176
indole derivatives 32, 83
– asymmetric Friedel–Crafts alkylation 83
– – with nitroalkenes using zinc catalysts 83
– electron-withdrawing/electron-donating substituents 32
indoles 26, 29, 31, 33, 39, 40–46, 47, 51, 53, 57, 58, 63–66, 69, 70, 72, 73, 78, 79, 82, 84–88, 91, 106, 112, 115–118, 122, 125, 127–131, 138, 151–155, 225, 228–230
– N-allyl-substituted 31
– anti-alkylation products 85
– anti-syn-isomers 41
– asymmetric addition to chalcones 63
– asymmetric Friedel–Crafts alkylation 31, 39, 51, 70, 73, 82
– – asymmetric glyoxylate catalyzed (S)-BINOL-titanium complex 106
– – asymmetric organocatalytic intramolecular with α,β-unsaturated aldehydes 66
– – chiral Lewis acid-catalyzed 122
– – H alkylation 87, 91, 92
– – of 4-phenyl-3-enyl methyl carbonate 151
– – thiourea catalyzed trans-β-nitrostyrene 69
– – using metal salts with ligands 82
– – with α,β-unsaturated thioesters 39
– – with nitroacrylates 85
– – with nitroalkenes 70, 73
– asymmetric Friedel–Crafts reaction 116, 125, 129
– – α-dicarbonyl compounds 116
– – with α-aryl enamides 129
– – with imines 125
– based compounds precursor 225
– – β-indolyl nitroalkanes 225
– N-besyl reaction 185
– (R)-carvone coupling pathways 254
– catalytic asymmetric Friedel–Crafts reaction 42–47, 79, 84, 122, 131
– – catalytic enantioselective allylic alkylation with Pd chiral catalyst 153
– – chiral phosphoric acid-catalyzed with N-sulfonyl imines 138
– – chiral thiourea-catalyzed N-sulfonyl imines reaction 127
– – experimental procedure 42–47
– – with imines 122
– – with nitroalkenes 79
– – with tritylsulfenyl 131
– cinchona alkaloid catalyzed hydroxyalkylation 118
– – with ethyl 3,3,3-trifluropyruvate 118
– copper-catalyzed Friedel–Crafts alkylations 31, 112
– – with α′-phosphonate enones 31
– – with N-sulfonyl aldimines 125
– containing products 223
– – 2-aminomethyl indoles 223
– – indolizidine alkaloids 223
– – β-indolyl-propanoic acids 223
– – polycyclic indoles 223
– – pyrrolizine 223
– – pyrrolo[1,2-a]pyrazines 223
– – synthesis 224–240
– – tryptamine 223
– – tryptophan derivatives 223
– CPA-catalyzed asymmetric Friedel–Crafts reaction 125, 130
– – N-sulfonyl imines 128
– – with α-aryl enamides 130
– – with enecarbamates 130

- enantioselective Friedel–Crafts reactions 39, 117, 153
- – – ethyl glyoxylate catalyzed by (S)-BINOL-Ti complex 115
- – – hydrogen bonding catalyzed 116
- – – organocatalytic with carbonyl compounds 117
- – – with ethyl trifluoropyruvate 117
- Friedel–Crafts addition of α-ketoester 118
- – – enantioselective addition to β,γ-unsaturated α-ketoesters 44
- Friedel–Crafts alkylation 29, 40, 86
- – – catalytic alkylation with aldehydes/nitroalkenes 91
- – – enantiodiscriminating step 40
- – – enantioselective 225, 228, 230
- – – metal-catalyzed with simple α,β-unsaturated ketones 64
- – – metal-catalyzed intramolecular with enones 66
- – – organocatalytic 53, 57, 58
- – – organocatalytic with nitroalkenes 78
- – – organocatalytic with simple α,β-unsaturated ketones 65
- – – organocatalytic with trifluoroborate salts 58
- – – with aromatic nitroalkenes 88
- – 3-H addition pathway 55
- – hypothetic trajectory 40
- – Ir-catalyzed enantioselective allylation 155, 156
- – metal-catalyzed allylation 151
- – nucleophilicity 154
- – Pd-catalyzed allylic alkylation 151–154
- – – with alcohols 152
- – – intramolecular 229
- – regiochemical alkylation 151
- – regioselective allylation 152
- – Ru-catalyzed allylation 152
- – – with allyl alcohol 152
- – scandium-catalyzed alkylations 33
indolizidine alkaloids 241
- epi-leuconalam 241
- epi-tashiromine 241
- leuconolam 241
- razhinal 241
- rhazinilam 241
- synthesis 241–242
- tashiromine 241
indolyl-based alkaloids 150
- auxins 150
- BMS-378806 150
- unnatural potent HIV inhibitors 150

(S)-(3)-indolyl glycine 133
- CPA-catalyzed asymmetric synthesis 133
2-indolyl methanamine derivatives 231
β-indolyl-propanoic acid synthesis 224, 225
indolyl α,β-unsaturated carbonyls 66
- intramolecular ring-closing Friedel–Crafts alkylation 66
intramolecular Friedel–Crafts alkylation 66, 229, 285
- catalytic enantioselective 241
- ring-closing 250
  strategy 66
iodo-arenes 147
- bimetallic-catalyzed cross-coupling 147
iodobenzenes 147
- intra/intermolecular Pd-In-mediated cross-coupling reaction 147
- – with allyl acetates 147
5-iodoindole 226
- enantioselective alkylation 226
iridium-catalyzed hydroarylation mechanism 205
isomerization reaction 276

k

β-keto phosphonate adducts conversion 32
- amide products 34
- carboxylic acid derivatives 33
- into enone products 32
- into ketone products 32
Knoevenagel condensation 211

l

LAB, see linear alkyl benzenes
lamellar aluminosilicate materials, clays 275
large scale hydrocarbon processing compounds 278–286
- alkylanilines/related compounds 280, 281
- alkylphenol production 281
- dialkylated biaryls 280
- ethylbenzene and cumene 278, 279
- linear alkyl benzenes 279
less reactive nucleophiles 52
- chiral N-triflyl phosphoramide-catalyzed asymmetric synthesis 132
- enantioselective catalytic addition 52
(−)-leuconolam synthesis 243
Lewis acid catalyzed enantioselective reactions 80–90
- Friedel–Crafts reactions of furans and pyrroles 88
- – catalytic asymmetric Friedel–Crafts alkylation 88

– – zinc catalysts   88
– Friedel–Crafts reactions of indoles   81
– – aluminum catalysts   84
– – copper catalysts   84
– – zinc catalysts   81
Lewis acids (LAs)   2, 4, 9, 67, 68, 80, 90, 102, 277
– aluminum trichloride   2, 102
– assisted activation   28
– boron trifluoride   2
– catalyst complex   18
– center   84
– ferric chloride   2
– resin-supported indium synthesis   175
– Salen-based   179
– use of   102
– water-tolerant   193
– zinc chloride   2
ligand-metal catalytic complex   91
linear alkyl benzenes   279
– production process   279
– sulfonates precursors, LAB   279
(−)-lintetralin synthesis   257–259, 265, 266
LUMO-lowering activation strategy   49–51
– advantages   49, 50

m
MacMillan's imidazolidinone catalyst   226
mandelic acid derivatives synthesis, optically active   250
mesoporous silicas/aluminas   275
– MCM-41   275
– SBA-15   275
metal-based catalysis   60, 62
– asymmetric Friedel–Crafts reactions   60
– Friedel–Crafts alkylation of indoles/pyrroles   62
2-methoxyfuran   89, 92, 126, 248
– asymmetric Friedel–Crafts alkylation   89
– – catalytic alkylation with nitroalkenes   89, 92
– – CPA-catalyzed N-Boc imines   126
– – with transition state   89
5-methoxy-2-methyl indole   53
– organocatalytic alkylation   53
2-methylfuran   189, 190
– enantioselective ruthenium-catalyzed propargylation   189, 190
methylindole   71, 107, 192
– Brønsted acid catalyzed reaction   192
– Friedel–Crafts alkylation mechanism   107
– – thiourea catalyzed   71
– – with methyl 3,3,3-trifluoropyruvate   107
– – with methyl pyruvate   107

– – with p-chlorophenylglyoxal   107
– – working model   108
N-methylindoline   105
– asymmetric ethyl glyoxylate catalyzed Friedel–Crafts reaction   105
– – by a BINOL-titanium complex   105
N-methylpyrrole   30, 31
– Friedel–Crafts alkylation reaction   30
– – with α'-phosphonate enones   31
Michael-type addition reactions   17, 68, 258
– acceptors   28
– chelating α,β-unsaturated compounds   17
– diastereoselectivity   258
microwave-enhanced reductive amination   254
Mobil–Badger process   278
mono-amine alkaloid, tryptamine   74
monodentate systems   18
montmorillonite clay   277
– diagramatic representation   277
multicomponent tandem reactions   91
– catalytic approach   91

n
2-naphthols   76, 80
– asymmetric Friedel–Crafts alkylation   76
– – catalytic alkylation with nitroalkenes   80
nitroacrylate(s)   40–42, 47, 85
– asymmetric Friedel–Crafts alkylation   40
– – catalytic reaction procedure   47
– – with indoles   41
– complexes   85
– Z/E conversion   41
nitroalkene(s)   67–69, 71, 83, 86
– activation of   83, 86
– aromatic/aliphatic   86
– catalytic asymmetric Friedel–Crafts reactions   69
– catalytic enantioselective Friedel–Crafts alkylations   67, 68
– Friedel–Crafts additions, catalytic system   71
– organocatalytic enantioselective reactions   68
non-Michael acceptor alkenes   204
– enantioselective hydroarylation   204
norbornenes   204
– enantioselective hydroarylation   204–206
– hydroamination   206
N-nosyl-aziridine   171
– enantiospecific opening   171
π-nucleophiles   53
– electron-rich benzenes   53

- furans 53
- indoles 53
- thiophenes 53
nucleophilic allylic alkylation processes 11, 145, 146, 160
nucleophilic cross-coupling partners 147
nucleophilic reaction partners, *see* pyrrole derivatives

o

olefins 19
- Friedel–Crafts alkylation reaction 19
- – catalytic/enantioselective 19
one-pot aromatization reaction 36
- 4,7-dihydroindole alkylations 36
- – of α,β-unsaturated 2-acyl imidazoles 36
one-pot multi-catalysts cascade reactions 57
optically active chromanes synthesis 250, 251
organocatalytic approach, *see* LUMO-lowering activation strategy
organocatalytic domino reaction 55
- Friedel–Crafts alkylation 55
organocatalytic enantioselective Friedel–Crafts reaction 55, 68, 102, 224, 242
- Friedel–Crafts alkylation of indoles 68–72
- – bissulfonamide catalysts 72
- – phosphoric acid catalysts 73
- – synthetic applications 74
- – thiourea catalysts 69
- with trifluoroborate salts 55
organocatalytic Friedel–Crafts strategy 51
organocatalytic pyrroloindoline construction 59, 60
organocatalytic system 60
ortho-hydoxyalkylation, chiral aluminum-based complexes 6
oxa-Michael addition reaction 250
oxocarbenium ion 261
- intermolecular Friedel–Crafts alkylation 261
- precursor 261
oxygen atoms 32
- carbonyl 32
- phenolic, intramolecular attack 76
- phosphonyl 32
oxygenated compounds 168
- alcohols 168
- ethers 168

p

palladium-catalyzed cross-coupling reaction 211
partial life cycle assessment 273
Pechmann synthesis 284

pendant alkene, hydrometallation 207
phenols 6, 102, 171
- enantioselective chloral mediated ortho-hydroxyalkylation 102
- – by chiral alkoxyaluminum chloride 102
- ortho-hydoxyalkylation 6
- regio/stereoselective Friedel–Crafts alkylation 171
phenyl group 113
- analog, ligand bifunctional mode of interactions 84
- π-π stacking interaction 113
2-phenylindole, Friedel–Crafts reaction 225
phosphine ligand 205, 208
- effect on enantioselective hydroarylation 208
phosphodiesterase inhibitor compound 226
α'-phosphonate enones 31, 32, 45
- asymmetric Friedel–Crafts alkylations 31
- – catalytic reaction procedure 45
phosphoramidite ligands 208
phosphoric acid catalyst 73
- BINOL-derived 73
Pictet–Spengler (PS) reaction 119, 132, 135, 137, 228
- asymmetric organocatalyzed (+)-harmicine synthesis 229
- condensation 5
- cyclization 74, 134
- – substrate-controlled 134
- 3,5-di(*tert*-butyl)-4-hydroxytoluene (BHT) 137
platinum catalyst 214
- catalyzed enantioselective hydroarylation 214, 215, 217, 220
- – limitations 217
- – mechanism 214
polycyclic indoles 217, 228
- synthesis 228, 229
potential surface acidic sites 277
- on $AlCl_3$-treated silica 277
protein kinase C (PKC) inhibitor 209
- via Rh-catalyzed enantioselective hydroarylation 210
proton slide mechanism 71
proton-transfer processes 71, 78
pyridine-2-carbaldehyde 112
- enantioselective N,N-dimethylaniline mediated Friedel–Crafts reaction 112
- – by chiral aluminum complex 112
pyrrole 30, 35, 51, 89–91, 129, 135, 136
- N-acyliminium ions regio/enantioselective catalytic cyclization 135, 136

– asymmetric Friedel–Crafts alkylation   30, 52, 51, 90, 91
– – catalytic cycle   91
– – CPA-catalyzed N-acyl imines reaction   129
– – with nitroalkenes   90
– containing compounds synthesis   240–246
– 2,5-disubstituted   89
– Friedel–Crafts reaction   35, 172
– – enantioselective reaction with N-acyl imines   127
– nitrogen, Boc protection   37
– π-system   51
– sequence steps   89
pyrrole derivatives   127
– enantioselective Friedel–Crafts reaction   127
– – with N-acyl imines   127
pyrroles   92
– catalytic asymmetric Friedel–Crafts alkylation   92
– – with nitroalkenes   92
pyrrolo[1,2-a]pyrazines synthesis   242–244, 246

q
quinine-derived catalyst   76

r
(−)-razhinal synthesis   243
(−)-razhinilam synthesis   243
rhodium-catalyzed enantioselective reactions   220
– conversion   220
– cyclization   220
– hydroarylation   207, 208, 212, 213
– protocols   219
rhodium phosphoramidite complexes   208
Ru-catalyzed phenol allylic alkylation   149
– branched allyl carbonates   149

s
Salen ligand   88
– AlCl complex   88
scandium (III)-pybox complexes   38
– Friedel–Crafts reaction   38
– – bidentate substrates   38
– X-ray crystal structure data   38
scandium (III) triflate complex   45, 232
– preparation procedure   45, 46, 232
serotonin reuptake inhibitor   53, 226, 227
– BMS-594726 synthesis   226, 227, 235–237
solid acids, zeolites   271
solvent effect   24
stereoselectivity, origin   51
sterically-demanding partners   62
– α,β-unsaturated ketones   62
Strecker reaction   120, 258
styrene catalytic dehydrogenation   278
styrene oxide   169
– ring-opening   169
substrate-based chirality transfer   211
substrate-catalyst complex   29
– stereomodel   29
N-sulfonyl aldimines   119
– asymmetric Friedel–Crafts reactions   119
Suzuki–Miyaura cross-coupling reaction   242

t
(−)-talaumidin synthesis   261
(−)-tashiromine synthesis   242
N-TBS protected indoles   127, 128
– asymmetric N-Boc imines catalyzed Friedel–Crafts reaction   127, 128
– – CPA-catalyzed   128
1,1,2,2-tetrachloroethane (TTCE)   22, 24
– transition state   25
tetrahydro-β-carbolines (THBCs)   66, 159
– enantioselective synthesis   159, 160
– ring systems   132
tetrahydro-γ-carbolines (THGCs)   229, 230, 239
– synthesis   239
tetrahydrocarbazole derivatives synthesis   231
tetrahydropyrano indoles (THPIs)   66
thiourea catalysts   70, 76, 78
– double hydrogen bond interaction   70
– use   78
transition-state models   25, 105, 109
– stereo-controlling model   106
1,2,3-triazoles addition   76
– cinchona alkaloids catalysts   77
trifluoromethyl/difluoromethyl-containing compounds   133
– CPA-catalyzed asymmetric synthesis   133
trifluoropyruvate, carbonyl moiety   113
tris(oxazoline) ligands   23, 25
– hetero-/homo-   25
Trost's complex   149
– sulfonate catalyst   152
tryptamine   74, 75, 134
– analogs synthesis   225–228
– chiral aldehyde   134
– derivatives   74, 228
– Friedel–Crafts products   75
– melatonin analogs synthesis   226
– Pictet–Spengler reaction   228

tryptophan 25, 225
- β-substituted 25
- analogs synthesis 225–228
- derivatives 134, 228
- nitro-precursors 226
- transformation 226
N-Ts α-imino esters 121
- Cu(I)-catalyzed asymmetric Friedel–Crafts reaction 121

u

unactivated alkenes 203, 204
- enantioselective hydroarylation 204
α,β-unsaturated acyl compounds 32–39
- 2-acyl imidazoles 34
- acyl phosphonates 32
α,β-unsaturated 2-acyl imidazoles 34, 36, 37, 46
- catalytic asymmetric Friedel–Crafts reaction procedure 46
- divergent behavior 37
- Friedel–Crafts reaction 38
- scandium-catalyzed alkylations 34, 36
- – with indoles 34
- – with pyrrole 36
α,β-unsaturated 2-acyl N-iso-propylimidazole 35
- Friedel–Crafts reaction 35
α,β-unsaturated acyl phosphates 45
- catalytic asymmetric Friedel–Crafts reaction procedure 45, 46
α,β-unsaturated aldehydes 49, 50, 53
- aminocatalytic Michael addition 53
- aromatic/hetero-aromatic substrates 49
- – catalytic asymmetric Michael-type addition 49
- asymmetric Friedel–Crafts alkylation 55
- asymmetric Michael addition 50
- enantioselective Friedel–Crafts reactions 53
- nucleophilic indole 55
- organocatalysis, asymmetric aminocatalysis 50
- organocatalytic Domino reactions 55
- stereocontrolled Friedel–Crafts alkylation 49
α,β-unsaturated carbonyl compounds 49, 50
- aromatic/hetero-aromatic substrates 49

- – catalytic asymmetric Michael-type addition 49
- LUMO-lowering activation 50
- – via iminium ion formation 50
unsaturated α-ketoesters 25, 27, 28
- enantioselective Brønsted acid catalyzed indole addition 28
- square-planar geometry 27
α,β-unsaturated ketones 32, 49, 60
- aromatic/hetero-aromatic substrates 49
- – catalytic asymmetric Michael-type addition 49
- Horner–Emmons reaction 32
- organocatalysis 61–63
- organometallic catalysis 60
- stereoselective Friedel–Crafts alkylation 60
- – catalytic system 60
- stereoselective Michael addition 60
α,β-unsaturated system 39
- binding interaction 39
α,β-unsaturated thioesters 39, 47
- catalytic asymmetric Friedel–Crafts reaction procedure 47
- enantioselective Friedel–Crafts alkylation 39
urea-derived catalysts 228

v

Vilsmeier–Haack formylation 242
vinyl-benzyl stereocenters 146
- metal-catalyzed approaches 146
1-vinyl-tetrahydronaphthalenes synthesis 157
- via intramolecular Mo-catalyzed arenes allylic alkylation 157

w

Wilkinson's catalyst 242
Wittig olefination reaction 240, 258

x

X-ray structures 72

z

zeolite 276, 277
- catalysis 274
- properties 276
- ZSM5, schematic diagrams 276
zirconocene complex, chiral 102